W0042872

THE LOGIC OF CATEGORIES

BOSTON STUDIES IN THE PHILOSOPHY OF SCIENCE

EDITED BY ROBERT S. COHEN AND MARX W. WARTOFSKY

VOLUME 85

GYÖRGY TAMÁS

Department of Logic
of the Eötvös Loránd University, Budapest

THE LOGIC
OF
CATEGORIES

Edited by
ROBERT S. COHEN

D. REIDEL PUBLISHING COMPANY

A MEMBER OF THE KLUWER  ACADEMIC PUBLISHERS GROUP

DORDRECHT / BOSTON / LANCASTER / TOKYO

Library of Congress Cataloging-in-Publication Data

Tamás, György.
    The logic of categories.

(Boston studies in the philosophy of science ; v. 85)
Bibliography: p.
Includes index.
1. Categories (Philosophy)   I. Title.   II. Series.
Q174.B67   vol. 85   [BC172]   001'.Ols   [160]   85—31238
ISBN-13: 978-94-010-8502-1   e-ISBN-13: 978-94-009-4494-7
DOI: 10.1007/978-94-009-4494-7

Translated by Ildikó Berkes

Sold and distributed in the U.S.A. and Canada
by Kluwer Academic Publishers,
101 Philip Drive, Assinippi Park, Norwell, MA 02061, U.S.A.

Distributors for Albania, Bulgaria, Chinese People's Republic, Cuba,
Czechoslovakia, German Democratic Republic, Hungary, Korean People's
Republic, Mongolia, Poland, Romania, the U.S.S.R., Vietnam, and Yugoslavia
Kultura Hungarian Foreign Trading Company,
P.O.B. 149, H-1389 Budapest, Hungary

Distributors for all remaining countries
Kluwer Academic Publishers Group,
P.O. Box 322, 3300 AH Dordrecht, Holland

Original title: *Kategóriák logikája*
First edition published in 1975 by Akadémiai Kiadó, Budapest.
First English edition published in 1986 by Akadémiai Kiadó, Budapest in co-edition with
D. Reidel Publishing Company Dordrecht, Holland.

All Rights Reserved.
© 1986 by Akadémiai Kiadó, Budapest, Hungary.
Softcover reprint of hardcover 1st edition 1986
No part of the material protected by this copyright notice may be
reproduced or utilized in any form or by any means, electronic or mechanical, including
photocopying, recording or by any information storage and retrieval system, without written
permission from the copyright owner.

TABLE OF CONTENTS

CHAPTER FIVE

CHAPTER SIX

CHAPTER SEVEN

EDITORIAL PREFACE

György Tamás works in the philosophy of logic, that difficult interdisciplinary region wherein the notion of categories is both basic and subtle. To understand ways of thinking, to understand patterns of whatever is real, to recognize what is possible and to reject the nonsensical and the impossible in thought and in fact, is to comprehend the categories. This was a recurring motive of European thought from the earliest self-aware beginnings, and Tamás knows that history well, as his critical respect demonstrates. Ancient, medieval, and modern thinkers appear in this book, set forth in their own words; and likewise we see that Tamás has built upon the historians and commentators, upon the pioneering historical investigation of the categories by Trendelenburg a century ago and by Bochenski in our days.

Tamás has two principal goals here: to investigate the logic, which is to say the structure and the relations, of the philosophical categories; and to set forth the logic of thought which may then be based upon the critically established system of categories obtained by that investigation. Ancillary but of striking value is his style of historical relevance which enables the reader to engage in a discussion that is both analytically sharp and developmentally insightful. Furthermore, Tamás draws upon his contemporary colleagues with similar critical respect: Łukasiewicz, Quine, Patzig, Menne, Tavanets and others. Has he reached his goals? No, but he would be the first to say that this is the case, that the range of categories studied should be extended to include, for example, the categories of deontic logic, and the spectrum of propositions extended to those of past, present and future; *the theory of the categories always remains incomplete, the task is always to seek greater comprehensiveness.* This is a sensitive insight for Tamás who has written at another occasion that an essential fault of Ramon Lull in the *Ars Magna* and of Leibniz in the *Characteristica Universalis* was their illusion of managing to reach completion.

Dr. György Tamás completed his university studies in 1952; he had studied with Béla Fogarasi, Gyula Kornis, László Kalmár, and György Lukács, among others. After some years as a university teacher, he joined the Institute of Philosophy of the Hungarian Academy of Sciences where he worked for fifteen years. Since 1973 he has been at the Department of Logic of the Eötvös Loránd University in Budapest. Tamás has studied and lectured widely, in Austria, Greece, the U.S.S.R., Belgium, Rumania, the Federal Republic of Germany, and the German Democratic Republic. Among his works is *Die wissenschaftliche Definition* (Budapest 1964) as well as some 25 books and research papers. This year, Dr. Tamás edited a volume of logical investigations for the Hungarian Academy, published in German under the title *Studien zur Logik* (Budapest 1983); he contributed a further result of his work on the logic of relations and categories. We are grateful to him and to his editorial and academic associates for their collaboration in bringing this fine work on the categories into our *Boston Studies.*

We thank Peter McLaughlin for his translations of many passages from the German.

December 1983

ROBERT S. COHEN
Center for Philosophy and
History of Science,
Boston University

MARX W. WARTOFSKY
Department of Philosophy,
Baruch College,
The City University of New York

INTRODUCTION

Categories are usually divided into two groups: the most general terms used by the special sciences and those used in philosophy. The categories of biology, e.g., life, nutrition, multiplication, perception, etc., belong to the former group while the latter includes quantity, quality, relation, etc.

What categories are used in logic? In order to answer this question, let us first consider the major stages of historical development! The *Organon* begins with the study of categories. Aristotle ranks the following among them in this work of his: substance, quantity, quality, relation, place, time, position, state, action, and affection. He examined them primarily as predicates of propositions and never as their subjects because, according to his postulate, there exist no terms more general than these and consequently nothing can be predicated of them.

He paid special attention to the interrelations of the categories: We must not be disturbed because it may be argued that, though proposing to discuss the category of quality, we have included in it many relative terms. We did say that habits and dispositions were relative. In practically all such cases the genus is relative, the individual not.[1]

If it were but in the *Organon* that Aristotle treated categories, it would be more or less justifiable to say that this philosophical issue *par excellence* has been drawn into logical studies due to the lack of distinction between philosophy and logic. There is, however, a detailed analysis of categories in the *Metaphysics*, too, especially in Book Five. The comparison of these two works reveals two different approaches to the same problem rather than any repetition.

---

[1] Aristotle. Vol. I. Oxford, 1928. 11a20–24.

It is well-known that the central, or at least, one of the fundamental issues of traditional logic was the study of syllogisms. That is why the 'First Analytics' was considered as the most important part of the Aristotelian logic. The following warning should serve as a compensation for this one-sidedness:

> It is almost indifferent how many true or false perceptions there are in this study of predicables which develops into a 'curriculum' in the Book of Categories. For, regardless of all the details, this study gives the key to two extremely important characteristics of Aristotle's syllogistic. He discusses *only* the conclusions of universal and particular propositions and totally ignores singular propositions within the *system* of syllogistic, on the one hand, and he examines propositions of the subject-predicate type only, i.e. those asserting whether the class of things covered by a *subject* is totally or partially included in the class of things covered by a *predicate,* on the other... The formal perfection and the self-sufficiency of the Aristotelian syllogistic as a whole are based on this dual limitation manifest in the study of predicables. This very limitation, however, raises numerous problems and also results in various defects because the authority of this system which is wholly justified by its merits has made it extremely difficult for nearly two thousand years not only to overcome the limitations inherent in it, being by itself formally perfect and self-sufficient but at the same time *restricted,* but also to properly assess completely different types of propositions, and completely different kinds of inferences, too.[2]

As to the structure of the *Organon,* the first book is devoted to the study of philosophical categories while the other books are mainly concerned with logical categories.

In traditional logic it was, first of all, term, proposition, and inference which stood high among the categories. How do they relate to the Aristotelian categories? Two extreme points of view have developed in this respect. According to one of them, the traditional approach is a direct continuation of that of Aristotle. That is, the following correspondence can be found between them:

---

[2] Szalai, Sándor, The Major Problems of the Development of the 'Organon' and the Structure of the Aristotelian Syllogistic. Budapest. 1961, pp. XXXV–XXXVI.

Categories   —   Study of Terms
Hermeneutics —   Study of Propositions
Analytics   —   Study of Inferences

According to the other view, the above correspondence is completely unfounded. For, among others, term is not congruent with '*horos*' and neither is proposition with '*apophantikos*'. This objection is justified so far as it denies the identity of the above categories joined in pairs. Nevertheless, it is a mistake to emphasize their difference only. However different the syllogism of Aristotle and that of the scholastic logic may be, both are inferences all the same. We get a true picture only if the degree of difference is correctly stated.—The main structural change was that the study of categories gradually ceased to be an independent part of works on logic, to receive explicit mention in relation to general terms only.

In Kant's transcendental logic, again, more importance was attached to categories. Kant tried to define them as follows:

> The same function which gives unity to the various representations *in a judgment* also gives unity to the mere synthesis of various representations *in an intuition*; and this unity, in its most general expression, we entitle the pure concept of the understanding. The same understanding, through the same operations by which in concepts, by means of analytical unity, it produced the logical form of a judgment, also introduces a transcendental content into its representations, by means of the synthetic unity of the manifold in intuition in general. On this account we are entitled to call these representations pure concepts of the understanding, and to regard them as applying *a priori* to objects—a conclusion which general logic is not in a position to establish.
>
> In this manner there arise precisely the same number of pure concepts of the understanding which apply *a priori* to objects of intuition in general, as, in the preceding table, there have been found to be logical functions in all possible judgments. For these judgments specify the understanding completely, and yield an exhaustive inventory of its powers. These concepts we shall, with Aristotle, call *categories*, for our primary purpose is the same as his, although widely diverging from it in manner of execution.[3]

---

[3] Kant, *Kritik* p. 149; Kemp Smith, pp. 112–113.

Thus, in Kant's view categories are *a priori* forms of human understanding which are present in the mind, independently of experience. The basic terms, as materialism sees it, are abstracted from experience: they are concentrated summaries of human cognition.

Kant's theory of categories gained many adherents. In due time, however, its imperfections—upon which I shall dwell later in detailed analysis—became more and more apparent.

Hegel was one of the few who also attached great importance to the categories from the point of view of logic.

> Gegen die Kahlheit der bloss formellen Kategorien hat der Instinkt der gesunden Vernunft sich endlich so erstarkt gefühlt, dass er ihre Kenntnis mit Verachtung dem Gebiete einer Schullogik und Schulmetaphysik überlässt, zugleich mit der Missachtung des Werthes, den schon das Bewusstseyn dieser Fäden für sich hat . . . [4]

He thoroughly examined the development, interrelations, and mutual transitions of the categories.

Mathematical logicians relegate the study of categories to the field of philosophy, or rather, epistemology. It does not mean, however, that they do not treat such basic terms as existence, modality, relation, etc.

The position of Marxist logicians has strongly been affected by the changes in Marxist philosophy. Early in the thirties the opinion that the main task of philosophy and logic is the examination of categories came to prevail in the Soviet Union. As a counter-reaction to this approach, a new trend developed and replaced the former one. It regarded the analysis of categories as a break-away from practical demands, as something destined to its own end. Although categories have been studied again from the second half of the fifties onward, so far this has affected dialectical logic only.

We have seen that opinions greatly vary on the question what kinds of categories logic should be concerned with. For my part I agree with those who think that existence, or relation, though philosophical categories, belong to the subject matter of logic as well.

---

[4] Hegel, *Wissenschaft*, p. 29.

In most cases the structure of logical works has been adjusted to the categories of logic (with the exception of modal logic, for example). This structure has produced certain stereotypes. In the present work I make an attempt to expound the subject concerned by relying on philosophical categories. I do so in the conviction that this approach makes it possible to solve some long-discussed problems, to reveal certain connections hitherto ignored or neglected, and to overcome many limitations.

As to the definition of philosophical categories, I have proceeded from the principle that they cannot be traced back to terms more general than themselves. Therefore I define categories in relation to one another, bearing in mind Hegel's remark:

> ... in the belief that a definition in itself must appear clear and worked out and only has its regulator and touchstone in presupposed notions, or at least in ignorance of the fact that the sense as well as the necessary proof of a definition is to be found only in its development and in its proceeding from this (development) as its result. [5]

* * *

Chapter One of the present work is concerned with the categories of existence, value, and quality. Chapters Two and Three raise the question of intension and extension. Modality, time, and space are discussed in Chapter Four. Chapter Five examines the logic of relations. Chapter Six investigates such problems of inference as are related to more than one category. Finally, in Chapter Seven, I am concerned with categories hardly, or not at all, touched upon in logic.

The above order is the result of two factors. On the one hand, I have built my study on historical development, and on the other I have proceeded from the simple to the complicated. In most cases these two factors have coincided.

I have found the demonstration of historical antecedents necessary because I wanted (1) to avoid the appearance of raising fictional problems, (2) to save the reader the trouble of looking up the major views on the questions raised in case he, or she, wants to compare with them the solutions I suggest, and (3) to make the denial of mistaken views more convincing by revealing their sources.

[5] Hegel, *Encyclopädie*. p. 9.

A special emphasis has been laid on Aristotle's views in the historical surveys. It is partly because he was the first to form an opinion, usually decisive for several thousand years to come, on the majority of the questions at issue, and partly because so far no trend in logic could avoid taking up a position on his logic.

There was a time in the history of mathematical logic when it was thought to be justified and reasonable to ignore Aristotle. After the Second World War, however, many logicians, to mention only Łukasiewicz, Patzig, and Menne, have abandoned this view, not considering it beneath their dignity to analyse the Aristotelian logic. Taking this into consideration, I do not think that the study of the syllogism, for example, renders one hopelessly retarded!

CHAPTER ONE

1. EXISTENTIAL TERMS AND PROPOSITIONS

I

In order to correctly assess the role of the category of existence in logic, it is, first of all, reasonable to take into consideration the relevant philosophical conceptions. Parmenides thought there is only existence, and no non-existence. One of the still extant fragments from Gorgias' work *On Non-existence, or on Nature*, however, began as follows: At first, nothing exists. This statement has made Gorgias the forefather of nihilism.

The atomists were of the opinion that the atom is existence while the void is non-existence, but the void exists just as much as the atom.

> ... Leucippus and his associate Democritus say that the full and the empty are the elements, calling the one being and the other non-being—the full and solid being being, the empty non-being (whence they say being no more is than the empty); and they make these the material causes of things.[1]

As regards the interrelation of objective and subjective existence, Democritus maintained that sensible properties, though evolving under the influence of objective reality, are still not objective in themselves but exist in the mind only.[2]

I have selected the following from among the statements of Aristotle on our subject:

> Since the science of the philosopher treats of being *qua* being universally and not in respect of a part of it, and 'being' has many senses and is not used in one only, it follows that if the word is used equivocally and in virtue of nothing common to its various uses, being does not fall under one science

---

[1] Aristotle, *Metaphysica*. 985b5–10.
[2] Cf.: Sextus Empiricus, p. 135.

(for the meanings of an equivocal term do not form one genus); but if the word is used in virtue of something common, being will fall under one science.[3]

While Democritus had ranked, first of all, the sensible, empirical properties among those existing in consciousness, philosophers in the Middle Ages relegated to consciousness mainly the products of thought, concepts, numbers, etc. The expression 'ens rationis' corresponds to this approach. What exists in the mind is distinguished from what exists in reality (ens reale).

These problems were also examined from the viewpoint of psychic activity. The orientation of the mind toward the object was called intentio. Avicenna distinguished intentio prima and intentio secunda. By the former he meant orientation toward real objects and by the latter orientation toward the objects of the mind.

Thomas Aquinas wrote the following on this subject:

> Being is two-fold, being in thought (ens rationis) and being in nature. Being in thought is properly said of those intentions which reason produces (adinvenit) in things it considers, e.g. the intention of genus, species and the like, which are not found among natural objects, but are consequent on reason's consideration. This kind, viz. being in thought, is the proper subject-matter of logic.[4]

According to the interpretation of ens and intentio respectively, many schools of thought evolved in medieval logic and words ran high in their debate. There is no need here to dwell upon its details [see: Prantl, Geschichte der Logik im Abendlande. Vol III–IV. (Berlin, 1955)]. It all boils down to the following: it is the school drawing a distinct line between logic and ontology which became dominant. Logical investigations were confined exclusively to the intentio secunda. Inferences were regarded as purely formal relations of propositions which have nothing to do with really existing things.

The intentio theory took a new shape with Leibniz who distinguished rational and factual truth. Rational truth includes all logical and

---

[3] Aristotle, Metaphysica. 1060b32–3.
[4] Thomas Aquinas, In Metaphysicam Aristotelis Commentaria.—Quoted in: Bochenski, p. 177.: Ivo Thomas tr., p. 154.

mathematical truths while factual truth covers those of the natural sciences. The former are inevitable while the latter are accidental. Identity, contradiction, and the law of excluded middle are sufficient in order to reveal rational truth but factual truth requires the law of sufficient reason as well.

Descartes took up the position of subjective idealism:

> ... seeing that I could pretend that I had no body and that there was no world or place that I was in, but that I could not, for all that, pretend that I did not exist, and that, on the contrary, from the very fact that I thought of doubting the truth of other things, it followed very evidently and very certainly that I existed; while, on the other hand, if I had only ceased to think, although all the rest of what I had imagined had been true, I would have had no reason to believe that I existed; I thereby concluded that I was a substance, of which the whole essence or nature consists in thinking, and which, in order to exist, needs no place and depends on no material thing ...[5]

According to Gassendi the above argument corresponds to the following syllogism:

All who think exist.
I think.
————————————
Therefore I exist.

Descartes protested against this interpretation in his work *Responsiones*:

> ...neque etiam cum quis dicit *ego cogito, ergo sum sive existo*, existentiam ex cogitatione per syllogismum deducit, sed tanquam rem per se notam simplici mentis intuitu agnoscit... ea enim est natura nostrae mentis ut generales propositiones ex particularium cognitione efformet.[6]

Although Descartes' view allows for several kinds of interpretation it is not without reason that those propagating the priority of thought regarded him as one of their forerunners. Hume also shared this opinion:

---

[5] Descartes, *Discours de la méthode* in *Œuvres choisies*. p. 29; Sutcliffe tr. p. 54.
[6] Descartes, *Válogatott filozófiai művek* [*Selected Philosophical Works*]. (Budapest, 1961). p. 288.

All the objects of human reason or enquiry may naturally be divided into two kinds, to wit, *Relations of Ideas*, and *Matters of Fact*. Of the first kind are the sciences of Geometry, Algebra, and Arithmetic; and in short, every affirmation which is either intuitively or demonstratively certain. *That the square of the hypothenuse is equal to the square of the two sides*, is a proposition which expresses a relation between these figures. *That three times five is equal to the half of thirty*, expresses a relation between these numbers. Propositions of this kind are discoverable by the mere operation of thought, without dependence on what is anywhere existent in the universe. Though there never were a circle or triangle in nature, the truths demonstrated by Euclid would forever retain their certainty and evidence.[7]

'Pure logic' evolved from the Platonic ideas. Bolzano thought the objects of logic are self-existent notions and theses and their relations. The meaning of these theses which he called '*Satz an sich*' does not exist in reality, but nevertheless is forever present, regardless of the theses being actually conceived or asserted. Inference is the interrelation of these theses which are independent of reality.

In Lotze's opinion it is the eternal, timeless meanings and truths which fall within the domain of logic. They constitute the world of ideas which exist by holding true. — Meinong purged logic not only of the real but also of ideal objects. In order to examine them, he developed the object theory. In this way he transformed logic into a science of purely practical character.

Mill's approach is similar to this:

Existence in general, is a subject not for our science, but for metaphysics. To determine what things can be recognised as really existing, independently of our own sensible or other impressions, and in what meaning the term is, in that sense, predicated of them, belongs to the consideration of 'Things in themselves', from which, throughout this work, we have as much as possible kept aloof.[8]

In mathematical logic there are basically two different approaches. One is indifferent to the question of existence, probably upon the consideration that it is a category of ontology and metaphysics, in brief, of philosophy.

[7] Hume, *Enquiries* p. 25.
[8] Mill, *System* p. 604.

The other tries to emphasize that logic should use the concept of existence in a broader sense. According to Russell, for example, numbers, Homeric gods, relations, chimeras, and four-dimensional spaces all exist, for if they were not some kind of entities nothing could be asserted of them.[9]

This opinion can be accepted after all. But let us see the following statement:

> In der Logik hat im allgemeinen Existenz formal nur als logische Existenz, d.h. Widerspruchsfreiheit, Berechtigung.[10]

This approach isolates logical existence, that is, existence in correct thinking, from real existence.

In fact, does the acceptance of things existing in the mind necessarily imply the view that they are independent of objective reality? This question has already been answered by Engels:

> Like the idea of number, so the idea of figure is borrowed exclusively from the external world, and does not arise in the mind out of pure thought. There must have been things which had shape and whose shapes were compared before anyone could arrive at the idea of figure. Pure mathematics deals with the space forms and quantity relations of the real world—that is, with material which is very real indeed. The fact that this material appears in an extremely abstract form can only superficially conceal its origin from the external world.[11]

In the foregoing a fairly complex picture has been given to us. Let us begin with the most extreme position which wants to make logic totally free of existence. This attempt, however consistent, is essentially all but an illusion, for in logic, as we shall see, we run into the problem of existence every now and then.

According to the moderate view, logic should treat what exists in the mind. It is supported basically by two arguments: (a) the objectively existing things belong to the domain of the special sciences and philosophy respectively; (b) all the objectively existing things may exist in the mind but it does not hold true the other way round, and consequently the latter,

---

[9] Russell, 'Existence and being'. In: *Mind*. No. 10, 1901.
[10] Menne, (1954). p. 128.
[11] Engels, *Anti-Dühring* pp. 44–45; Eng. tr. p. 58.

2*

mind, has a wider scope. Let us reply to the second argument first. It is true
that in the mind there also exist things which do not exist in reality, but
eventually even the wildest phantasies are but distorted reflections of
reality. Accordingly—and it is already my answer to the first argument—
existence in logic is always imbued with objective existence. Thus, it is
useless to argue about the necessity for logic to be concerned with
objective reality because it does so *nolens-volens*.

II

Our terms can be distinguished according to their referring to things which
do or do not exist in reality. Traditional logic used mainly those terms
which were existential concepts. In general, however, S and P are
designations devoid of any knowledge of existence. The only reservation is
that they must not be replaced by anything else than terms. Therefore
terms indefinite from the point of view of existence are usually represented
by an empty circle:

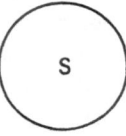

Fig. 1.

The term 'Martian' is, for example, indefinite because at the present
stage of our knowledge we are not able to decide whether or not such
beings exist. The following relation can be predicated of the term
concerned: S is identical with S.

Even if we are able to make up our mind on the question of existence, a
term remains formally indefinite as long as its existence is not stated.

> A sentence is a significant portion of speech, some parts of which have an
> independent meaning, that is to say, as an utterance, though not as the
> expression of any positive judgement. Let me explain. The word 'human'
> has meaning, but does not constitute a proposition, either positive or

negative. It is only when other words are added that the whole will form an affirmation or denial.[12]

Consequently, if existence is to be taken into consideration, it must be indicated. In this way, first of all, the following two versions are arrived at: (1) *neS*; (2) *eS* [*ne* = non-existential term, *e* = existential term]. They can be represented by circles, one with shade-lines indicating non-existence, the other with a dot in it indicating existence.

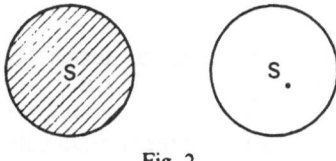

Fig. 2.

When a term is definite from the point of view of existence, identity means the following: *neS* is identical with *neS*; *eS* is identical with *eS*. A further relation which can be stated is: *neS* contradicts *eS*.

Aristotle was already aware of the necessity to speak of terms reflecting the non-existent. It is indicated by his frequent use of the 'goat-stag' as an example. He realized as well that the value of propositions is affected by the fact whether they refer to existent or non-existent things.

> 'Socrates is ill' is the contrary of 'Socrates is well', but not even of such composite expressions is it true to say that one of the pair must always be true and the other false. For if Socrates exists, one will be true and the other false, but if he does not exist, both will be false; for neither 'Socrates is ill' nor 'Socrates is well' is true, if Socrates does not exist at all.[13]

However, he did not consider such terms in his syllogistic.

The operation by which the existence of the phenomena reflected in the terms is postulated was called 'appellation' in the Middle Ages. Petrus Hispanus defined appellation as the usage of terms referring to existing

---

[12] *The Works of Aristotle.* 1928. 16b26–30.
[13] *Op. cit.* 13b14–19.

things since a term which means something non-existent does not name
anything, e.g., 'Caesar' or 'Antichrist'.[14]

The problem of how existential terms affect the validity of inferences
was raised in the 14th century. Vinzenz Ferrer argued as follows: the
inference that all men run, consequently Socrates runs, holds true only if
Socrates lives. A dead, or a not-at-all existing Socrates cannot run.—
Paulus Venetus referred to the following example: all men are living
beings, hence an individual man is a living being. This inference holds true
only if men exist.[15]

According to Bacon non-existential terms have a purely adverse
influence.

> The *idols* imposed by words on the understanding are of two kinds. They
> are either names of things which do not exist (for as there are things left
> unnamed through lack of observation, so likewise are there names which
> result from fantastic suppositions and to which nothing in reality
> corresponds), or they are names of things which exist, but yet confused and
> ill defined and hastily and irregularly derived from realities. Of the former
> kind are Fortune, the Prime Mover, Planetary Orbits, Element of Fire, and
> like fictions which owe their origin to false and idle theories. And this class
> of *idols* is more easily expelled, because to get rid of them it is only necessary
> that all theories should be steadily rejected and dismissed as obsolete.[16]

As opposed to this, Herbart thought propositions including non-
existential terms to be true if the existence of an ideal reality were accepted.
Consequently, he found the following conversion correct: the rage of the
Homeric gods is frightening—some frightening phenomena are the rage of
the Homeric gods.[17]

Mathematical logic examines existential terms (i.e. concepts concerning
existence) from an extensional point of view. Accordingly, the existential
terms are treated as a not-empty class, and non-existential ones as an
empty class. Since the problems of classes are discussed in our Chapter 2,
Sect. 5 'The Extension of Terms', I shall analyse this approach there.

---

[14] Petrus Hispanus, *Summulae Logicales*. pp. 10–11.
[15] Cf.: Bochenski, 1956. pp. 257–259.
[16] Bacon, *Selection* p. 342 (*The New Organon*, 1(LX)).
[17] Herbart, p. 53.

III

Aristotle, among others, wrote the following about existential propositions:

> We admit that of composite expressions those are contradictory each to each which have the verb 'to be' in its positive and negative form respectively. Thus the contradictory of the proposition 'man is' is 'man is not'...[18]

Elsewhere in the same book[19] it becomes obvious that in his view the propositions concerned have two components.

Boethius still followed Aristotle in putting the predicate before the subject in the proposition. This order was reversed in medieval logic. This modification did not change the essence of the Aristotelian approach since the order of subject and predicate is indifferent in existential propositions. It is the same, for example, whether we say 'man is' or 'is man'. The terms of existential propositions can be simply converted.

Kant did not include existential propositions into his table of propositions. It is one of the reasons why the majority of logicians in modern age dealt with this kind of proposition only incidentally, or not at all. Tavanets was right when commenting:

> Logicians denying the specific nature of existential propositions obviously fail to understand what an important role this kind of proposition plays in the cognition of reality. In fact, the knowledge that some object, phenomenon, etc., exist in reality is of paramount importance for science. Let us remember the propositions stating that there exist radioactive elements in reality, there is nuclear energy, there are cosmic rays, and various micro-organisms causing diseases, etc.[20]

Now let us turn to some of those who have shown some interest in this subject. Some assumed the position that existential propositions have but one component. Herbart was of the opinion that the propositions under discussion are, after all, propositions without a subject (*impersonalia*),

---

[18] Aristotle. 1928. 21a38–21b1.
[19] *Op. cit.*, 19b5–35.
[20] Tavanets, *Proposition* p. 85.

e.g., 'it snows', 'it rains'. It is widely known, however, that propositions without a subject are but abridged propositions where the subject is not expressed in a linguistic form, so it denies in no way the role of the logical subject.

The counterpole of Herbart's approach was represented, among others, by Brentano. He maintained that the formula '*S* is' does not refer to the connection of *S* with the verb denoting existence but merely to *S* itself. Erdmann regarded existential propositions as subject propositions (*Subjektsurteile*). The following objection can be made: by not considering 'is' in the proposition 'man is' as a predicate, we degrade it into an unnecessary addition. And if 'is' can be ignored in this case, 'man' is no longer the subject of a proposition but a mere term.

Both Herbart and Erdmann have fallen into the error of blurring the distinction between proposition and term. They speak of proposition but, as the content of their message reveals, they mean some term by it. Relying on this, I find it an error to describe existential propositions as having one component only.

As we have seen, Aristotle interpreted one of the two components of the existential proposition as the subject and the other as the predicate. As opposed to this, Karinski wrote the following:

> The so-called copula expressed by the auxiliary verb 'is' is essentially the assertion of existence. However, it is not the existence of the object that is asserted by it since its existence has already been taken for granted in advance, or has been accepted conditionally. It asserts the existence of a relation of the object to some other reality or phenomenon. If I say 'The whale is a mammal' by the word 'is' I evidently indicate existence just as much as by the proposition 'There are microbes spreading disease'. What the copula of the former proposition states, however, is not the existence of the subject as in the latter proposition—for the existence of the subject is taken for granted here—, but the existence of a certain connection between the object concerned, which is already known to be existing, and a certain process of feeding offspring.[21]

Existential propositions consist of *subject* and *copula* accordingly. But the function of the copula is to express a connection between terms. All

---

[21] Tavanets, *Proposition* p. 48.

that can be said of existential propositions is that there the copula coincides with the predicate.

Karinski's view has been criticized by Tavanets from the following aspect:

> This is a misinterpretation of the nature of the copula because it cannot be applied to existential propositions unless they are regarded as propositions without a predicate. For in this case such propositions as 'There are microbes spreading disease', 'There are alleles', etc., are proved to be without a predicate.
>
> However, by accepting the existence of propositions without a predicate, our fundamental thesis which we have proceeded from, becomes untenable, i.e. that every simple proposition consists of three components, namely subject, predicate, and copula.[22]

The author has given no reason either here or elsewhere why *every* simple proposition *has to have* a three-component structure. He has not pointed out, either, to what extent there is an independent copula in the existential propositions. His fundamental thesis is disproved by the existential propositions themselves.

Mathematical logic examines both existential propositions and terms from the point of view of extension. Accordingly, existential propositions are treated as quantified propositions.

> The so-called *existential quantifier* '$(\exists x)$' corresponds to the words 'there is something x such that'. Application of '$(\exists x)$' to the expression:
>
> (1) x is a book. x is boring
>
> in the fashion:
>
> (2) $(\exists x)(x$ is a book. x is boring$)$
>
> is called *existential quantification* of (1).[23]

---

[22] *Loc. cit.*
[23] Quine, p. 83.

This view will be discussed at length in Chapter 3, Sect. 7 'The Extension of Propositions'.

Existence has been taken as a predicate in the propositions discussed so far. However, it can also be seen as a subject, e.g., 'existence is a category of philosophy'. Or, 'the existence of Pluto has been proved'.

Existence can refer to propositions as well. Let us proceed from a proposition *p*. This designation implies that it can be replaced by propositions only. Now let us consider as an example the proposition 'the Martians like music'. This proposition is indefinite from existential point of view, since—at the present stage of our knowledge—we are unable to decide whether or not such a relation exists.

If the existence of some relation is to be considered, it must be indicated. There are, first of all, two versions: (1) *nep* (it does not exist that *p*), (2) *ep* (it exists that *p*). The basic relations of existentially definite propositions are: (1) *nep* is identical with *nep*, (2) *nep* contradicts *ep*, (3) *ep* contradicts *nep*, (4) *ep* is identical with *ep*. These four relations can be reduced to two: (a) the propositions are compatible in cases (1) and (4); (b) they are incompatible in cases (2) and (3).

As our survey has revealed, unjustly little attention has been paid to existential propositions in the history of logic. If they were treated at all, it was mostly with the intention to trace them back to propositions of some other type. I share Aristotle's view in this question: they constitute another kind of proposition in its own right.

2. THE TRUTH VALUE OF TERMS AND PROPOSITIONS

I

As to the nature of truth, two conflicting views developed in ancient Greece. One of them regarded truth as a reflection. The following statement can be read in one of the fragments by Heraclitus: the greatest virtue is thinking, wisdom consists in telling the truth and in acting always in accordance with nature.

Protagoras, on the other hand, held that man is the measure of all things. Consequently he thought one can make two opposite assertions of all things with equal right, and in fact, there is no truth.

Aristotle rejected this relativistic approach.

> From the same opinion proceeds the doctrine of Protagoras, and both doctrines must be alike true or alike untrue. For on the one hand, if all opinions and appearances are true, all statements must be at the same time true and false. For many men hold beliefs in which they conflict with one another, and think those mistaken who have not the same opinions as themselves; so that the same thing must both be and not be. And on the other hand, if this is so, all opinions must be true; for those who are mistaken and those who are right are opposed to one another in their opinions...[24]

He formulated his opinion in the affirmative as well:

> ... he who thinks the separated to be separated and the combined to be combined has the truth, while he whose thought is in a state contrary to that of the objects is in error. This being so, when is what is called truth or falsity present, and when is it not? We must consider what we mean by these terms. It is not because we think truly that you are pale, that you *are* pale, but because you are pale we who say this have the truth.[25]

Early Stoics still admitted that truth is a reflection. Their successors, however, abandoned this view. It became especially obvious with regard to compound propositions. According to late Stoics, whether they are true or not depends on the value of the simple propositions which constitute them rather than on the content expressed in them.

Skeptics went even further. The founder of the school, Pyrrho, taught that one should abstain from all judgement because any proposition can equally be proved and disproved. The truth of propositions depends on a consensus among men only and not on reality.—Sextus Empiricus pointed out that the basic argument of scepticism to which everything else can be traced back is the principle of relativism. In his view a skeptic

---

[24] Aristotle, *Metaphysica.* 1009a5–13.
[25] *Op. cit.* 1051b3–9.

should not even try to formulate true propositions for they do not exist anyway, but rather he should warn people against the dangers of dogmatism.

Scholasticism at its initial stage approached the issue of truth from a theological point of view. Compared to this, the theory of double truth which developed later was progressive. There are two kinds of truth according to this theory: a theological and a philosophical. It was Occam who went further than anybody else in this field by propounding not only the existence of two kinds of truth but also their potential contradiction to one another. As a conclusion to this doctrine, the position of philosophy may be true even if it contradicts the theological approach. Nevertheless, all the participants in this controversy over the theory of truth were equally far from considering truth as a reflection.

The theory of double truth gradually lost its progressive character in the modern age, and in fact, turned into the opposite. As soon as natural sciences began to develop, this theory ceased to function as an opposition to theology but strove for a compromise between theology and philosophy. At this stage only those propagated double truth who did not dare openly and unequivocally to break with theology.

In the modern age it was, first of all, Bacon who professed the principle of reflection.[26] Descartes, on the other hand, wrote the following about the logical rules established by himself:

> The first was to accept nothing as true which I did not evidently know to be such, that is to say, scrupulously to avoid precipitance and prejudice, and in the judgments I passed to include nothing additional to what had presented itself to my mind so clearly and so distinctly that I could have no occasion for doubting it.[27]

That is, there is no need to compare our knowledge with the external objects, for knowledge itself involves the criteria of its being true. It is sufficient if we are convinced of the truth of our knowledge.[28]

---

[26] Bacon, *Selection* p. 333 (*The New Organon*, 1(XIX)).
[27] Descartes, *Œuvres Choisies.* Tome Premier (Paris) p. 16. Kemp Smith tr. from *Discourse on Method*, Part II. in *Descartes' Philosophical Writings*, p. 129.
[28] Spinoza, p. 70.

Hume confronted the reflection theory by rejecting any connection between knowledge and the external world.

> By what argument can it be proved, that the perceptions of the mind must be caused by external objects, entirely different from them, though resembling them (if that be possible) and could not arise either from the energy of the mind itself, or from the suggestion of some invisible and unknown spirit, or from some other cause still more unknown to us? It is acknowledged, that, in fact, many of these perceptions arise not from anything external, as in dreams, madness, and other diseases. And nothing can be more inexplicable than the manner, in which body should so operate upon mind as ever to convey an image of itself to a substance, supposed of so different, and even contrary a nature.
>
> It is a question of fact, whether the perceptions of the senses be produced by external objects, resembling them: how shall this question be determined? By experience surely; as all other questions of a like nature. But here experience is, and must be entirely silent. The mind has never anything present to it but the perceptions, and cannot possibly reach any experience of their connexion with objects. The supposition of such a connexion is, therefore, without any foundation in reasoning.[29]

As opposed to skeptical and agnostic views, Marxism maintains that both cognition and truth are reflections of reality. Without the existence of objective things, no image could develop in our mind. Cognition, however, depends also on the mind and not on the objective reality only. The state and the development of the mind do affect the evolution of our knowledge.

In other respects cognition and reality are identical and different at the same time. As Marx has pointed out, their identity consists in the fact that the mental is but the material transferred and translated in the human mind. On the other hand, the fact that our knowledge gives only a relatively true image of reality makes their difference apparent.

---

[29] Hume, *Enquiries* pp. 152–153.

II

Aristotle wrote the following about the truth value of terms in his *Metaphysics*:

> A false *account* is the account of non-existent objects, in so far as it is false. Hence every account is false when applied to something other than that of which it is true; e.g. the account of a circle is false when applied to a triangle. In a sense there is one account of each thing, i.e. the account of its essence, but in a sense there are many, since the thing itself and the thing itself with an attribute are in a sense the same, e.g. Socrates and musical Socrates (a false account is not the account of anything, except in a qualified sense). Hence Antisthenes was too simple-minded when he claimed that nothing could be described except by the account proper to it,—one predicate to one subject; from which the conclusion used to be drawn that there could be no contradiction, and almost that there could be no error. But it is possible to describe each thing not only by the account of itself, but also by that of something else. This may be done altogether falsely indeed, but there is also a way in which it may be done truly; e.g. eight may be described as a double number by the use of the definition of two.[30]

In contrast to the above said, the following can be read in the *Hermeneutics*:

> ... truth and falsity imply combination and separation. Nouns and verbs, provided nothing is added, are like thoughts without combination or separation; 'man' and 'white', as isolated terms, are not yet either true or false. In proof of this, consider the word 'goat-stag'. It has significance, but there is no truth or falsity about it, unless 'is' or 'is not' is added, either in the present or in some other tense.[31]

Bacon held the following view:

> The *idols* and false notions which are now in possession of the human understanding, and have taken deep roots therein, not only so beset men's minds that truth can hardly find entrance, but even after entrance obtained, they will again in the very instauration of the sciences meet and trouble us,

[30] Aristotle, *Metaphysica*. 1024b28–39
[31] Aristotle. 1928. 16a11–18

unless men being forewarned of the danger fortify themselves as far as may be against their assaults.[32]

And now let us make a big jump in time. Lenin quoted and complimented Engels's view on this question:

'... From the moment we turn to our own use these objects, according to the qualities we perceive in them, we put to an infallible test the correctness or otherwise of our sense-perceptions. If these perceptions have been wrong, then our estimate of the use to which an object can be turned must also be wrong, and our attempt must fail. But if we succeed in accomplishing our aim, if we find that the object does agree with our idea of it, and does answer the purpose we intended for it, then that is positive proof that our perceptions of it and of its qualities, so far, agree with reality outside ourselves...

Thus, the materialist theory, the theory of the reflection of objects by our mind, is here presented with absolute clarity: things exist outside us. Our perceptions and ideas are their images. Verification of these images, differentiation between true and false images, is given by practice.[33]

How do contemporary Marxist philosophers relate to this view? Let us begin with Schaff's opinion:

This question arises because the classics of Marxism and Leninism describe even terms, or notions, as true or false. This is, however, only an apparent difficulty: the analysis of the text concerned reveals that the classics of Marxism and Leninism use terms, or notions, in a broader sense as propositions, or cognitive thoughts, and accept them as true only if they correspond to objective reality. Thus, in such cases terms, or notions, should be understood not in their narrow but broader sense which coincides with the mental form of cognition, i.e. the proposition. In this interpretation they can, of course, be true or false. Besides, the classics of Marxism and Leninism speak of the truth of notions, or sense-perceptions, in their relation to the objects, that is, whether the given sense-perceptions make the formulation of true propositions possible or not.[34]

Now let us listen to the objections:

[32] Bacon, *Selection* p. 335. (*The New Organon*, I(XXXVIII)).
[33] Lenin, 1952. pp. 105–106.
[34] Schaff, 1955. p. 14.

In our opinion Schaff is mistaken in confining truth (and of course falsehood too) to the proposition. Truth is primarily an epistemological category. It is a property of all ideal forms of reflection in as much as they convey a picture which corresponds to objective reality. Truth is "the correct reflection of the external world in the consciousness of humans..." We believe that Schaff's mistake consists in reducing truth to *one* of the reflection forms. Objectively speaking, we are dealing here (just as with Kant) with an unjustified separation of the sensual and the rational levels of knowledge, which credits only the rational level of knowledge—and furthermore within this rational level only one of its forms, namely the proposition—with truth. But in this way the unified process of knowledge is torn apart, and sensual knowledge as well as concepts are denied in principle the ability to convey truth, that is, a correct picture of reality. This is the standpoint of rationalism with all of its one-sidedness for epistemology.[35]

It might be of interest to see another argument for and against respectively. Klaus argued as follows:

When dealing with propositions, we mentioned that truth and falsehood are attributed only to propositions. Does the problem of truth not apply to concepts? Of course the problem exists, but not in the sense that we can ascribe truth or falsehood to the concepts themselves. We cannot even without further ado assert that concepts are reflections or pictures of reality, for some of the examples we have used have shown us that there are concepts to which nothing in reality corresponds.[36]

Then, the author proceeds to the examination of the relation between extension and intension and concludes:

Figuratively speaking, we can denote a term as true, if it has an extension, and as false if this is not the case. We can also speak of a term as picturing something or as picturing nothing. In both cases however we are dealing with incorrect modes of expression, which we want to avoid completely from now on. It is more correct to speak of terms which are directed towards reality and of those for which this is not the case. Human

---

[35] Händel–Kneist, p. 28.
[36] Klaus, p. 145.

imagination can produce terms at will, sensible ones as well as senseless ones.[37]

Kondakov, having quoted the above opinion, made the following comment:

> Such an interpretation of notions, or terms, gives the impression that they are pure products of thought which merely 'are, or are not orientated' to reality. Then, however, the question arises where they have come from. As to the assertion that imagination is able to create notions which have no sense, let me stress that they are usually reflections of reality but in a distorted form.[38]

Tavanets has written the following critical remarks on Klaus's approach:

> Some adherents to the view that notions, or terms, cannot be said to be true or false maintain that terms can never be treated as reflections of reality since there are terms (e.g. 'god', etc.) to which nothing corresponds in reality ... This reasoning is not satisfactory. There are not only terms but also propositions to which nothing corresponds in reality, and yet we find it possible to speak of propositions as reflections of reality and also to describe this or that proposition as true or false ... The fact that there are terms, or notions, to which nothing corresponds in reality does not disprove but, on the contrary, supports the thesis that the characterization of 'being true or false' can be applied to terms as well. Thus, if some maintain that the object of such a term as 'god' exists in reality while in fact it does not, then the term 'god' whose intension includes the property 'it exists in reality' cannot be described as anything but false.[39]

Marxist logicians generally agree that terms are reflections. It is with regard to the truth value of terms that their opinions vary. For my part, I agree with the above argument and believe that terms, just as much as propositions, can be true or false.

---

[37] *Op. cit.*, pp. 146–147.
[38] Kondakov p. 272.
[39] Tavanets, *On the truth value of terms* Voprosy filosofii 1959, No. 12.

III

What is, after all, the criterion for a term to be true or false? Kondakov assumed the following position:

> If the intension of a term rightly reflects the essential properties of the objects of the objective world, the term concerned is true. If, however, the intension of a term does not correspond to reality this term is false.[40]

What is the relation between the true and the false, on the one hand, and between the existent and the non-existent, on the other? Aristotle expressed the following views at various places:

> Again, 'being' and 'is' mean that a statement is true, 'not being' that it is not true but false...[41]
>
> But since that which *is* in the sense of being true, or *is not* in the sense of being false, depends on combination and separation...[42]
>
> As regards the '*being*' that answers to truth and the 'non-being' that answers to falsity, in one case there is truth if the subject and the attribute are really combined, and falsity if they are not combined; in the other case, if the object is existent it exists in a particular way, and if it does not exist in this way it does not exist at all.[43]

Elsewhere, however, the following can be read:

> To say of what is that it is not, or of what is not that it is, is false, while to say of what is that it is, and of what is not that it is not, is true; so that he who says of anything that it is, or that it is not, will say either what is true or what is false...[44]

The first three quotations, as I interpret them, imply that, e.g., if *something* does not exist, it is false. The fourth quotation, however, suggests that if something *is said* to be non-existent, this statement can be either true or false. In my view the latter holds true and the former does not, for what does not exist can also be true or false.

---

[40] Kondakov, p. 351.
[41] Aristotle, *Metaphysica*. 1017a32–33.
[42] *Op. cit.*, 1027b19–21.
[43] *Op. cit.*, 1051b33–35.
[44] *Op. cit.*, 1011b26–30.

It is, of course, different to evaluate something or an assertion about something. The reason why the former can be true or false is not because the latter implies this possibility. In order to give the grounds for my statement of the former, it is necessary to define the criterion for the evaluation of terms.

What is needed for that has been known in logic for a long time. The only reason it has not been used for the purpose indicated is that even the possibility of evaluating terms was denied. Let us start from Fogarasi's arguments:

> ι The elements, attributes, traits of an object, which differentiate it from other objects are called the properties of the object in logic. To the properties of the object correspond, as components of the term, the properties or characteristics of the term. The usual logical terminology does not differentiate precisely between properties and characteristics. We suggest, in accordance with the standpoint of materialism, that one ought to differentiate terminologically between the properties of the objects and the characteristics of the terms.[45]

Tavanets was right in pointing out in his article that the criterion for the evaluation of terms is to be found in the relation of the properties we attribute to things and their actual properties. If the properties we attach to a term are in accordance with the properties of the given object, the term concerned is true, and if there is no correspondence it is false (like, e.g., Bacon's idols).

A false term may result from the mixing up, misuse, misinterpretation, etc. of terms. Its typical manifestations are the following: the use of the same term in several meanings, the identification of different terms, the unjustified extension of the validity of a term, etc. The term 'element' is, for example, false when applied to bronze.

As we have seen above, existential terms may be not only true but also false. The value of some term depends on its correspondence to the object expressed by it and not on the fact whether it refers to existent things or not. This criterion being accepted, it is not difficult to recognize that terms referring to the non-existent can also be true. The term 'mermaid', for

---

[45] Fogarasi, p. 122.

3*

example, is true if applied to the imaginary creatures denoted by this word. It is false, however, if applied, say, to goblins.

It is no accident that those who regard the non-existent as false illustrate their arguments mainly with mythological creatures. Indeed, they are 'felt' to be false under all circumstances even if eventually they also have evolved on the basis of reality. Yet the imaginary number, the rhomboid, etc. are just as fictive. Should they also be described as false?

Let us compare Quine's position with what has been said above.

> It is the peculiarity of a statement to be true or false. It is the peculiarity of a term, on the other hand, to be *true of* many objects, or one, or none, and false of the rest. The term 'Greek' is true of each Greek, and the term 'wicked' is true of each wicked individual, and nothing else. The term 'satellite of the earth' is true of each satellite of the earth and nothing else, hence true of but one object, the moon. The term 'centaur' is true of each centaur and nothing else, hence true of nothing at all, there being no centaurs.
>
> In place of the clumsy phrase 'is true of' we may also say 'denotes', in the best sense of this rapidly deteriorating word. But I prefer here to resist the temptation of good usage. 'Denotes' is so current in the sense of 'designates', or 'names', that its use in connection, say, with the word 'wicked' would cause readers to look beyond the wicked people to some unique entity, a quality of wickedness or a class of the wicked, as named object. The phrase 'is true of' is less open to misunderstanding; clearly 'wicked' is true not of the quality of wickedness, nor of the class of wicked persons, but of each wicked person individually.[46]

This approach fundamentally differs from what I consider correct. Still, the above quotation could be turned back from being upside down to up again with little modification. First of all, the expression 'is true of' should be understood as reflection. Then, it would become obvious at once that there is nothing like the Great Wall of China between terms and propositions for the latter are also true *of* a certain connection.

---

[46] Quine, p. 65.

IV

As we have seen, whether or not terms have a value is still an issue under discussion. The value of propositions, on the contrary, is unequivocally accepted. Aristotle established this view, which has remained dominant ever since:

> Every sentence has meaning, not as being the natural means by which a physical faculty is realized, but, as we have said, by convention. Yet every sentence is not a proposition; only such are propositions as have in them either truth or falsity. Thus a prayer is a sentence, but is neither true nor false.[47]

Aristotle abandoned this view but once:

> Everything must either be or not be, whether in the present or in the future, but it is not always possible to distinguish and state determinately which of these alternatives must necessarily come about. Let me illustrate. A sea-fight must either take place tomorrow or not, but it is not necessary that it should take place to-morrow, neither is it necessary that it should not take place, yet it is necessary that it either should or should not take place to-morrow. Since propositions correspond with facts, it is evident that when in future events there is a real alternative, and a potentiality in contrary directions, the corresponding affirmation and denial have the same character.[48]

That is, the disjunctive proposition at issue is necessarily true but none of the alternatives is definitely true.

Thus, the value of certain propositions can be stated definitely while that of others cannot. The latter are denoted by several expressions in logic. They say, for example, that the proposition 'It will rain tomorrow' can be false and can be true. Or let us take immediate inferences. If one of the propositions is true in a subcontrary relation, the value of the other cannot be decided. To make the terminology uniform, from now on I shall speak of propositions of indefinite and definite value respectively and shall designate the former by the distinctive sign t/f.—Since the present work is

[47] Aristotle. 1928. 17a1–5.
[48] Op. cit., 19a28–35.

confined to two-valued logic, there is no need here to discuss approaches to polyvalence.

To decide whether indefinite propositions have a value or not is a typical borderline case. Opinions vary on this question. Aristotle and others answered 'yes'. For those, however, who did not admit the existence of indefinite propositions this question had no sense at all.

Indefinite propositions, at least in this formulation, are not admitted in mathematical logic, either. Let us put the question like this: has the open formula a value or not? Quine wrote the following:

> Open sentences are neither true nor false, but they may, like terms ..., be said to be *true of* and *false of* various objects. The open sentence 'x is a book' may, like the term 'book' itself, be said to be true of each book and false of everything else; and 'x is a book. x is boring' may be said to be true of each boring book and false of everything else. '$x = x$' and 'x is a man $\supset x$ is mortal' are true of everything. In general, to say that an open sentence is true of a given object is to say that the open sentence becomes a true statement when 'x' is reinterpreted as a name of that object.[49]

Accordingly, if I understand it right, open formulas have a potential value. This interpretation is supported by the following comparison.

> What have propositions and logical functions in common and what is the difference between them?
>
> They have in common the potential to be true or false, i.e. their logical value can be t or f. The logical value of one part of the propositions is t, and that of the rest is f. The logical values of logical functions can vary according to the values of their variables but the choice is not bigger: the only alternatives are the values t and f.
>
> And what can the values of the variables be?
>
> The elements of any not-empty set.[50]

Relying on these findings, I think that indefinite propositions also have a value.

---

[49] Quine, pp. 90–91.
[50] Varga, pp. 151–152.

Aristotle identified the logical evaluation of propositions with inference. This operation deduces the truth of a proposition from one or more other propositions. The Stoics founded their investigations on compound propositions, and decided their values not by inference but according to the values of the simple propositions they are composed of. However, to state the value of simple propositions, they also had recourse to inference.

What is the difference between these two methods? Let us take as an example the formula 'if *p,* then *q*'. The Aristotelian method defines the value of the *consequent* by the help of the antecedent as an *external* means, while the Stoic method evaluates the *proposition as a whole*, based on the *internal* factors. What both operations have in common is that a simple proposition in itself cannot be evaluated by them. On the one hand, inference would require at least one more proposition. On the other, to apply the Stoic method of evaluation, the proposition under discussion ought to be broken down into simple ones.

These views survived into, and were expounded at great length in the that neither true nor false propositions can be made about things and their relations because they are noncognizable. The right thing to do is to refrain from any judgement.

These views survived into, and were expounded at great length in, the Middle Ages. As is well-known, the Aristotelian approach prevailed.

Bacon, on the contrary, was of the following opinion:

> There are and can be only two ways of searching into and discovering truth. The one flies from the senses and particulars to the most general axioms, and from these principles, the truth of which it takes for settled and immovable, proceeds to judgement and to the discovery of middle axioms. And this way is now in fashion. The other derives axioms from the senses and particulars, rising by a gradual and unbroken ascent, so that it arrives at the most general axioms last of all. This is the true way, but as yet untried.[51]

Descartes, in turn, proposed this solution:

> Now by method I intend to signify rules which are certain and easy and such that whosoever will observe them accurately will never assume what is

[51] Bacon, *Selection* p. 333 (*The New Organon,* I(XIX)).

false as true, or uselessly waste his mental efforts, but gradually and steadily advancing in knowledge will attain to a true understanding of all those things which lie within his powers. . . .

But if our method rightly explains how by making use of intuition we can avoid falling into the contrary error, and how by way of deduction we can reach to a knowledge that takes in all things, nothing else, it seems to me, is needed to render our knowledge complete, since, as already remarked, no knowledge can be acquired save by way either of intuition or of deduction. There can be no question of extending the method so as to show how these two operations ought to be performed, since they are the simplest of all mental operations, and primary. If our understanding were not of itself qualified to perform them, it would be unable to comprehend any of the precepts prescribed by the method, however easy. As to those other (syllogistic) mental operations, which dialectic (relying on the aid of these prior ones) labours to direct, they are here useless or rather positively harmful; for nothing can be added to the pure light of reason which does not in some way obscure it.[52]

Although the novelty of both Bacon's and Descartes's ideas goes unchallenged, their reflections proved to be unsatisfactory at least in two respects. On the one hand, both of them were one-sided, the former overemphasizing empiricism, the latter rationalism. On the other, they failed to make their ideas concrete with adequate methods.

These shortcomings must have played a role in making the following approach prevail in traditional logic:

Many logicians have also divided propositions according as they are *true* or *false*, and it might well seem to be a distinction of importance. Nevertheless, it is wholly beyond the province of the logician to consider whether a proposition is true or not in itself; all that he has to determine is the comparative truth of propositions—that is, whether one proposition is true when another is. Strictly speaking, logic has nothing to do with a proposition by itself; it is only in converting or transmuting certain propositions into certain others that the work of reasoning consists, and the truth of the conclusion is only so far in question as it follows from the truth of what we shall call the premises. It is the duty of the special sciences each in its own sphere to determine what are true propositions and what are false,

---

[52] Descartes, *Œuvres* pp. 16–17; Kemp Smith tr. pp. 15–16.

and logic would be but another name for the whole of knowledge could it take this duty on itself.[53]

Mathematical logic has preserved the basic principle of the Stoics, even if in a thoroughly elaborated and advanced form:

> Logic, like every science, lays claim to truth, but to a truth essentially different from that of a particular scientific discipline. Truth itself is the subject matter of logic, which accordingly is the study of the most general laws and relations of truth. This definition should not be misunderstood. The investigation of the truth or falsehood of the proposition, 'This ceiling is white' is not the task of logic. The investigation of such propositions is the task of particular sciences. Logic deals rather with the most general relations of truth, which are independent of the concrete object. For instance, the logical inference: if all S are M, and if all M are P, then all S are P, is one such most general truth relation.[54]

A few remarks are still to be made. To say only that mathematical logic deals with the determination of the value of compound propositions might lead to misunderstanding. Let me add, on the one hand, that not all compound propositions are treated therein (e.g. 'p because q'), and on the other, certain simple propositions are evaluated, too (e.g., propositions 'All S is P' are true if S is an empty class).

In the following I shall try to demonstrate that:

(1)   simple propositions can also be evaluated;
(2)   the criteria for the evaluation of compound propositions are questionable;
(3)   both simple and compound propositions can be evaluated by a uniform method;
(4)   in many cases it is more reasonable to use evaluation than the usual inferences.

[53] Jevons, p. 70.
[54] Klaus, p. 9.

V

Existential propositions can be evaluated according to the diagrams of
Figure 2:

TABLE 1

	(1)	(2)
(1) there is S	f	t
(2) there is no S	t	f

The pairing of S and P leads to the following combinations:

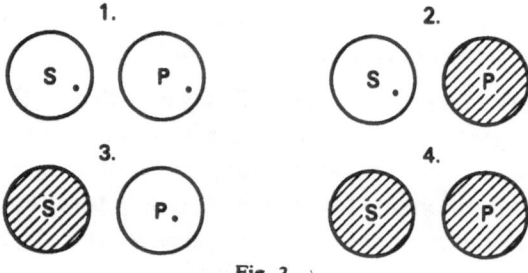

Fig. 3.

These combinations help to evaluate the following interrelations of
existential propositions:

TABLE 2

		(1)	(2)	(3)	(4)
(1)	there is S or there is no S	t	t	t	t
(2)	there is S or there is P	t	t	t	f
(3)	there is S or there is no P	t	t	f	t
(4)	there is S	t	t	f	f
(5)	there is no S or there is P	t	f	t	t
(6)	there is P	t	f	t	f
(7)	there is S and there is P or there is no S and there is no P	t	f	f	t
(8)	there is S and there is P	t	f	f	f
(9)	there is no S or there is no P	f	t	t	t

		(1)	(2)	(3)	(4)
(10)	there is no S and there is P or there is S and there is no P	f	t	t	f
(11)	there is no P	f	t	f	t
(12)	there is S and there is no P	f	t	f	f
(13)	there is no S	f	f	t	t
(14)	there is no S and there is P	f	f	t	f
(15)	there is no S and there is no P	f	f	f	t
(16)	there is no S and there is S	f	f	f	f

I shall base the evaluation of all propositions with the structure 'S is P' on the relations of the above terms. Traditional logic usually distinguished the following kinds of term:

(1)   Terms whose extension coincides are identical, e.g. pig and swine.

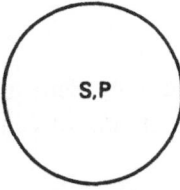

Fig. 4.

(2)   The relation of inferior and superior terms is characterized by the fact that the extension of the former is included in, but does not amount to, that of the latter, e.g. iron—metal.

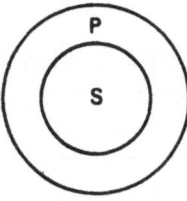

Fig. 5.

(3)   Terms whose extension is partially identical are in a crossing relation, e.g. table—brown.

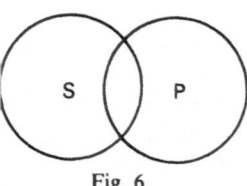

Fig. 6.

(4)   Co-ordinated terms are typically incompatible with one
        another within the same genus, e.g. slave society—feudal
        society.

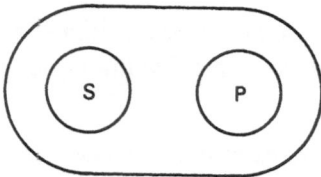

Fig. 7.

(5)   Terms which mutually deny the whole extension of one
        another are in a relation of contradiction, e.g. guilty—not-
        guilty.

Fig. 8.

(6)   Terms which differ most within the same genus are opposites,
        e.g. guilty—innocent.

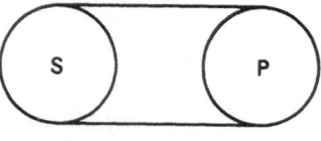

Fig. 9.

Euler's diagrams above are empty—not so much to illustrate how uncertain and indefinite our knowledge is but rather because it was taken for granted in traditional logic that they stand for existential terms.

As a rule, the above relations were divided into two groups: the first three relations were considered as compatible, and the latter three as incompatible. The negations of these relations were usually not touched upon, with the exception of nonidentity used in a sense synonymous with difference.—The relations discussed above imply the following operations: identification, sub- and superordination, crossing, co-ordination, contradiction, and contrariety.

Gergonne assumed a position different from the usual approach in his work *Essai de dialectique rationnelle*. He distinguished five basic relations between terms: identity, inferiority, superiority, crossing, and exclusion. These relations can be described as follows by combining propositions:

(1)   All A is B, and all B is A.
(2)   All A is B, but not all B is A.
(3)   All B is A, but some A is not B.
(4)   Some B is not A, or some A is not B.
(5)   No A is B, and no B is A.

That is to say, he made the following modifications as compared to the traditional approach: he regarded inferiority and superiority as independent term relations while he described co-ordination, contradiction, and contrariety as special cases of exclusion which are at a level different from that of the other term relations.

Mathematical logicians who take the empty classes into consideration as well have added three more to the above five relations. Venn illustrated all these relations as follows:

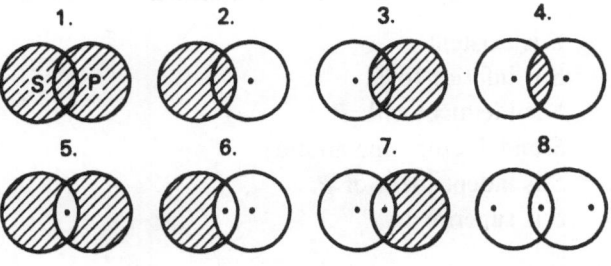

Fig. 10.

I interpret the relations illustrated above as follows:

(1)   *Sne, Pne.*
(2)   *Sne, Pe.*
(3)   *Se, Pne.*
(4)   *Se* is alien to *Pe.*
(5)   *Se* is identical with *Pe.*
(6)   *Se* is inferior to *Pe.*
(7)   *Se* is superior to *Pe.*
(8)   *Se* and *Pe* cross one another.

Relying on these findings, some have come to the following conclusion:

> Our presentation omits the discussion of the so-called *contrary* opposition of terms. In traditional logic two terms are called contrary if the greatest difference exists between them within the framework of a higher term. Some examples of contrary terms are: good and bad, empty and full, etc. Contrary terms are thus formed by dividing up a class of things into subclasses in such a way that the subclasses are arranged in a series according to a particular ordering principle. We then call the two terms which occupy the end positions of the row, opposite terms. Since the respective ordering principle is *intensional*, the contrary opposition has no place in formal logic. There are many diverging opinions on the adequate characterization of the contrary contradiction.[55]

If we are confined to the diagrams in Figure 10, it is disputable whether the contrary is really illustrated. Let us, however, make the following restriction: *S* refers to something existent and is not identical with the universe. The following diagrams correspond to the above restriction (see Figure 11):

These diagrams can be interpreted as follows:

(1)   *P* is existent.
(2)   *S* is inferior to *P*.
(3)   *S* is identical with *P*.
(4)   *S* and *P* cross one another.
(5)   *S* is independent of *P*.
(6)   *S* is superior to *P*.

[55] *Op. cit.*, p. 159.

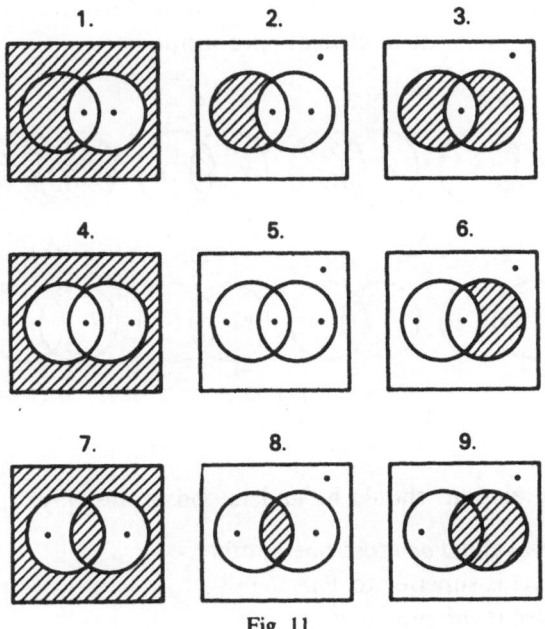

Fig. 11.

(7)   *S* contradicts *P*.

(8)   *S* is contrary to *P*.

(9)   *P* is non-existent.

All the fourth diagram in Figure 10 allows us to state is that *S* is alien to *P* (it illustrates the contrary but only implicitly). If, however, the universe is also considered, it becomes possible to tell contradictory and contrary apart. In this way the relation of contrariety need not be excluded from formal logic but must be put into its own place within it. The components of this relation typically exclude one another but allow for the possibility of some third component.

The interpretation of Figure 10 reveals that the relations of non-existential terms, as opposed to existential ones, are not completely illustrated. Since this would lead to confusion in the interpretation of propositions, it is reasonable to do away with it at this early stage. Emptiness will be denoted by *y*. Three factors (existent, non-existent, empty)[56] may have no more than 27 variations. However, the following 12 are already sufficient to evaluate propositions:

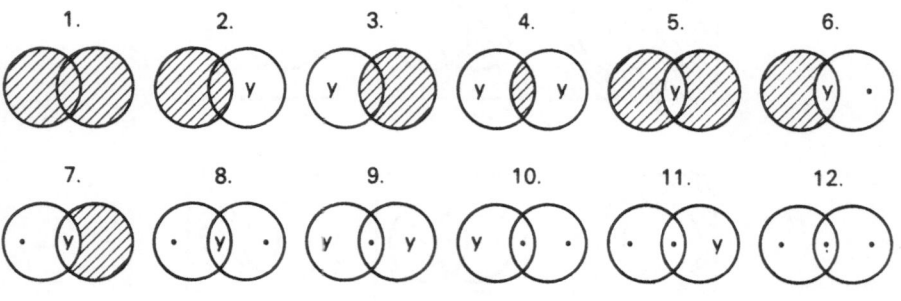

Fig. 12.

The above relations should be understood in this way:

(1)   *Sne* and *Pne* cross one another.
(2)   *Sne* is superior to *Pne*.
(3)   *Sne* is inferior to *Pne*.
(4)   *Sne* is identical with *Pne*.
(5)   *Sne* is alien to *Pne*.
(6)   *Sne* is alien to *Pe*.
(7)   *Se* is alien to *Pne*.
(8)   *Se* is alien to *Pe*.
(9)   *Se* is identical with *Pe*.
(10)   *Se* is inferior to *Pe*.
(11)   *Se* is superior to *Pe*.
(12)   *Se* and *Pe* cross one another.

[56] The distinction between non-existent and empty will be explained later.

3. THE QUALITY OF TERMS
AND PROPOSITIONS

I

Terms may be positive and negative. Aristotle raised the following question in this respect:

> ...the expressions 'it is a not-white log' and 'it is not a white log' do not imply one another's truth. For if 'it is a not-white log', it must be a log: but that which is not a white log need not be a log at all. Therefore it is clear that 'it is not-good' is not the denial of 'it is good'. If then every single statement may truly be said to be either an affirmation or a negation, if it is not a negation clearly it must in a sense be an affirmation. But every affirmation has a corresponding negation. The negation then of 'it is not-good' is 'it is not not-good'. The relation of these statements to one another is as follows. Let A stand for 'to be good', B for 'not to be good', let C stand for 'to be not-good' and be placed under B, and let D stand for 'not to be not-good' and be placed under A. Then either A or B will belong to everything, but they will never belong to the same thing; and either C or D will belong to everything, but they will never belong to the same thing. And B must belong to everything to which C belongs. For if it is true to say 'it is not-white', it is true also to say 'it is not white': for it is impossible that a thing should simultaneously be white and be not-white, or be a not-white log and be a white log; consequently if the affirmation does not belong, the denial must belong. But C does not always belong to B: for what is not a log at all, cannot be a not-white log either. On the other hand D belongs to everything to which A belongs. For either C or D belongs to everything to which A belongs. But since a thing cannot be simultaneously not-white and white, D must belong to everything to which A belongs. For of that which is white it is true to say that it is not not-white. But A is not true of all D. For of that which is not a log at all it is not true to say A, viz. that it is a white log. Consequently D is true, but A is not true, i.e. that it is a white log. It is clear also that A and C cannot together belong to the same thing, and that B and D may possibly belong to the same thing.[57]

---

[57] Aristotle. 1928. 51b28–52a14.

4

Let us illustrate, first of all, the example of the log:

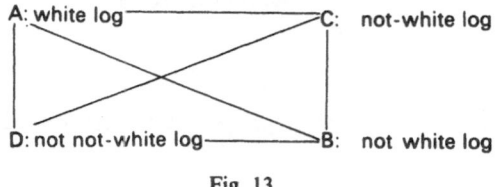

Fig. 13.

Aristotle established the following interrelations: [58]

(1)   *A* and *B* contradict one another.
(2)   *C* and *D* contradict one another.
(3)   'if *C*, then *B*' does not hold true the other way round.
(4)   'if *A*, then *D*' does not hold true the other way round.
(5)   *A* and *C* are contrary to one another.
(6)   *B* and *D* are compatible.

The truth of the first two relations requires no further explanation (e.g.,
The birch is a white log—The birch is not a white log). As to the third one:
The pine is a not-white log → The pine is not a white log. But: The pine is
not a white grass (*t*).—The pine is a not-white grass (*f*). That is, Aristotle
meant by a not-white log that, e.g. in the case of a pine, its whiteness, and
not its being a log, is negated. The third and the fourth relations mean
basically the same.

Why did Aristotle regard the relation of *A* and *C* as contrary? Let us see
a commentary on this question:

> In Aristotle's view the affirmation that something is *not-equal*, or *not-grateful*, or *not-peaceful*, presupposes the existence of something unequal,
> or ungrateful, or unpeaceful respectively. The affirmation that something is
> not equal, not grateful, not peaceful does not, however, presuppose
> anything like this as he saw it. Accordingly, we could assert, for example, of
> a *cliff* that it is not grateful for it, indeed, does not have the attribute of
> gratefulness, but it should not be said that it is not-grateful because it would
> imply that it belongs to the things which have the attribute of ungrateful-
> ness, and this is obviously not true. According to the above view, '*S* est *P*'

---

[58] *Op. cit.*, 55b2–56b2.

and 'S non est P' are contradictory, and the propositions 'S est P' and 'S est non-P' are contrary, i.e. with any S and P in mind, one of the two components of the first pair of propositions should be true while even both components of the second pair could be false.[59]

I would like to add the following: Aristotle proceeded from the fact that B and C differ from one another. If B is a negation of A, then C cannot be the negation of A because an affirmation has but one negation. He did not deny, however, that C is opposed to A. Upon these considerations he found the explanation that C is contrary to A reasonable enough.

Can this problem be solved without giving up the distinction between B and C? As Aristotle interpreted it, B may belong to anything except white logs. C implies that the universe discussed consists of logs only, with the white ones excluded. The difference between B and C arises from the choice of the universe to be discussed. B contradicts A in a universe without limitations while C does so within a limited universe.

In the foregoing our example was a compound term ('white log'), but simple terms behave in the same way, too. Not white is to be understood as anything except white. Not-white, in turn, implies the exclusion of white from an adequate limited universe (e.g. colour). Thus, Aristotle was right in asserting that there is a difference between not good and not-good.

Theophrastus is the first to be mentioned among the post-Aristotelian logicians. He maintained that e.g. 'not white log' and 'not-white log' mean the same thing because the extension of both terms covers anything but white logs. Following this basic principle, he modified cases (3)–(6) in the following way:

(3)   B and C are equipollent.
(4)   A and D are equipollent.
(5)   A and C are contradictory.
(6)   B and D are contradictory.

In cases (5) and (6) the relation of contradiction is stated by postulating the equipollence of, rather than a distinction between, B and C.

It was Theophrastus's approach which came to prevail in the history of logic. Let us see one of its manifestations in modern times:

[59] Aristotelés, *Organon.* (Budapest 1961). p. 428.

4*

The fourth principal division of names, is into *positive* and *negative*. Positive, as *man, tree, good*; negative, as *not-man, not-tree, not-good*. To every positive concrete name, a corresponding negative one might be framed. After giving a name to any one thing, or to any plurality of things, we might create a second name which should be a name of all things whatever, except that particular thing or things. These negative names are employed whenever we have occasion to speak collectively of all things other than some thing or class of things. When the positive name is connotative, the corresponding negative name is connotative likewise; but in a peculiar way, connoting not the presence but the absence of an attribute. Thus, not-white denotes all things whatever except white things; and connotes the attribute of not possessing whiteness.[60]

Some, however, reckoned with a limited universe as well:

There is a *relation of contradiction* between two terms if one of them generalizes things, occurrences, and relations which have a definite (A) essential attribute in common while those which lack this essential attribute (not-A) belong to the other. For example, 'working' and 'not-working', 'progressive' and 'not-progressive', 'superstructure phenomenon' and 'not-superstructure phenomenon', etc. are in such a relation. Terms in contradiction can be described also as follows: two terms contradict one another if one of them generalizes a certain group of things included in the extension of a general term while the other generalizes all the rest of things which belong to this general term.[61]

Accordingly, traditional logic illustrated it as follows:

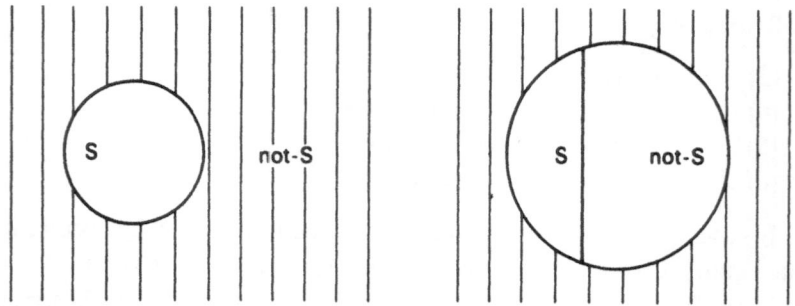

Fig. 14.

[60] Mill, p. 41.
[61] *Logika* (Budapest 1956). p. 101.

Mathematical logic interprets the terms concerned in this way: if the given term is denoted by '*A*', its logical negation is a complementary term denoted by '*A*'. Complementary terms completely exclude one another. In this formulation the designation 'complementary' does not conform to the interpretation 'exclusive'. In my opinion it should be put as follows: the terms at issue are contradictory to one another (they exclude one another) but complementary in their relation to the universe (they complement each other).

Finally, let us examine Fogarasi's position:

> The negative terms have no independent meaning, but they are not nonsensical as the anti-dialectically oriented traditional logic and identity metaphysics maintain; rather they must always be interpreted in relation to a positive term. *Non-Euclidean geometry* taken in a purely formal sense can refer to various geometries, but in the development of science its concrete sense was the absolute geometry founded by Bolyai and Lobachevsky and developed by Riemann.[62]

Fogarasi was right so far that negative terms should not be assessed in an abstract way only. One should strive to reveal their concrete meaning as well. Such negative terms, however, must be used as complementary to, and not instead of, or at the expense of, abstract negative terms.

The following figure illustrates the relation of the existentially indefinite positive and negative terms:

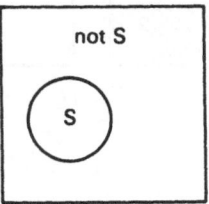

Fig. 15.

The relation of *S* and not-*S* is a special case of '*S* is alien to *P*', and therefore it can be illustrated by diagram 4. The relation of 'not *S*' and 'not-*S*', in turn, can be represented as follows:

---

[62] Fogarasi, pp. 147–148.

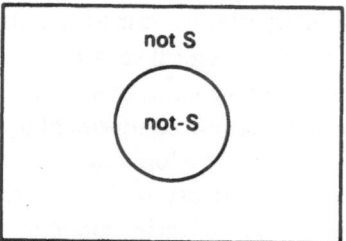

Fig. 16.

Bearing existence in mind, one arrives at the following variations:

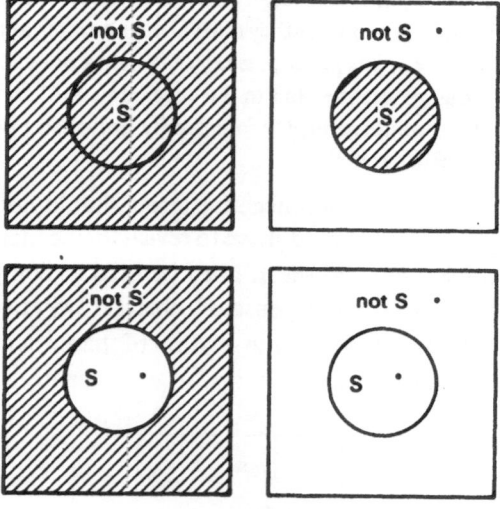

Fig. 17.

II

This is how Aristotle defined the quality of propositions: "An affirmation is a positive assertion of something about something, a denial a negative assertion."[63] This definition covers only such propositions as '*S* is *P*' and '*S* is not *P*'.

[63] Aristotle. 1928 17a25–27.

In order to go beyond this narrow scope, first of all, the relation between reality and the quality of propositions has to be made clear. Aristotle wrote about it the following way:

> For it is manifest that the circumstances are not influenced by the fact of an affirmation or denial on the part of anyone. For events will not take place or fail to take place because it was stated that they would or would not take place, nor is this any more the case if the prediction dates back ten thousand years or any other space of time.[64]

In his view speech is the expression of thought, and thought refers to things.

> Now if the spoken word corresponds with the judgement of the mind, and if, in thought, that judgement is contrary of another, which pronounces a contrary fact, in the way, for instance, in which the judgement 'every man is just' pronounces a contrary to that pronounced by the judgement 'every man is unjust', the same must hold good with regard to spoken affirmations.[65]

Theophrastus and Eudemus abandoned this view. They ceased to relate the affirmative and negative propositions to reality and studied them as purely formal relations. Negative propositions were seen as negations of some affirmative *proposition* rather than as assertions about the lack of some relation. They separated logical investigation from what was called metaphysics and tried to relate it to grammar.

The Stoics went even further on this way. Having quoted the Stoics' approach to this issue, Bochenski made the following remark:

> This text shows something to which many passages bear witness, that the Stoics constructed their logic not merely formally, but quite formalistically. This was blamed by Apuleius and Galen, who said that the Stoics were only interested in linguistic form.[66]

The Stoic view had a great revival in mathematical logic:

---

[64] *Op. cit.*, 18b39–19a1.
[65] *Op. cit.*, 23a32–35.
[66] Bochenski, p. 133; Ivo Thomas tr., p. 116.

The peculiarity of *statements* which sets them apart from other linguistic forms is that they admit of truth and falsity, and may hence be significantly affirmed or denied. To deny a statement is to affirm another statement, known as the *negation* or *contradictory* of the first.[67]

Fogarasi took the following position:

Affirmative and negative propositions make reference to reality. That is their original and essential function. An affirmative proposition expresses a situation which exists in reality. A negative proposition expresses the *non-existence* of a situation in reality. Affirmative and negative propositions not only express something about *reality*, they *affirm* or *negate* something about reality. Thus they are activities of consciousness, special activities of thought. The idealistic theory of the proposition interprets this fact falsely. According to idealistic logic it is merely *connections of representations* (or connections of concepts) about which affirmative or negative propositions affirm or negate something. This conception completely distorts the sense of the proposition. Let us take an example. Radium is a radioactive element. According to idealistic logic I affirm in this proposition the connection of the notion of radium with the notion (or concept) of a radioactive element. *Natural science* does not however maintain about the notion of radium that it is a 'radioactive-element-notion'; rather it asserts about radium, a part of the objective world, that radioactivity is its objective property independent of any representation.

In certain cases I do in fact make an assertion about the *concept* of radium instead of about radium itself: for instance, when I deal not with physics but with concept formation in physics, with logic. In the same way with an affirmative proposition we *emphasize* in certain cases for particular reasons the aspect of affirmation, of conscious affirmation. When Galileo ascertained that the earth moves, he expressed an objective situation with an affirmative proposition. But when he said: '*It does so move*', he meant: I *assert* and insist that the earth moves.[68]

I agree with the above criticism of the idealistic view but I do not share Fogarasi's opinion of how to establish an objective basis. Let us first see the definition of affirmative propositions as the expressions of some relation which exists in reality. An example to the contrary is: furies have

<hr />

[67] Quine, p. 1.
[68] Fogarasi, pp. 196–197.

snake-like hair. It is an affirmative proposition but, as far as I know, there is no such relation objectively.—The second definition is: negative propositions suggest that a certain relation does not exist in reality. An example to the contrary is: it is not raining outside. This relation exists in reality, and yet is expressed by a negative proposition. That is to say, negative propositions can just as well refer to what exists in reality as to what does not.

If propositions expressed real relations only, materialist philosophy would have an easy job. In fact, however, they express fictive, often even fantastic, relations as well. What is their objective basis? It obviously cannot be some concrete real relation but only a relation in general which objectively exists. The proposition 'Furies have snake-like hair' can be made only because there is a relation at all. Relations as such can only be reflected by man while certain concrete particular ones can even be created by him. The fact that two things (e.g. table and island) are not related can be generalized, too. Consequently, the objective basis of affirmative propositions is a relation in general while that of negative ones is a nonrelation.

According to the prevailing view:

> Every proposition (let it be about things or their relations) affirms or denies something of something.[69]

Let us denote any one proposition as usual by p. Since any proposition can be substituted for this designation, p is indefinite as far as the quality of the proposition is concerned.

An objection to what has been said above can be the following: if p is replaced by some proposition, it must be either affirmative, or negative. The example 'If I walk in the street, I see people' is evidence to the contrary. It is, no doubt, an affirmative proposition. At the same time it is a compound proposition, that is, it can be broken down into simple ones. And it is generally accepted that the antecedent in the above example, which should be regarded as a proposition, is neither affirmed nor denied. Thus, in a certain context the quality of a proposition can be indefinite.

[69] *Logika*. (Budapest 1956). p. 139.

The simplest form of affirmation and negation respectively is the existential proposition: (1) there is S, (2) there is no S. The latter can also be described as: there is not S. We get two more versions by including the negative term: (3) there is not-S, (4) there is no not-S. Aristotle commented on these combinations as follows:

> Every affirmation, then, and every denial, will consist of a noun and a verb, either definite or indefinite. There can be no affirmation or denial without a verb; for the expression 'is', 'will be', 'was', 'is coming to be', and the like are verbs according to our definition, since besides their specific meaning they convey the notion of time. Thus the primary affirmation and denial are as follows: 'man is', 'man is not'. Next to these, there are the propositions: 'not-man is', 'not-man is not'.[70]

Let us compare these two versions: there are not white logs; there are not-white logs. No further explanation is needed to prove they are different. In this way it has been reaffirmed that, in contrast to the opinion of Theophrastus and his followers, there is a difference whether or not a hyphen is used after the negative particle.

Let us determine the values of the propositions concerned, relying on Figure 17:

TABLE 3

	(1)	(2)	(3)	(4)
(1)	f	f	t	t
(2)	t	t	f	f
(3)	f	t	f	t
(4)	t	f	t	f

Figure 17, being composed of S-es only, is not sufficient for the evaluation of the proposition 'S is P'. Therefore I use here the diagrams of Figure 12 as a basis. For simplicity's sake I proceed from cases (5)–(8). S and P are alien in these cases, so the proposition at issue is false, e.g. 'white is black' (diagram 8). [Cf. p. 34]

---

[70] Aristotle. 1928. 19b10–16.

In case of identity (9) or subordination, or inferiority (10), the value will be *t*, e.g. 'a fox is a fox'. Diagrams 11 and 12 reveal that *S* can just as well be *P* as not *P*, hence the value of the proposition under discussion is *t/f*, e.g. 'the astronauts are men'. The same values are successively arrived at in cases (3), (4), (1), and (2) as well because the relations concerned are identical. There is no need to dwell upon the evaluation of '*S* is not *P*' for it is but the negation of '*S* is *P*'.

TABLE 4

	(1)	(2)	(3)	(4)	(5)	(6)	(7)	(8)	(9)	(10)	(11)	(12)
(1) *S* is *P*	*t/f*	*t/f*	*t*	*t*	*f*	*f*	*f*	*f*	*t*	*t*	*t/f*	*t/f*
(2) *S* is not *P*	*f/t*	*f/t*	*f*	*f*	*t*	*t*	*t*	*t*	*f*	*f*	*f/t*	*f/t*

III

Aristotle made the following statement about the copula:

> When the verb 'is' is used as a third element in the sentence, there can be positive and negative propositions of two sorts. Thus in the sentence 'man is just' the verb 'is' is used as a third element, call it verb or noun, which you will. Four propositions, therefore, instead of two can be formed with these materials. Two of the four, as regards their affirmation and denial, correspond in their logical sequence with the propositions which deal with a condition of privation; the other two do not correspond with these. I mean that the verb 'is' is added either to the term 'just' or to the term 'not-just', and two negative propositions are formed in the same way. Thus we have the four propositions. Reference to the subjoined table will make matters clear:
>
> (A) Affirmation.   Man is just. ╲   ╱(B) Denial.   Man is not just.
> (D) Denial.   Man is not not-just. ╱ ╲(C) Affirmation.   Man is not-just.
>
> Here 'is' and 'is not' are added either to 'just' or to 'not-just'.[71]

While Aristotle regarded the copula as a component equal in rank with the other components of a proposition, Boethius introduced the following

---

[71] *Op. cit.* 19b18–30.

distinction: the subject and the predicate are the terms of a proposition; the copula, however, is not a term but only an indication of the quality of the proposition *(significatio qualitatis)*.

In this way he became the forerunner of the medieval view (e.g. Shyreswood) according to which the proposition '*S* is *P*' has two components *(de secundo adiacente)*. Buridan wrote the following on this subject:

> When form and matter are here spoken of, by the matter of a proposition or consequence is understood merely the categorematic terms, i.e. the subject and predicate . . ., all else, we say, belongs to the form. Hence we say that the copula, both of the categorical and of the hypothetical proposition, belongs to the form of the proposition.[72]

Petrus Hispanus and his followers, however, adopted Aristotle's approach and spoke of propositions as having three components *(de tertio adiacente)*. Walter Burleigh, in turn, laid great stress on the objective nature of the copula.[73]

Kant continued the tradition of Boethius:

> In categorical judgements the subject and predicate constitute the matter of the judgement; the form through which the relation (of agreement or opposition) between subject and predicate is determined and expressed is called the *copula*.[74]

This strict separation of the form and the material (content) of propositions is the manifestation of the same metaphysical method which Kant adopted to distinguish between phenomena and what he called the *Ding-an-sich*. This approach, when consistently applied, leads to the postulation of an absolute pure form and an absolute pure content. In reality, the form and the content of a proposition constitute a unity, that is, these factors mutually exclude, and at the same time pervade, one another. It is not to say that I do not find it justified to look beyond the content of a proposition, but I accept it within this unity only.

---

[72] Buridan, *Tractatus consequentiarum magistri*. Chapter VII. in: Bochenski p. 181; Ivo Thomas tr. p. 158.
[73] Cf.: Prantl, pp. 42 and 303.
[74] Kant, *Schriften* p. 535.

Sigwart went even further in Kant's footsteps:

> *The copula is not the vehicle but rather the object of negation;* there is no negative copula, there is only a negated copula.[75]

This conception has at least two imperfections: (1) it develops further Kant's subjectivist view that the only function of the negative copula is to prevent fallacies, (2) it denies the equality of the affirmative and the negative copula by ascribing an independent role to the former only.

The copula plays no role in today's mathematical logic.

Fogarasi took up a nominalist position in this question:

> The usual formula for the composition of a proposition *(S est P)* raises the question of the logical sense of the copula *est*, although this does not at all affect the essence of this structure. As is well known, the copula plays a part only in certain language systems: the Hungarian language, for instance, has no copula. It follows that the copula is not a necessary logical element of a proposition, but rather merely a form of expression in certain languages. Consequently, the connection subject-predicate *(S-P)* remains as the basic form of logical propositions and statements.[76]

The assertion that the copula is an indispensable element of all propositions is, no doubt, false. Any existential proposition is an example to the contrary. It is equally mistaken, however, to state that no proposition requires the copula as an indispensable element. This aversion to the copula is partly due to traditional logic which identified it almost exclusively with 'is'. It was a narrow-minded attitude, indeed. What is better after all: to reject the copula, or to understand it in a broader sense? In my opinion it should be defined as follows: *the copula is the link between, and one of, the components of a proposition.* This definition refers to both simple and compound propositions. One should start with the most abstract form of the copula, namely 'is' which is the copula indicating quality, and then proceed to the discussion of more and more concrete copulas corresponding to the specific relations of the propositions, e.g. *'S is necessarily P', 'S is identical with P', 'p or q'*, etc.

[75] Sigwart, p. 123.
[76] Fogarasi, p. 185.

One more thing to remark: it is not conservatism which makes me stick
to the copula. I want to keep it, on the one hand because I see it as a
manifestation of the relations reflected in propositions; and on the other
hand, in this way the continuity with the old logic can be maintained while
the ways to surpass it are also made clear.

IV

What kind of relation is there between the copula and the existence of
things reflected in the terms of propositions? With regard to this, the
following can be read in the *Hermeneutics*:

> Verbs in and by themselves are substantial and have significance, for he
> who uses such expressions arrests the hearer's mind, and fixes his attention;
> but they do not, as they stand, express any judgement, either positive or
> negative. For neither are 'to be' and 'not to be' and the particle 'being'
> significant of any fact, unless something is added; for they do not
> themselves indicate anything, but imply a copulation, of which we cannot
> form a conception apart from the things coupled.[77]

That is, the copula expresses the relations, and not the existence, of the
things reflected in the terms.

Kant treated this subject in relation to his disproof of the ontological
proof of God's existence:

> *'Being'* is obviously not a real predicate; that is, it is not a concept of
> something which could be added to the concept of a thing. It is merely the
> positing of a thing, or of certain determinations, as existing in themselves.
> Logically, it is merely the copula of a judgment. The proposition, 'God is
> omnipotent', contains two concepts, each of which has its object—God and
> omnipotence. The small word 'is' adds no new predicate, but only serves to
> posit the predicate *in its relation* to the subject. If, now, we take the subject
> (God) with all its predicates (among which is omnipotence), and say 'God
> is', or 'There is a God', we attach no new predicate to the concept of God,
> but only posit the subject in itself with all its predicates, and indeed posit it
> as being an *object* that stands in relation to my *concept*. The content of both
> must be one and the same; nothing can have been added to the concept,

---

[77] Aristotle. 1928. 16b19–26.

which expressed merely what is possible, by my thinking its object (through the expression 'it is') as given absolutely ...

By whatever and by however many predicates we may think a thing—even if we completely determine it—we do not make the least addition to the thing when we further declare that this thing *is*. Otherwise, it would not be exactly the same thing exists, but something more than we had thought in the concept; and we could not, therefore, say that the exact object of my concept exists.[78]

Kant was right in assuming that 'is' *as copula* does not imply whether or not the things reflected in the subject exist. In the proposition 'God is omnipotent', 'is' refers to the statement of the relation concerned and not to the existence of the subject itself. Consequently, it does not follow from the affirmation 'God is omnipotent' that God exists.

Kant accepted 'is' *only* as copula. According to this interpretation, existential propositions should be regarded as nonsense, since it is the very function of such propositions to affirm, or deny, the existence or the non-existence of the things expressed in the subject. If the verb expressing existence cannot be a predicate, existential propositions have no justification. But what should be done then with such statements like e.g. 'the German Democratic Republic exists', 'phlogiston does not exist'?

Hegel wrote the following:

> The predicate which is attached to the subject should, however, also *belong* to it, that is, be in and for itself identical with it. Through this significance of *attachment*, the *subjective* meaning of judgement and the indifferent, outer subsistence of subject and predicate are sublated again: this action *is* good; the *copula* indicates that the predicate belongs to the *being* of the subject and is not merely externally combined with it. In the *grammatical* sense, that subjective relationship in which one starts from the indifferent externality of the subject and predicate has its complete validity; for it is *words* that are here externally combined.[79]

Supposing that the subject refers to existing things, one has to agree with Hegel. In general, however, there is no guarantee for what Hegel has relegated to the field of grammar. In my view the external relation treated

---

[78] Kant. *Kritik* pp. 655–656; Kemp Smith, pp. 505.
[79] Hegel, *Wissenschaft* II. p. 69; A. V. Miller tr. p. 626.

by Hegel is of not purely grammatical character but the starting point of logical investigation into the question under discussion.

According to Drobisch, such propositions as 'God is just', 'the soul is not mortal' do not state the existence of God and the soul respectively, just as there is no reference whatsoever to the existence of the subject in the following propositions: 'furies have snake-like hair', 'ghosts appear at night'. All that is stated in these propositions is: if the subject is postulated, some given predicate goes with it.[80]

Überweg made the following remark: whoever does not want to accept the existence of the subject in propositions as self-evident has to express himself in conditionals, e.g. 'if God exists'. This provision can be omitted only if obviously fictive beings (e.g. Zeus, the Sphinx) are spoken of.[81]

Let us now look at a version of this approach expressed a hundred years later.

> When it is affirmed, or denied, that some property belongs to some object our proposition reflects also whether or not the object of the proposition exists in reality.
>
> Such simple propositions as e.g. 'There are cosmic rays', 'Mermaids do not exist in reality', etc. directly affirm (or deny) whether or not the object of the proposition exists in reality. In other simple propositions it is known in advance whether the object of the proposition exists in reality or not. Consequently, what is directly affirmed (or denied) in these propositions is not the fact itself that the object of the proposition exists in reality but that it exists at a certain place or at a certain time.[82]

I share the view that propositions reflect reality in some way. It does not follow, however, that they include knowledge of existence and non-existence respectively without further ado. What is to be done, for example, with this proposition: 'the Martians go in for sports'? If the author knows in advance whether or not Martians exist, I do envy him for that. If, however, he does not know, which I find very probable, the statement in the last paragraph of the quotation does not hold true.

---

[80] Cf.: Drobisch, p. 60.
[81] Cf.: Überweg, p. 150.
[82] Tavanets (*The Proposition and its Kinds*) pp. 33–34.

Mill assumed a correct position on this issue:

> It is apt to be supposed that the copula is something more than a mere sign of predication; that it also signifies existence. In the proposition, Socrates is just, it may seem to be implied not only that the quality *just* can be affirmed of Socrates, but moreover that Socrates *is*, that is to say, exists. This, however, only shows that there is an ambiguity in the word *is*; a word which not only performs the function of the copula in affirmations, but has also a meaning of its own, in virtue of which it may itself be made the predicate of a proposition. That the employment of it as a copula does not necessarily include the affirmation of existence, appears from such a proposition as this, 'A centaur is a fiction of the poets'; where it cannot possibly be implied that a centaur exists, since the proposition itself expressly asserts that the thing has no real existence.[83]

Sigwart was right in pointing out the inconsistency of logicians asserting that an affirmative copula implies also the assertion of the existence of the subject for then they ought to admit that a negative copula implies the non-existence of the subject. Accordingly 'Socrates is not ill' would mean at the same time that there is no Socrates. This kind of reasoning cannot be accepted, and consequently the former is also untenable.[84]

Nevertheless, it becomes clear at another point that he wanted to separate the copula from the subject in order to state that the copula is "a component of the predicate".[85] In this way he succeeded in opposing one extreme with another.

So we have come to the conclusion that the copula does not automatically imply the existence of the subject. Therefore the non-existent must also be reckoned with both in propositions and inferences. This is the way to meet the requirement to investigate the given problem in general. This being accomplished, we may proceed to the special treatment of such cases where the existence, or non-existence, of the things reflected in the terms are indicated. The simplest cases are the following:

(1)   There is an S which is P.

(2)   There is an S which is not P.

---

[83] Mill, pp. 78–79.
[84] Sigwart, Logik. pp. 123–124.
[85] *Op. cit.*, p. 92 ('...das Verbum 'Sein' bildet einen Bestandtheil des Prädikats').

(3)   There is no S which is P.
(4)   There is no S which is not P.

Let us compare these formulas: 'there is such a case when S is P'; 'there is such an S which is P'. One can see that existence refers to the proposition as a whole in the first case and to the term in the second. Propositions (1)–(4) acquire the following values according to the diagrams of Figure 12:

TABLE 5

	(1)	(2)	(3)	(4)	(5)	(6)	(7)	(8)	(9)	(10)	(11)	(12)
(1)	*f*	*f*	*f*	*f*	*f*	*f*	*f*	*f*	*t*	*t*	*t*	*t*
(2)	*f*	*f*	*f*	*f*	*f*	*f*	*t*	*t*	*f*	*f*	*t*	*t*
(3)	*t*	*t*	*t*	*t*	*t*	*t*	*t*	*t*	*f*	*f*	*f*	*f*
(4)	*t*	*t*	*t*	*t*	*t*	*t*	*f*	*f*	*t*	*t*	*f*	*f*

V

As far as the relation of quality and existence is concerned, let us proceed from the following statement:

> Now it is possible both to affirm and to deny the presence of something which is present or of something which is not, and since these same affirmations and denials are possible with reference to those times which lie outside the present, it would be possible to contradict any affirmation or denial. Thus it is plain that every affirmation has an opposite denial, and similarly every denial an opposite affirmation.[86]

The above text allows for two interpretations at least. First, existence acts as a term (in the order of the quotation): (1) there is a man, (2) there is no man, (3) there is a devil, (4) there is no devil. Otherwise, existence refers to propositions:

(1)   There is a rule that man is a living being.
(2)   There is no such rule that man is a living being.
(3)   There is a rule that glass is a good heat-conductor.
(4)   There is no such rule that glass is a good heat-conductor.

[86] Aristotle. 1928 17a27–32.

Both interpretations are logically relevant.—It should be noted here that two meanings of 'is' have to be distinguished. In one case it refers to present tense: there is now. In the other it implies an existing phenomenon which is not related to a certain time. This was what Aristotle had in mind writing 'with reference to those times which lie outside the present'. When using 'is' without any attribute, I have the broader sense of the word in mind.

Let us compare affirmative and negative propositions, on the one hand, and existential and non-existential ones, on the other. Let us take the following two propositions as an example: (a) it is not raining, (b) there is no such case that it is raining. If existence is not confined to the present state of affairs but implies the existence of a relation in general, there is a difference between these two propositions. The first one is t in certain cases and f in other cases. The second proposition, however, is always f.

Their interrelation can be described as follows. If (b) is t then (a) is also t but it is not true the other way round. Accordingly, the formulas (c) p and (d) there is such a case that p must be distinguished, too. The latter being t, the former is also t, but it does not hold true the other way round.

Let us now study the interrelation of quality and value. Kant approached this problem in this way:

> Owing to the general desire for knowledge, negative judgments, that is, those which are such not merely as regards their form but also as regards their content, are not held in any very high esteem. They are regarded rather as the jealous enemies of our unceasing endeavour to extend our knowledge, and it almost requires an apology to win for them even tolerance, not to say favour and high repute.
>
> As far as *logical* form is concerned, we can make negative any proposition we like; but in respect to the content of our knowledge in general, which is either extended or limited by a judgment, the task peculiar to negative judgments is that of *rejecting error*.[87]

If Kant had meant to describe as negative only those propositions which prevent fallacies, his view would have been limited but not indefensible.

---

[87] Kant, *Kritik* p. 740; Kemp Smith, p. 574.

Let us, however, confront his actual opinion with this question: what fallacy is prevented by the proposition 'man is not a living being'?

The issue under discussion can be approached under another aspect, too. Couturat argues as follows: to say a sentence is true means to affirm the sentence itself. In other words, to affirm a sentence is the same as to state the truth of it. He called this thesis the principle of affirmation.[88]

Fogarasi opposed this view:

> The relation of affirmative and negative propositions to one another is often confused with the relation of true and false propositions, as if an affirmative proposition were the the form of expression of a true proposition, and a negative proposition were that of a false proposition. In reality, the affirmative proposition, just as the negative proposition, can be true or false. Metal is a good insulator. This is a false affirmative proposition. Metal is not a good conductor. This is a false negative proposition. Metal is not an insulator; a true negative proposition. *We see that the affirmative or negative quality of a proposition does not depend on its character as true or false*...[89]

The first part of the above statement can be accepted: if it is true that it is raining, then it is raining. It seems to hold true also in a converted form: if it is raining, then it is true that it is raining. In this case, however, the antecedent was understood *nolens-volens* as: if it is, indeed, raining. Let us formulate this conditional proposition as follows: if I state that it is raining, then it is true that it is raining. Now it is more difficult to prove that the antecedent really entails the consequent. However, it becomes even more difficult to do so in the following case: if I state that glass is a heat-conductor, then it is true that glass is a heat-conductor. Relying on the above said, I agree with those who reject the embroilment of affirmative and true propositions.

VI

Now let us turn to propositions with positive and negative terms respectively.

> An affirmation is the statement of a fact with regard to a subject, and this subject is either a noun or that which has no name...for I stated that the

---

[88] Couturat, *L'algèbre de la logique*. (Paris 1905).
[89] Fogarasi, *Logik* pp. 195–196.

expression 'not-man' was not a noun, in the proper sense of the word, but an indefinite noun, denoting as it does in a certain sense a single thing. Similarly the expression 'does not enjoy health' is not a verb proper, but an indefinite verb...

(A) Affirmation. Man is just.   (B) Denial. Man is not just.

(D) Denial. Man is not not-just. (C) Affirmation. Man is not-just.

...there are moreover two other pairs, if a term be conjoined with 'not-man', the latter forming a kind of subject. Thus:

(A)' Not-man is just.   (B)' Not-man is not just.

(D)' Not-man is not not-just.   (C)' Not-man is not-just.[90]

Let me list the variations treated here in traditional formulation:

(1) S is P.
(2) S is not P.
(3) S is not-P.
(4) S is not not-P.
(5)   Not-S is P.
(6)   Not-S is not P.
(7)   Not-S is not-P.
(8)   Not-S is not not-P.

Further on, Aristotle suggested that negative terms are indefinite. This statement holds true so far as the extension of such terms, indeed, requires further explanation. If, however, such a term is taken in general, it is definite in so far as it is a *negative* term.

Boethius introduced the following designations: (1) simple affirmation (*affirmatio simplex*), (2) simple negation (*negatio simplex*), (3) infinite affirmation (*affirmatio infinita*), (4) infinite negation (*negatio infinita*). He failed to deal with cases (5)–(8).

The view that negation must refer to the copula in negative propositions (*in propositione negativa negatio afficere debet copulam*) came to prevail in scholasticism. That is, negation should be understood as referring to the

[90] Aristotle. 1928. 19b5–6, 7–10, 28–30, 39–44.

proposition as a whole and not to the terms only. This is how the eight cases described by Aristotle were reduced to (1)–(2).

In modern times some went to the other extreme.

> Some logicians, among whom may be mentioned Hobbes, state this distinction differently; they recognize only one form of copula, *is*, and attach the negative sign to the predicate. 'Caesar is dead', and 'Caesar is not dead', according to these writers, are propositions agreeing not in the subject and predicate, but in the subject only. They do not consider 'dead', but 'not dead', to be the predicate of the second proposition, and they accordingly define a negative proposition to be one in which the predicate is a negative name.[91]

Wolff and Reimarus returned to the approach of Boethius who had distinguished simple and infinite propositions. Kant divided propositions according to their quality as follows: (a) affirmative (*S* is *P*), (b) negative (*S* is not *P*), (c) infinite (*S* is not-*P*). He wrote in his formal logic:

> According to the principle of the exclusion of every third element (*exclusi tertii*), the sphere of one concept relative to another is either exclusive or inclusive.—Since logic deals only with the form of judgements and not with the concepts as far as their content is concerned, thus the differentiation of infinite judgements from negative judgements does not belong to this science.[92]

Elsewhere he gave the following answer to the problem under discussion:

> ... *infinite judgments* must, in transcendental logic, be distinguished from those that are *affirmative*, although in general logic they are rightly classed with them, and do not constitute a separate member of the division. General logic abstracts from all content of the predicate (even though it be negative); it enquires only whether the predicate be ascribed to the subject or opposed to it. But transcendental logic also considers what may be the worth or content of a logical affirmation that is thus made by means of a merely negative predicate, and what is thereby achieved in the way of addition to our total knowledge.[93]

---

[91] Mill, p. 80.
[92] Kant, *Schriften* p. 535.
[93] Kant, *Kritik* p. 143; Kemp Smith tr., p. 108.

As compared to his forerunners, Kant took a big step forward. He recognized that there is a qualitative difference between (1)–(2), on the one hand and (3), on the other, because the latter also takes into account the intension of terms. If Kant had been consistent he should have introduced two groups according to this distinction. Then, however, he would have been confronted with a double problem. On the one hand, the system of his table of propositions where each group consisted of three versions would have been destroyed. On the other, he should have given a reason why there was but one version in the second group, namely S is not-P. Kant took the line of least resistance and merged the two groups. Consequently, the qualitative difference he had correctly recognized was again pushed into the background and the propositions concerned levelled off.

Hegel, on the contrary, laid a special emphasis on their relation of subordination:

> The negatively-infinite judgment, in which the subject has no relation whatever to the predicate, gets its place in the Formal Logic solely as a nonsensical curiosity. But the infinite judgment is not really a mere casual form adopted by subjective thought. It exhibits the proximate result of the dialectical process in the immediate judgments preceding (the positive and simply-negative), and distinctly displays their finitude and untruth. Crime may be quoted as an objective instance of the negatively-infinite judgment. The person committing a crime, such as a theft, does not, as in a suit about civil rights, merely deny the particular right of another person to some one definite thing. He denies the right of that person in general, and therefore he is not merely forced to restore what he has stolen, but is punished in addition, because he has violated law as law, i.e. law in general.[94]

This view was fiercely opposed by formal logicians.

> In opposition to attempts to interpret all negative propositions as if a predicate, not-B, were being attributed to a subject, stands the dominant tradition that the negation applies to the *copula*; and one speaks therefore of the affirmative and negative quality of the copula. This doctrine is correct, at least in the sense that the negation is not in the elements of the proposition but rather in the way they are related to one another. It is

[94] Hegel, *System*. Erster Teil pp. 374–375; Wallace tr. pp. 306–307.

however incorrect to suppose two opposing kinds of copula, an affirmative and a negative.[95]

Let us see another, less fierce, reaction:

Kant, when classifying the propositions according to their quality, includes also those which are called *infinite* (*limitative*). The copula of infinite propositions is affirmative, and their predicate is negative. E.g. 'this book is not black'—hence it can be of any other colour. This kind of proposition is called infinite because, with the exception of one (black), an infinite number of predicates can be related to its subject. The infinite is limited by this single predicate only. There is, however, no justification for this kind of proposition to exist in its own right *beside* affirmative and negative propositions because it is an affirmative proposition itself. In fact, it is but a subdivision of affirmative propositions where the predicate is a contradictory term. Kant needed the class of infinite propositions for he had recognized that even simple affirmative propositions have a dialectical nature.[96]

I agree that infinite propositions must not be regarded as co-ordinated ones. It would be, however, a mistake to consider infinite propositions merely as one of the versions of affirmative propositions. The former are related to the latter as the *concrete* to the *abstract*. That is why the dialectical character of affirmative propositions does not make the discussion of infinite propositions superfluous.

---

[95] Sigwart, p. 122.
[96] *Logika.* (Budapest 1956.) pp. 151–153.

CHAPTER TWO

4. ABSTRACT AND CONCRETE

I

To begin with, the operation of abstraction is to be analysed.

> As the mathematician investigates abstractions (for before beginning his investigation he strips off all the sensible qualities, e.g. weight and lightness, hardness and its contrary, and also heat and cold and the other sensible contrarieties, and leaves only the quantitative and continuous, sometimes in one, sometimes in two, sometimes in three dimensions, and the attributes of these *qua* quantitative and continuous, and does not consider them in any other respect...)[1]

The significance of abstraction is, first of all, to help to find one's way in the complicated network of things and phenomena. Without this method it would have been impossible to break away from the undifferentiated unity of phenomena. It is by abstraction that e.g. the notion, or term, of number could develop. If we had not been able to apply this method we would still count on our fingers or the things themselves.

The results achieved by abstraction have supported the view that it is the method which makes scientific exactness possible. This opinion goes back to Aristotle:

> And in proportion as we are dealing with things which are prior in definition and simpler, our knowledge has more accuracy, i.e. simplicity. Therefore a science which abstracts from spatial magnitude is more precise than one which takes it into account; and a science is most precise if it abstracts from movement... Each question will be best investigated in this way—by setting up by an act of separation what is not separate, as the arithmetician and the geometer do. For a man *qua* man is one indivisible thing, and the arithmetician supposed one indivisible thing, and then

[1] Aristotle, *Metaphysica*. Oxford 1954. 1061a29–36.

59

considered whether any attribute belongs to a man *qua* indivisible. But the
geometer treats him neither *qua* man nor *qua* indivisible, but as solid. For
evidently the properties which would have belonged to him even if
perchance he had not been indivisible, can belong to him even apart from
these attributes (sc. indivisibility and humanity).[2]

No doubt it is easier to have a more exact idea of the simple than of the
complicated. It does not mean, however, that only the simple can be
studied in an exact way. Marx, for example, has studied capitalism in its
concrete complexity. And still, could one deny the accuracy of his results?
Thus, the difference between abstract and concrete is not that the former is
accurate while the other is not, or less so, but that the abstract can lead to
exact results more quickly.

It was a widely held view in the Middle Ages—mainly due to Duns
Scotus and his disciples—that abstraction is the demonstration of the
essence. In my opinion it should be accepted in general that the separation
of any property is abstraction. Abstraction involves the separation of the
essentials as an important but still special aspect.

What arguments can support the view that abstraction is the separation
of the essentials only? One can argue that the abstraction of inessential
properties is unnecessary. It may go unchallenged that one of the basic
functions of abstraction is to separate the essentials. It is, however, a
mistake to understand abstraction in such a narrow sense, if for nothing
else because it is not easy to decide whether a property is essential or not.
In order to answer this question, the property concerned has to be
examined. If the separation necessary for this is not abstraction, what
should then be the name of this operation?

Another argument is based on the idea that abstraction cannot include
all the properties of a certain object because they are of infinite number.
No doubt that a rabbit, for example, has infinitely many properties and we
cannot learn them all. But is it really indispensable to know them all in
order that the notion, or term, of rabbit could include all the characteristic
features of this object? Let us not forget that there are innumerable
rabbits. The notion, or term, of rabbit, however, comprises all rabbits
without our knowing them. Consequently, if a notion including all rabbits

---

[2] *Op. cit.*, 1078a7–28.

can exist I see no reason why a notion, or term, comprising all the characteristics of rabbit could not exist.

Kant argues as follows:

> In logic the term *abstraction* is not always used correctly. We must not say: abstract *something* (*abstrahere aliquid*) but rather abstract *from something* (*abstrahere ab aliquo*). If for instance, taking a purple cloth, I think only of the red colour: then I abstract from the cloth; if I also abstract from the colour and think of the purple cloth simply as a material stuff: then I abstract from several more determinations and my concept has thus become even more abstract. For the more differences of things are omitted from a concept or the more determinations of a thing have been abstracted from: the more abstract the concept is. Abstract concepts ought therefore really to be called *abstracting* concepts (*conceptus abstrahentes*) that is, concepts in which a number of abstractions occur. Thus for instance, the concept of *body* is really not an abstract concept; for I cannot abstract from the body itself; otherwise I would not have the concept of it. But I must indeed abstract from the size, the colour, the hardness of fluidity: in short from all special determinations of particular bodies.—The *most abstract* concept is the one that has nothing in common with any different concept. This is the concept of *something*; for whatever is different from it is *nothing* and has nothing in common with the something.[3]

According to Kant the right thing to do is to abstract from something and not to abstract something. It is a metaphysical contraposition. We always abstract something from something.

Traditional logic interpreted this 'something' as property. Opinions about the nature of properties do not differ in logic, only the wordings of definitions do. Let us proceed from the following formulation.

> *Property* is all what things and occurrences have, or have not, in common, any indicator or aspect of a thing or occurrence through which one can learn, define, or describe the thing or occurrence concerned. Every object and every phenomenon one thinks of has the most different properties.[4]

[3] Kant, *Schriften* pp. 525–526.
[4] Kondakov, pp. 290–291.

Based on the above definition, I think abstraction is the separation of any property.

Traditional logic usually makes a distinction between two kinds of abstraction: generalizing abstraction and isolating abstraction. As a result of a generalizing abstraction, some property of the object (e.g. white) is separated from the others. Isolating abstraction, on the contrary, isolates the given property not from the others only, but also from the object itself (e.g. whiteness).

Traditional logic usually enlists the following operations as the fundaments of thought: identification and differentiation, analysis and synthesis, generalization and limitation, and abstraction. As the above list has revealed, among the operations concerned it is only abstraction which is condemned to loneliness in traditional logic.

Hegel was the first to explicitly call for proceeding from the abstract towards the concrete:

> Abstraction, therefore, is a *sundering* of the concrete and an *isolating* of its determinations; through it only *single* properties and moments are seized; for its product must contain what it is itself. But the difference between this individuality of its products and the Notion's individuality is that in the former the individual as *content* and the universal as *form*, are distinct from one another—just because the former is not present as absolute form, as the Notion itself, or the latter is not present as the totality of form. However, this more detailed consideration shows that the abstract product itself is a unity of the individual content and the abstract universality, and is therefore a *concrete* and the opposite of what it aims to be.[5]

This ascent from the abstract to the concrete has been demonstrated by Marx in *Capital* in a classical way. Starting from the simplest abstractions—labour, value, etc.—, he went on systematically to the more complicated and more concrete. The first volume deals with the abstract general forms of the capitalist production process. The second volume turns to the analysis of the process of circulation, indispensable for

---

[5] Hegel, *Wissenschaft* p. 61; A. V. Miller tr., pp. 619–620.

capitalist economy. Finally, the third volume examines the relation of capitalist production as the unity of production and circulation.

It seems to be reasonable to use the word 'concretization' to denote the above operation.

> Thales of Miletus was already aware of the electric properties of bodies. The notion of electron as a distinct thing has been conceived relatively late, nevertheless. How could one change over from the properties to the thing itself? First, the discrete character of electric properties had been revealed as early as in Faraday's experiments. Then Thomson, and especially Millikan, defined the concrete value of elementary charge as a result of their experiments. Thomson attached a globular volume to this charge and defined its radius. In addition, a definite mass was attributed to the electrical charge within this volume. Later on spin and other attributes have been ascribed to the charge. This is how the notion, or term, of the thing—the electron—has developed. This process can be called concretization.[6]

While abstraction separates certain properties and aspects of the object, concretization strives for an overall demonstration of the given object. Concretization has a function in thought similar to synthesis because both of them aim at completeness, or totality. The only difference is that by synthesis the parts of the object concerned are united while by concretization its properties are integrated.

The more concrete is the way things and occurrences are reflected, the more authentic is the idea we get of them. Let me illustrate this thesis with an example from a special science.

> ... the pictures of tropical fish in colour, reproduced with extreme care in scientific works, are mainly false, or rather unnatural. Almost invariably, dead or captured animals are used as models, rather than observations of fish in their usual environment and in their normal mood. An underwater photographer can do valuable research in this field. The only trouble is that even colour film is affected by the light-absorption of water. Not more than ten metres deep in the sea the red, and soon also the yellow, rays disappear from the spectrum, so that shots made in greater depth display blue and green tonalities only. To make really authentic colour photos of fish would require an underwater flashlight which does not distort the tonalities of

---

[6] Uemov, (translated from the Russian) 1966. p. 139.

colour. Even in this case, however, one should take into account the age of
the animal (for, very often, colours also fade with the years), the season of
the year (whether or not it is a mating-season), and even the hour of the day
(for many fish change their colours at night). In addition, the specifics of the
ground and the temperature of the water, which also affect the colour of a
fish, ought to be considered, not to speak of the mood of the animal at the
moment (whether it is full or hungry, tired or rested, calm or alarmed)! It
would probably reveal that there are very many fish which have no 'normal'
colour at all, the colour being but a reflection of their living conditions just
like facial expression and mimicry with man.[7]

II

As to the relation of abstract and concrete, I proceed from the following
statement of Aristotle:

> ... if attributes do not exist apart from their substances (e.g. a 'mobile' or a
> 'pale'), pale is prior to the pale man in definition, but not in substantiality.
> For it cannot exist separately but is always along with the concrete thing;
> and by a concrete thing I mean the pale man. Therefore it is plain that
> neither is the result of abstraction prior nor that which is produced by
> adding determinants posterior; for it is by adding a determinant to pale that
> we speak of the pale man.[8]

In order to interpret the above text, one has to consider the attitude of
Aristotle to Plato's theory of ideas. Plato, as is well known, was of the
opinion that every idea, even the idea of whiteness, for example, exists
independently. Aristotle, on the contrary, argued that properties, or
attributes, cannot exist in themselves but only in objects. Thus, both
whiteness and white exist only in a combination with some object.

And what is the relation between whiteness and white? This question is
partially answered in the *Categories*:

> ... both the name and the definition of the predicate must be predicable of
> the subject. For instance, 'man' is predicated of the individual man. Now in
> this case the name of the species 'man' is applied to the individual, for we use

---

[7] Hass, p. 198.
[8] Aristotle, *Metaphysica*. 1077b5–11.

the term 'man' in describing the individual; and the definition of 'man' will also be predicated of the individual man, for the individual man is both man and animal. Thus, both the name and the definition of the species are predicable of the individual.

With regard, on the other hand, to those things which are present in a subject, it is generally the case that neither their name nor their definition is predicable of that in which they are present. Though, however, the definition is never predicable, there is nothing in certain cases to prevent the name being used. For instance, 'white' being present in a body is predicated of that in which it is present, for a body is called white: the definition, however, of the colour 'white' is never predicable of the body.[9]

For Aristotle, white by definition meant whiteness. That is to say, the abstract is to be interpreted as the essence of something.

Being of a content character, these categories failed to arouse special interest both in ancient and medieval logic. It was not before Duns Scotus that the expressions *abstractum* and *concretum* have gradually come to stay in the literature.

Let us see now the typical views held in modern times. Mill wrote the following:

> A concrete name is a name which stands for a thing; an abstract name is a name which stands for an attribute of a thing. Thus *John*, *the sea*, *this table*, are names of things. *White*, also, is a name of a thing, or rather of things. Whiteness, again, is the name of a quality or attribute of those things. Man is a name of many things; humanity is a name of an attribute of those things. *Old* is a name of things; *old age* is a name of one of their attributes.[10]

Mill himself was also aware of the problematic character of this distinction, which is why he returned to this question:

> It may be objected to our definition of an abstract name, that not only the names which we have called abstract, but adjectives, which we have placed in the concrete class, are names of attributes; that *white*, for example, is as much the name of the colour as *whiteness* is. But (as before remarked) a word ought to be considered as the name of that which we intend to be

---

[9] Aristotle. 1928, 2a19–34.
[10] Mill, p. 29.

understood by it when we put it to its principal use, that is, when we employ it in predication. When we say snow is white, milk is white, linen is white, we do not mean it to be understood that snow, or linen, or milk, is a colour. We mean that they are things having the colour. The reverse is the case with the word whiteness; what we affirm to *be* whiteness is not snow, but the colour of snow. Whiteness, therefore, is the name of the colour exclusively: white is a name of all things whatever having the colour; a name, not of the quality whiteness, but of every white object.[11]

The term 'white' is the result of a generalizing abstraction as some property, or quality, is abstracted from the others, but not from the object itself. The term 'whiteness', on the other hand, is the result of an isolating abstraction as the given property is abstracted not from the other properties only but also from the object itself. This difference, however, must not be pushed beyond the range of its validity. The fact that we do not abstract from the object in the first case does not mean that the term 'white' reflects some object.

Fogarasi offered the following opinion:

> When we speak of abstract terms, we mean by this that the terms abstract from the individual varying, accidental properties of phenomena and emphasize a particular common trait or a number of common traits fixing them in thought and expressing them in a determinate manner in language. Thus, for instance, 'beauty' is an abstract term. In reality there are beautiful people, beautiful landscapes, beautiful works of art, but there is no 'beauty'. The abstract term 'beauty' is expressed by the root word '*beau*' and the additional syllable '*té*' or some other suffix. On the other hand, we speak of the "beautiful Helen" or say: "The Sistine Madonna is one of Raphael's most beautiful paintings." The beautiful is a concrete quality or property of individual people or objects that we apprehend with our senses and express in language with an adjective.[12]

Two widely held views are manifest in the above quotation. One of them could be summed up like this: concrete is what is real, and abstract is what is in our mind. In other words, concrete is what has an objective existence while abstract is what does not exist objectively. What follows from

[11] *Op. cit.*, p. 30.
[12] Fogarasi, *Logik.* p. 138.

accepting this statement? Let us mention but one thing. To proceed from concrete to abstract is to proceed from existent to non-existent. Is that really the function of abstraction?

In my opinion, a concrete term, or notion, refers to an object while an abstract term refers to a property, or quality. Both concrete and abstract terms can be existential and non-existential respectively: (1) a concrete existential term: man, (2) an abstract existential term: white, beautiful, (3) a concrete non-existential term: absolutely black body, Zeus, (4) an abstract non-existential term: snake-haired.

III

The other view evident in the quotation from Fogarasi requires a somewhat more detailed analysis. It has become a common belief in the logic of modern times, mainly under the influence of Locke, that all notions, or terms, are abstract. Here is what Kant wrote on the subject:

> The expressions *abstract* and *concrete* apply not so much to the concepts in themselves—for every concept is an abstract concept—but rather only to their *use*. And this use can in turn have various degrees;—accordingly as one treats the concept now more, now less abstractly or concretely; that is, as one omits or attaches now more, now less determinations. With abstract use a concept approaches the highest genus, with concrete use on the other hand it approaches the individual.[13]

Hegel, on the contrary, held the following view:

> No complaint is oftener made against the notion than that it is *abstract*. Of course it is abstract, if abstract means that the medium in which the notion exists is thought in general and not the sensible thing in its empirical concreteness. It is abstract also, because the notion falls short of the idea. To this extent the subjective notion is still formal. This however does not mean that it ought to have or receive another content than its own. It is itself the absolute form, and so is all specific character, but as that character is in its truth. Although it be abstract therefore, it is the concrete, concrete altogether, the subject as such.[14]

---

[13] Kant, *Schriften* p. 530.
[14] Hegel, *Encyclopädie* p. 160; Wallace tr., p. 295.

6

What are terms, or notions, like by nature: are they abstract or concrete? To answer this question, let us start with analysis of the relation between experience and terms. The prevailing opinion is that experience is concrete. This statement is usually taken as evident, and consequently no explanation is offered. If, nevertheless, some explanation is needed the following argument is referred to: as a result of experience our mind forms a notion of the things and occurrences. This notion is a more or less vivid, and by all means manifold, reflection of reality. That is why experience is often also called vivid observation.

Notions, or terms, on the contrary, abstract from the complexity of reality and are but its pale and gray reflection.

> Every one will readily allow, that there is a considerable difference between the perceptions of the mind, when a man feels the pain of excessive heat, or the pleasure of moderate warmth, and when he afterwards recalls to his memory this sensation, or anticipates it by imagination. These faculties may mimic or copy the perceptions of the senses; but they never can entirely reach the force and vivacity of the original sentiment. The utmost we say of them, even when they operate with greatest vigour, is, that they represent their object in so lively a manner, that we could *almost* say we feel or see it: But, except the mind be disordered by disease or madness, they never can arrive at such a pitch of vivacity, or to render these perceptions altogether undistinguishable. All the colours of poetry, however splendid, can never paint natural objects in such a manner as to make the description be taken for a real landscape. The most lively thought is still inferior to the dullest sensation.[15]

Let us first dissect the assertion that experience is concrete. Psychology and epistemology distinguish between sensation and perception. Sensation is an aspect of experience which reflects certain properties, or qualitites, of an object. Sensation makes possible the cognition of colours, sounds, tastes, etc. Our senses in action do abstraction by selecting some property from among the others. If, for example, we touched something in the dark and found it wet, we had an abstract experience.—Perception is another aspect of experience which grasps an object as a whole. What we perceive in the course of perception are not certain stimuli as during

---

[15] Hume, *Enquiries* p. 17.

sensation, but we rather form an overall picture of an object under the joint influence of several stimuli. For example, such an overall picture can be developed at a wine-tasting.

Thus, experience is not only concrete but abstract as well. And what about notions, or terms?

> It is customary in classifying terms to differentiate between 'abstract' and 'concrete' terms. As a matter of fact however, almost all so-called concrete terms are likewise abstract. Examples:
> (a) In Arabic the tails of donkeys, lions, horses, etc. are designated by different words.
> (b) We know now that the natural numbers are abstraction classes of numerically equal sets. Beside them there still remain remnants of more primitive levels of abstraction. In some primitive cultures the number 3 is designated in connection with sheep by a different term than it is in connection with stones. But this sort of thing is also still the case in the German language. We speak of two lovers as a *couple* (*Liebespaar*) but of two oxen as a *yoke* (*Joch*), even though from the abstraction standpoint of numbers—but only in this regard—both mean the same thing. Or we speak of *three score* (*ein Schock*) eggs; and we use the term *dozen* only for particular kinds of things.[16]

The following questions arise with regard to the above. (1) Can one of these doctrines be proved by examples? (2) Do these examples reveal why they are abstract terms? (3) If not all, but almost all so-called concrete terms are abstract, which of them are not?

It is mainly the following arguments which are used to prove the abstract character of terms: No concept, or term, can be as vivid as looking at the things themselves. Thought abstracts from the concrete complexity of reality. Thought disregards the external, accessory, and accidental properties of things and grasps only what is essential in them. While forming a notion, one disregards the qualities which are characteristic of certain individual objects only, and looks for what is common in them.

First of all, the following question arises: is it a criterion for the concreteness of knowledge that it should be vivid? First, vivacity is not an

---

[16] Klaus, p. 140.

6*

exact criterion. Second, some knowledge is concrete in so far as it gives an exhaustive reflection of the object concerned.

Further, if we start from the supposition that a notion, or term, is the reflection of the essentials of things it follows that we should regard all terms as being abstract, since the separation of essential properties postulates abstraction. In the first chapter of the present work, I tried to demonstrate that a term may include all properties of an object. If so, it is not justified any more to consider terms, or notions, as abstract.

Finally, is the singular more concrete than the general?

> The general law of the change of form of motion is much more concrete than any single 'concrete' example of it.[17]

Thought is unable to form an entirely true notion of the things at once. The mind gradually develops a real image of things and occurrences. It is the ability to abstract which makes it possible for man to disentangle himself from the entangled net of accidental and secondary circumstances. By abstraction one can arrive at a deeper and more thorough knowledge of reality than by experience only. The trouble starts when abstract terms, or notions, are thought to be the end-station of cognition. In reality, they are but a starting point to proceed from in the course of thinking.

> ... The concrete is concrete because it is the concentration of many determinations, hence unity of the diverse. It appears in the process of thinking, therefore, as a process of concentration, as a result, not as a point of departure, even though it is the point of departure in reality and hence also the point of departure for observation (*Anschauung*) and conception ...... the method of rising from the abstract to the concrete is only the way in which thought appropriates the concrete, reproduces it as the concrete in the mind.... the concrete totality is a totality of thoughts, concrete in thought, in fact a product of thinking and comprehending; but not in any way a product of the concept which thinks and generates itself outside or above observation and conception; a product, rather, of the working-up of observation and conception into concepts.[18]

Thus, notions, or terms, are not only abstract but also concrete.

---

[17] Engels, *Dialektik der Natur.* p. 236; Clemens Dutt tr., p. 295.
[18] Marx, *Einleitung* p. 632. Nicolaus tr., p. 101.

IV

Now let us see what relations there are between objects and properties in a proposition. Aristotle has stated the following in this respect: "An affirmation is a positive assertion of something about something, a denial a negative assertion."[19]

This is a rather abstract statement which allows for various interpretations. One of them is, for instance, the negative thesis of nominalists: it is impossible to affirm a thing of a thing (*res de re praedicari non potest*). They tried to justify this view with the following argument: the predicate of a proposition cannot be singular but general only. The general, however, is but a name, consequently it cannot be a thing. I think it is needless to give a detailed criticism of this view for I have already proved in what precedes that things may also be general.

The attributive theory has developed in modern logic. According to this, the relation of the subject and the predicate is to be seen as a relationship between an object and its property.

> The subject in the logical sense is the object about which we say something. The predicate expresses the attributes of the object, i.e. its properties and relations. Radium is a radioactive element.[20]

So, the essence of a proposition is to affirm or deny whether a property belongs to the object. The attributive relation is directly manifest, for example, in the statement: a tree is green. There are, however, propositions, for example 'a tree is a plant', where this relation is present in a hidden form only. In order to reveal the relation, the proposition has to be reformulated, for example, like this: 'a tree has the attributes of plants'. From this consideration some have arrived at the following conclusion:

> The form of simple propositions reflects the objectively existing relation of the properties and the object. Consequently, all simple propositions are of attributive character, that is, they either affirm, or deny the fact that some property belongs to the object concerned.[21]

---

[19] Aristotle. 1928. 17a25–26.
[20] Fogarasi, p. 187.
[21] Tavanets, p. 32.

The attributive relation is usually interpreted from the point of view of subordination or identity. A relation of subordination is manifest in the well-known axiom: the property of the property of an object is also the property of this object, and what is contradictory to the property of an object is also contradictory to this object (*nota notae...*). For example: it is a property of sodium that it is a metal. It is the property of metal that it is an element. Hence it is a property of sodium, too, that it is an element. This axiom is the attributive version of '*dictum de omni et nullo*'.

It is the following relation which is usually referred to in the other case: the proposition '*S* is *P*' implies that *S* objects are identical with all objects which have *P* properties, that is, they have this very *P* property in common. The proposition 'metals are flexible' expresses the identity of metals with objects which have the property of flexibility. To make this relation obvious, it is practical to reformulate the proposition in our example like this: metals are flexible objects.

Finally, there is also a view which reckons with several kinds of relations between objects and properties.

> The relationship of the subject and the predicate mediated by the copula expresses *as content* that the object of the proposition has a certain property (or properties). The term 'property' is used here in its widest possible sense. It may mean the quality (attribute) of the object, e.g. 'a rose is red'. It may express the essence of the object, especially in definitions, e.g. 'by endocrine secretion we mean the development and excretion of matters in the blood—hormones—which have a specific chemical effect on the tissues of the organism'. It may imply that the object belongs to a class, e.g. 'a fox is one of the canidae'.[22]

The attributive theory holds true so far as there are, indeed, propositions whose function is to express the relation of objects and properties. To accept nothing but this relation in simple propositions is, however, the basic mistake of this view. As a result, it tacitly shares the above opinion of nominalists.

Some (e.g. Ziehen) add what is called a substantial relation to the attributive one. In this interpretation the subject expresses the property

[22] *Logika*. (Budapest 1956). p. 149.

and the predicate the object, è.g. 'what takes nourishment is a living being'. Since such propositions occur relatively rarely this relation is seldom mentioned in the literature of logic.

This one-sidedness can be avoided by taking into consideration the following variants. For simplicity's sake let us choose propositions composed of two terms (S and P):

(I)	S object is P object	e.g.	all apples are fruit
(II)	S object is P property	e.g.	some roses are red
(III)	S property is P object	e.g.	some red are roses
(IV)	S property is P property	e.g.	all lemon-yellow are yellow

5. THE EXTENSION OF TERMS

A

I

Socrates and Plato agreed that terms should express the general. They disagreed, however, on the following: Socrates did not separate general from singular while Plato did so, and even rigidly. Aristotle was highly critical of Plato's approach and argued that the general is comprised by the singular.

Aristotle, on the other hand, accepted general and singular as equal in rank.

> Some things are universal, others individual. By the term 'universal' I mean that which is of such a nature as to be predicated of many subjects, by 'individual' that which is not thus predicated. Thus 'man' is a universal, 'Callias' an individual. Our propositions necessarily sometimes concern a universal subject, sometimes an individual.[23]

On the other hand, he acted in a onesided way by pushing the general into the foreground: "...every science is of universals and not of *infimae species*..."[24]

[23] Aristotle. 1928. 17a38–17b2.
[24] Aristotle, *Metaphysica*. 1059b26.

Traditional logic followed in Aristotle's footsteps also in this respect: if there is no indication for its extension, the term at issue is to be considered as a general one.

Let us denote a term at random by S. Theoretically S can be empty and not empty respectively. So I shall consider S as the sign of a term with an indefinite extension. Terms with a definite extension can then be expressed by taking of all S, some S, etc.

But is it necessary to indicate extension even if a concrete term is substituted for S? Mill gave the following answer:

> A general name is familiarly defined, a name which is capable of being truly affirmed, in the same sense, of each of an indefinite number of things. An individual or singular name is a name which is only capable of being truly affirmed, in the same sense, of one thing.
>
> Thus, *man* is capable of being truly affirmed of John, George, Mary, and other persons without assignable limit; and it is affirmed of all of them in the same sense; for the word man expresses certain qualities, and when we predicate it of those persons, we assert that they all possess those qualities. But *John* is only capable of being truly affirmed of one single person, at least in the same sense. For, though there are many persons who bear that name, it is not conferred upon them to indicate any qualities, or anything which belongs to them in common; and cannot be said to be affirmed of them in any *sense* at all, consequently not in the same sense.[25]

Mill is right so far as we want to denote a certain person by the name John. John is, however, general so far as it refers to all men of this name. If this general aspect is disregarded it leads to numerous misunderstandings and confusion in everyday life. Such declarations in the press as, for example, 'I am not identical with the John Smith who has misappropriated public funds', try to avoid these very misunderstandings.

Thus, the name 'John' can be equally understood as singular and general respectively. That is why:

> The singular term 'Jones' is ambiguous in that it might be used in different contexts to name any of various persons, but it is still a singular term in that it purports in any particular context to name one and only one person. The same is true even of pronouns such as 'I' and 'thou'; these again

[25] Mill, p. 28.

are singular terms, but merely happen to be highly ambiguous pending determination through the context or other circumstances attending any given use of them. The same may be said of 'the man', or more clearly 'the President', 'the cellar'; these phrases (unlike 'man', 'president', and 'cellar' themselves) are singular terms, but the one and only one object to which they purport to refer in any given use depends on attendant circumstances for its determination.[26]

Klaus has called attention to the following:

> One speaks, for instance, of *general terms*, but the concept of the general itself is usually rather vague. As we have already ascertained, all terms are in themselves general except for those which denote a particular individual. The general can be taken in quite a number of different ways; for instance, one can denote all generic and specific terms as general since they apply to many individuals and not merely to one.
>
> One can also denote the so-called plural terms as general terms, for instance, such terms as 'the students at the Philosophy Department', which can be considered general terms because they apply to a majority [plurality] of objects. But one could with just as much justification from a different point of view call the term 'the students at the Philosophy Department' a singular term, because taken as a whole, they constitute one individual within the Philosophy Department.
>
> Finally, any universal term can be understood as a general term, for instance, the term 'all spruce trees', since not only a number of objects is meant here, but also at the same time every single such object.[27]

After the above considerations, I find it necessary to indicate extension in order to avoid misunderstanding and to guarantee accuracy. All terms where extension is not indicated will be considered as indefinite terms in what follows.

It was essentially the Aristotelian approach that traditional logic followed in the extensive classification of terms. Let us look at some typical departures from this view:

[26] Quine, pp. 203–204.
[27] Klaus, pp. 161–162.

... there is no such thing as abstract or general ideas, properly speaking; but that all general ideas are, in reality, particular ones, attached to a general term, which recalls, upon occasion, other particular ones, that resemble, in certain circumstances, the idea, present to the mind. Thus when the term Horse is pronounced, we immediately figure to ourselves the idea of a black or a white animal, of a particular size or figure. But as that term is also usually applied to animals of other colours, figures and sizes, these ideas, though not actually present to the imagination, are easily recalled; and our reasoning and conclusion proceed in the same way, as if they were actually present. If this be admitted (as seems reasonable) it follows that all the ideas of quantity, upon which mathematicians reason, are nothing but particular, and such as are suggested by the senses and imagination... [28]

Hume was right in accepting particular terms, yet it was mistake to regard them as contrary to general terms.

Kant took the following position as regards the classification of terms:

It is a mere tautology to talk about general or common concepts;—a mistake which is based on an incorrect division of concepts into *general*, *particular*, and *singular*. Not the concepts themselves—only *their use* can be so divided. [29]

This view was based on the supposition that all notions, or terms, are general by nature. This position will be criticized in detail later on.

In order to sum up and complete the versions discussed above let us have a look at the following figure:

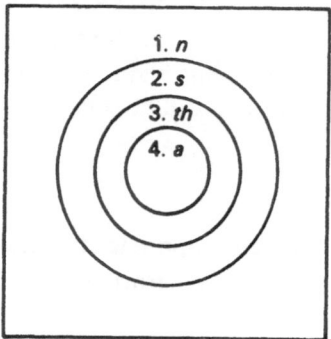

Fig. 18.

[28] Hume, *Enquiries* p. 158.
[29] Kant, *Schriften* p. 521.

Rings and circles must be distinguished in this figure. A ring marked off by two circular lines represents a definite quantifier. A circle, however, includes all quantifiers indicated on its territory.

The meaning of the signs is: n = nothing (empty), s = some (particular), th = this (singular), a = all (general).

Ring 2 represents 'only some' while ring 3 denotes 'only this'. Rings 2 and 4 together mean 'some but not only this' which equally excludes both the singular and the empty. Rings 2 and 3 together represent 'some but not all', and I mean by it what rules out both the general and the empty. The universe (u) must be taken into account as well. Since four elements can have 16 combinations, they can be summed up as follows (o = or):

0	—	not u	2 o 3	—	s, but not a
1	—	n	2 o 4	—	s, but not only th
2	—	only s	3 o 4	—	th
3	—	only th	1 o 2 o 3	—	not a
4	—	a	1 o 2 o 4	—	not only-th
1 o 2	—	not th	1 o 3 o 4	—	not only-s
1 o 3	—	only th o n	2 o 3 o 4	—	s
1 o 4	—	a o n	1 o 2 o 3 o 4	—	u

With persons in mind, the basic operators are: (1) nobody, (2) some, somebody, (3) I, you, he, etc., (4) everybody, anybody.

II

It is a common attitude in philosophical and logical manuals that the discussion of universals (a) concerns the way of existence of general terms and (b) is confined to medieval philosophy.[30] In my opinion—partially in agreement with the views of certain authors to be mentioned later—both of the above assertions are of restricted value. On the one hand, I try to prove that this discussion has covered numerous fundamental problems of logic and on the other, I want to document that it started as early as in ancient logic and has been going on without significant interruptions ever since.

[30] Cf.: Eisler, p. 1584 ('Universalien').

This controversy goes back to Socrates who started from the fact that
the material world is liable to change. But one cannot obtain firm scientific
knowledge of something which keeps on changing. Therefore cognition
should strive for grasping what is general and permanent. This aim can be
reached by the development of notions, or terms, since they allow us to
register the permanent.

Terms, however, are usually neither definite nor unambiguous.
Therefore one of the most important functions of cognition is to elucidate
terms. This demand can be met by a constant effort to give an exact
definition of every term. The most significant means of definition is
induction. "... for two things may be fairly ascribed to Socrates—
inductive arguments and universal definition, both of which are concerned
with the starting-point of science:—but Socrates did not make the
universals or the definitions exist apart ..."[31]

The category of the universal played an important part in Plato's
theory.

> For, having in his youth first become familiar with Cratylus and with the
> Heraclitean doctrines (that all sensible things are ever in a state of flux and
> there is no knowledge about them), these views he held even in later years.
> Socrates, however, was busying himself about ethical matters and
> neglecting the world of nature as a whole but seeking the universal in these
> ethical matters, and fixed thought for the first time on definitions; Plato
> accepted his teaching, but held that the problem applied not to sensible
> things but to entities of another kind—for this reason, that the common
> definition could not be a definition of any sensible thing, as they were always
> changing. Things of this other sort, then, he called Ideas, and sensible
> things, he said, were all named after these...[32]

Aristotle had several objections to this theory of ideas. According to
Plato each thing, species, and genus had a separate idea. This view is
(logically) indefensible because (a) it leads to an unnecessary doubling of
the world, (b) Plato is unable to prove the independent existence of even
one single idea, (c) if not only the individual things but even genera and
species have ideas, there will be more ideas than actually existing things.

[31] Aristotle, *Metaphysica*. 1078b28–31.
[32] *Op. cit.*, 987a33–987b9.

While Plato held that the universal (the idea) exists outside the individual, Aristotle was of the opinion that the universal (the form) is in the individual things themselves.

Lenin made the following remark:

> In Aristotle, objective logic is *everywhere confused* with subjective logic and, moreover, in such a way that everywhere objective logic is *visible*. There is no doubt as to the objectivity of cognition. There is a naive faith in the power of reason, in the force, power, objective truth of cognition. And a naive *confusion*, a helplessly pitiful confusion in the *dialectics* of the universal and the particular—of the concept and the sensuously perceptible reality of individual objects, things, phenomena.[33]

As a counterpoint to the objective interpretation of the universal, there developed a subjectivist approach in ancient idealistic philosophy. It can be detected in an embryonic state as early as the Sophists. Gorgias proceeded from the supposition that people can communicate their thoughts by means of words and signs only. However, words and signs respectively are not identical with the things they stand for. Therefore individuals can attach any meaning to words they like.

Similar views were professed by other Sophists as well. Diogenes Laertius recorded of Protagoras that in disputes he paid attention to words only, not caring about their meaning.[34]

It was the Stoics who further developed this conception and studied logic in its relation to rhetoric and grammar. The relation of language and thought stood in the centre of their investigation. While Aristotle considered thoughts to be prior to linguistic expressions, the Stoics took the latter as a starting-point and subordinated thought content to them. What the followers of the Aristotelian and the Stoic approach agreed about was that logic should treat universal relations and not particular thought contents. According to the followers of Aristotle, however, the universal as a form of spiritual nature is in the things themselves, while the Stoics held the view that the universal is independent of the individual and becomes manifest in linguistic expressions.

---

[33] Lenin, *Philosophical Notebooks*. p. 368.
[34] *Leben und Meinungen berühmter Philosophen*. Vol. II. (Berlin 1955). p. 186.

Both the objective and the subjective approaches were maintained by several schools of thought from the 3rd century B.C. to the 7th century A.D. The former was represented by the Platonists and the Peripatetics and the latter by philosophers who belonged to the middle and late Stoa respectively. Let me mention here only the Neoplatonist Porphyry who put the following questions in his commentary on the *Organon*: does the universal exist in, or outside of, things? Is it in, or outside of, the mind? Is it corporeal or incorporeal? Having posed these question, Porphyry became one of the forerunners of the medieval controversy over universals.

The characteristic feature of the period of transition to scholasticism is the endeavour to reconcile conflicting views. According to Erigena there is a close interdependence between the names of things and the things themselves. The names of things are not external to them but correspond to their nature. Cognition leads us from the names to the things themselves.—Ibn-Sina (Avicenna) supposed that the universal existed in three ways: (a) in a divine sense, (b) as the essence of an individual thing, (c) in a human sense.

Views on the question of whether or not universals existed in reality were divided into two main schools of thought in medieval philosophy. The realists were of the opinion that universal notions, or terms, had the same existence as individual phenomena. This school had taken up the position of objective idealism. As to the interpretation of the way of existence of universals, however, the realists held different views. In this respect two typical schools of thought had developed, those of extreme and moderate realism.

The extremists among the realists claimed that the universal was independent of individual phenomena. In their view there existed not only square or round tables, yellow or brown tables, etc. in reality, but also— irrespective of them—the table as such. That is, there exist both individual and universal things but the universal, as it were, is prior to the individual (*universale ante rem*). This conception is a revival of Plato's theory of ideas.

Based on this principle, Anselm had developed an ontological argument for God's existence: if something is absolutely perfect, it exists. His contemporary, Gaunilo, had pointed out that, accordingly, whatever was most perfect of its kind (e.g. an island) had to exist. Anselm replied that his argument only concerned a being most perfect in all respects. However, he could not resolve the tautology of this argument whereby God exists because he is perfect, but only what exists can be perfect.

The moderate realists (Albertus Magnus, Thomas Aquinas, etc.), on the other hand, believed that the universal existed in individual phenomena (*universale in re*), and not independently of them. This concept was also idealistic since it viewed the universal as ethereal.

The nominalists confronted realism by claiming that the universal concepts were but names, which are preceded by the individuals (*universale post rem*). Nominalists disagreed among themselves just as much as realists did. The early nominalists (e.g. Roscellinus, Duns Scotus), assumed that universals, though not existing objectively, were nevertheless based on reality, and words were not arbitrary inventions but natural formations.

There were certain elements of materialism in this theory. As Marx has pointed out:

> Nominalism, the *first form* of materialism, is chiefly found among the *English* schoolmen.[35]

Referring to this quotation, some Marxist philosophers have identified nominalism with materialism.

In my opinion this identification is mistaken. Not every nominalist was a materialist, and not every materialist was a nominalist. At that time, it is true, materialism often appeared in the guise of nominalism. In the Middle Ages materialist philosophy was persecuted to such an extent that it could not be openly stated. Nominalism offered the most suitable ideological framework for the expression of the materialist position.

Late nominalism differs significantly from its early formulation. First of all, in late nominalism universals are thought to be independent of reality

---

[35] Marx/Engels, *Die heilige Familie*. p. 257; Eng. tr. cited from *Collected Works 4* (New York, n.d., c. 1978) p. 127. (The German text gives 'Materialisten' where the English has 'schoolmen'—Ed.).

and exclusively of a spiritual nature, free of materialist elements and every notion is imbued with subjective idealism.

Thus, similarly to realism, a distinction can be made between moderate (early) and extreme (late) nominalism.

Other schools of subjectivism also can be mentioned here, for example, conceptualism. According to its followers (e.g. Occam) the universal exists in concepts. The terminists (e.g. Buridan) took up the position that universals can be concepts just as well as words. Abélard is thought to be the founder of sermonism. In his view the universal exists in propositions.

In the history of the controversy over universals three distinct periods can be seen. The first period (11–12th c.) was dominated by Platonic realism, the second (13th c.) and the third (14–15th c.) witnessed the prevailing influence of Aristotelian realism and nominalism respectively. While realism was to support the official theological school of thought, nominalism had taken a relatively dissenting stand. Thus, the doctrine of the Holy Trinity was attacked, for example, by Roscellinus who claimed that existing substances are necessarily unique.

III

The above trends can be detected in modern logic as well. As a counterreaction to scholastic logic—among the first—the Platonic academy (Plato, Bessarion, Ficino) was founded in Florence in the 15th century. Later the position of Platonic realism was taken up even by what is called the Cambridge school (More, Cudworth).

Leibniz, too, was an adherent of realism and had developed the following theory: first of all, a whole system of categories should be drawn up from which true propositions could be derived by an operation called *calculus ratiocinator*. Any other truth could be deduced and proved from, and on the basis of, these fundamental propositions. It is not difficult to see a similarity with the idea of Platonic classification.

An essentially new approach was introduced by Hegel who studied the question concerned from the point of view of dialectics. Contrary to his predecessors, he did not separate the universal from the individual but insisted on their being inseparable. I do not analyse his concept at this point for it will be discussed at length later.

It must be mentioned, however, that Hegel could not get away from Plato's objective idealism.

> We add a remark upon the account of the origin and formation of notions which is usually given in the Logic of Understanding. It is not *we* who frame the notions. The notion is not something which is originated at all. No doubt the notion is not mere Being, or the immediate: it involves mediation, but the mediation lies in itself. In other words, the notion is what is mediated through itself and with itself. It is a mistake to imagine that the objects which form the content of our mental ideas come first and that our subjective agency then supervenes, and by the aforesaid operation of abstraction, and by colligating the points possessed in common by the objects, frames notions of them. Rather the notion is the genuine first; and things are what they are through the action of the notion, immanent in them, and revealing itself in them.[36]

As compared to Hegel's dialectical view, the metaphysical approach of so-called pure logic (Lotze, Meinong, Husserl, Pauler) represented a great regression. This trend was also basically Platonic. According to Lotze the world of ideas is composed of eternal, timeless truths. The continuance of ideas has its validity, as opposed to the existence of things. The ideal sphere is the unchanged archetype of the ever changing reality. Logic should deal with these very pure forms.

The modern renaissance of Aristotelian realism started in the 19th century. The first of its representatives to be mentioned is Überweg who had taken a materialist position. In his opinion general notions, or terms, correspond to the objective essence of things. Furthermore, propositions conform to objective relations, inferences to objective rules, and the system to the totality of objective existence.

Just as in the Middle Ages, nominalism was interwoven with materialist views in certain philosophers' thinking, so also at the beginning of modern times. Hobbes, for example, "systematized Bacon's materialism" and at the same time considered logic as a formal study of calculation with notions, or terms. He believed that propositions are the complexes of signs

---

[36] Hegel, *System* (1929) pp. 360–361; Wallace tr. pp. 293–294.

denoting concepts and drew a definite line between two kinds of truth in propositions. A proposition is logically true if it can be formally deduced from the axioms of the given system, and it is empirically true so far as it is based on the meaning of certain signs.

Soon, however, the tide turned: again, the philosophers representing subjective idealism took the leading role in propounding nominalism. In this respect let me refer to nominalism interwoven with positivism (Dühring, Schuppe, Laas, Mach). Laas denied that logical propositions contain any knowledge and considered them to be of tautological character. Although Mach did not challenge the cognitive content of these propositions, he thought that the primary function of logic is to ensure the economy of thinking. If some proposition is recognized to be a logical conclusion of another, it saves us the study of the former.

These trends are discernible in mathematical logic as well. Let us begin with one of the forerunners of this logic, Bolzano. In his opinion notions, or terms, have a timelessly valid essence, irrespective of their being conceived or pronounced (*Vorstellung-an-sich*). The objects of logic are the self-sufficing terms and propositions and their relations.

Frege, too, adopted the approach of objective idealism. He meant by thought not the subjective activity of thinking but its objective content which may become public property.[37] According to Frege cognition is:

> ... an activity that does not create what is known but grasps what is already there. The picture of grasping is very well suited to elucidate the matter. If I grasp a pencil, many different events take place in my body: nerves are stimulated, changes occur in the tension and pressure of muscles, tendons, and bones, the circulation of the blood is altered. But the totality of these events neither is the pencil nor creates the pencil; the pencil exists independently of them. And it is essential for grasping that something be there which is grasped; the internal changes alone are not the grasping. In the same way, that which we grasp with the mind also exists independently of this activity, independently of the ideas and their alterations that are a

[37] Frege, '*Über Sinn und Bedeutung*'. p. 32.

part of this grasping or accompany it; and it is neither identical with the totality of these events nor created by it as a part of our own mental life.[38]

It is H. Scholz who has represented the position of Platonism most definitively among recent philosophers by claiming that the objects, and also the field of work, of mathematics exist by themselves (*an-sich*) as do the Platonic ideas.[39]

Among the representatives of moderate realism, first of all, Boole should be mentioned. I share Bochenski's opinion that there is a great similarity between the relation of Boole to Leibniz, on the one hand, and that of Aristotle to Plato, on the other.[40] Later this trend has become clear most of all in Church's views. In his opinion logic is related to ontology by the existential operator.[41]

Russell has explicated his main ideas partly in the work he had written together with Whitehead,[42] and partly in his book *Our Knowledge of the External World* (London 1914). Russell himself has classified his theory as realism: logical atomism, the only school of philosophy he wants to defend, has had a strong influence on what he calls neorealism.[43]

Indeed, Russell has postulated the existence of the external world but excluded this question from logic. He considered the task of logic to be the revelation of the simplest—as he put it, atomic—facts, and the registration of the same in elementary propositions. If two, or more, elements are connected by simple logical relations (conjunction, implication, etc.) we obtain what he called molecular propositions.

There are, however, universal propositions which have to be distinguished from the former. Generalization proceeds from the recognition that there are infinitely many things, for if there were but a finite number of them all universal propositions could be composed out of elementary ones and they would become common molecular propositions. Whoever considers, for example, the proposition 'all men are mortal' as a

---

[38] Frege, *Grundgesetze* p. XXIV; Montgomery Furth tr., pp. 23–24.
[39] Scholz–Hasenjaeger, p. 1.
[40] Bochenski, p. 347.
[41] Church, 'Ontological Commitments'. p. 1013.
[42] Russell and Whitehead, *Principia Mathematica*.
[43] Russell, *Our Knowledge of the External World*. p. 14.

proposition of the subject-predicate type postulates that, apart from individual men, there exists also all men. The class of men, however, does not exist, being but a logical fiction. To avoid mistaken philosophical speculations, the proposition concerned should be reformulated as follows: it is true of every individual that, if he is a man, he is mortal. This reasoning shows the obvious traits of nominalism and positivism.

While Russell has still preserved certain objective elements by admitting the existence of individual things, however restricted, his disciple, Wittgenstein, on the contrary, held the view that the world consists of empirical facts, and not of things. The similarity to the way Neo-Kantians tried to get rid of Kant's *Ding-an-sich* is rather striking.[44]

Wittgenstein's views started the development of neo-positivist logic. The important representatives of this trend are the Vienna Circle (Carnap, Neurath, Reichenbach) and the Polish school (Łukasiewicz, Łesniewski, Tarski). Let me mention from among what has been written on the subject after the Second World War 'Steps towards a Constructive Nominalism' by Goodman and Quine (*Journal of Symbolic Logic,* 1947/12) and *Methods of Logic* by Quine.

Finally, let us see a materialist interpretation:

> What is the basis for the formation of general terms? This basis is to be seen in the fact that in reality there are not merely individual things, as nominalism maintains, but also classes of things (species, genera) ...
>
> Humans form concepts, and the basis of the concepts is the *existence* of classes. It was not Linnaeus who formed the class of vertebrate animals. It existed independently of his scientific research. Rather, Linnaeus formed the scientific concept of the class of vertebrates. The nominalist position denies the objective existence of classes in reality and sees them as merely subjective products of human mental activity. Present day nominalism is almost always associated with idealistic philosophies. Its social function is quite different now than it was at the time of the controversy over universals. At that time it retained materialistic traits, in as much as it stood in confrontation with the Platonic idealist conception of the pre-existence of general concepts and ideas and turned the attention of science from metaphysical speculation on universals to the study of concrete individual things. Today however it has become an impediment to the progress of

[44] Wittgenstein, *Tractatus.*

human knowledge. For instance, the denial of the existence of social classes and the attribution of this category to merely subjective human formulation and stipulation serves in fact to disguise the real social relations.[45]

IV

The controversy over universals may very well command interest even nowadays not only because it has not come to an end, but also because it touches upon several problems of logic directly or indirectly. Let us first of all see the theory of terms. There the controversy concerns the relation of general and singular terms, generalization and limitation, division and classification. It has an influence on the interpretation of universal propositions. It plays an important role in the assessment of sub- and superordinating inferences, inductive and deductive inferences, and proof respectively.

From the above historical survey, primarily, the following conclusion can be derived: the conflict between realism and nominalism is one of the main forms the confrontation of objective and subjective idealism can take. Nominalists argue that universals exist in human consciousness only, as opposed to realists who take them as something independent of the mind, though of spiritual nature in some way. Eclectic views are also quite common.

As to the interpretation of universals, it is between materialism and idealism, and not between realism and nominalism that the main dividing line can be drawn. According to materialism, the universal exists in material reality. There are certain cases where materialism and idealism may come closer to each other. Under the force of circumstances, as we have seen, materialist views were propagated in the guise of nominalism. On the other hand, there are materialist elements in realism, too, e.g. with Aristotle. It is, however, not a consistent materialism by any means. The efforts to reconcile materialist views either with realism, or with nominalism, have resulted in eclecticism, and, in the long run, in idealism.

From the point of view of methodology, it is the metaphysical and dialectical attitudes that are opposed to one another. Metaphysical thought separates the universal and the individual and does so mainly in

[45] Klaus, pp. 143–144.

two ways. On the one hand, it is only the individual and not the universal which is admitted to exist in reality. That is, the former may be objective as well, while the latter is only subjective. Thus, the categories concerned are established at two different levels, and this operation even in itself results in their separation. On the other hand, both of them are dealt with at the same (subjective or objective) level, but in a way as though they were categories independent of each other.

Hegel has pointed out that what is indeed universal cannot be separated from the individual:

> If e.g. we take Caius, Titus, Sempronius, and the other inhabitants of a town or country, the fact that all of them are men is not merely something which they have in common, but their universal or kind, without which these individuals would not be at all. The case is very different with that superficial generality falsely so called, which really means only what attaches, or is common, to all the individuals. It has been remarked, for example, that men, in contradistinction from the lower animals, possess in common the appendage of ear-lobes. It is evident, however, that the absence of these ear-lobes in one man or another would not affect the rest of his being, character, or capacities: whereas it would be nonsense to suppose that Caius, without being a man, would still be brave, learned, etc.[46]

Formal logic concentrates on the abstract universal. As the term itself implies, one has to disregard the individual objects, and their specific properties, and to concentrate only on what they have in common in order to arrive at the abstract universal. This interpretation of the universal, among others, has also contributed to the development of the view according to which general notions, or terms, are but names.

The speculations of dialectical logic are concerned with the concrete universal which coincides with the substance of things. Substance is the total amount of properties a certain object or phenomenon cannot do without, as the properties distinguish it from other objects and phenomena. The universal interpreted in this way becomes an essential condition for the existence of individual objects.

Now let us compare the above tendencies from another point of view. The fundamental contradiction in ancient logic was between the

---

[46] Hegel, *System* 1929 pp. 377–378; Wallace tr. p. 309.

Aristotelian and the Stoic logic. The followers of both schools kept on emphasizing, first of all, what separated them from one another. Nevertheless, a reconciliation of conflicting views was also attempted. Among others, Apuleius (2nd c.), and later Boethius (6th c.), tried to reach this goal.

The nominalists in the Middle Ages labelled the Aristotelian system as old logic (*logica vetus*), and that of the Stoics as new logic (*logica nova*). It requires no special explanation that the indication of their having been developed at different times was but of secondary importance. In fact, it was meant to emphasize that the former was an outmoded way of thinking as compared to the latter.

Some called previous logic as a whole (i.e. both *logica vetus* and *logica nova* included) antique logic (*logica antiqua*) and termed nominalism modern logic (*logica moderna*). But as soon as nominalism had become differentiated, Duns Scotus and his followers were also ranked among the old (*antiqui*), and nothing but the logic of Occam and his adherents was considered modern (*moderni*).

The antithesis of the old and the new logic, however, was not absolute even with their most extreme representatives. The modern were drawing upon the Aristotelian logic, and the old logic became more and more imbued with certain elements of what was called new logic. Similarly to ancient times, this period again witnessed some efforts to reconcile these two kinds of logic. It was Petrus Hispanus who had attained this objective in the most effective way. The first six chapters of his work *Summulae logicales* covered the old logic, and the seventh chapter dealt with the new one. This work was used as a basic reference book for centuries.

It was in the 19th century that mathematical logic started to develop. Soon it was thought to be the only modern form of logic. According to the representatives of this view, there was traditional logic on the one hand, and modern formal logic on the other. It would be a mistake to assume that only traditional formal logic was considered old. Dialectical logic seen merely as one of the trends of the last century was also included in it.

In fact, since mathematical logic has developed, such personalities as Boole and Schröder are remembered, if at all, with some reverence only. Bochenski still ranks among the modern those who adopted the approach of Frege and Russell respectively. Nowadays the following distinction prevails: two-term propositions, and predicate logic respectively, are called classical, and whatever goes beyond them is non-classical.

What lessons can be drawn from the above survey? The attitude characterized by a fetishism of the new has been followed in its historical outline. This position is false in at least two aspects. On the one hand, what is new may be a regression as compared to the old. On the other, those who had introduced these distinctions invariably meant by 'new' a logic of a nominalistic character. They do not find this adjective adequate, for example, for the Port-Royal logic, though it was supposed to be new in its own time as compared to Occam's theory.

Some see merely a historical interest in the old. But was it not exactly a return to ancient times which made the development of the renaissance possible? Evolution can best be illustrated by a spiral symbolizing the reappearance of the old on a higher level. New and old are relative concepts. New becomes old, and old may revive.

The term old logic was designed, first of all, to discredit realism. It is easy to understand that materialism was also relegated there. But is it indeed the right expression? No, it is not—for on the one hand, realism and nominalism are trends equal in rank within idealistic logic, and on the other, both of them are equally old and equally new respectively, as illustrated by our historical survey.

There seem to be two alternatives: either to accept one of the two, or to strive for their reconciliation. Both solutions contain elements of truth, especially as far as the criticism of the other is concerned. However, the basic question—the relation of the universal and the individual—has not been satisfactorily answered by any of them. Consequently, the above alternatives are unacceptable. I agree with those who believe that this problem can be solved only within the framework of materialist dialectical logic.

B

I

As far as the degree of universality is concerned, there is a difference between *specific terms* and *generic terms*. Plato was the first to discuss their relation in detail. He has demonstrated how one arrives from specific terms to more and more general generic terms, and finally to categories, in the course of forming notions. He regarded categories not only as a result, but also as a starting-point proceeding from which one can reveal the whole range of relevant specific terms by division.

From among Aristotle's statements on this subject let me cite first the one about the ambiguity of the word 'genus':

> The term 'race' or 'genus' is used if generation of things which have the same form is continuous, e.g. 'while the race of men lasts' means 'while the generation of them goes on continuously'. . . There is genus in the sense in which 'plane' is the genus of plane figures and 'solid' of solids; for each of the figures is in the one case a plane of such and such a kind, and in the other a solid of such and such a kind; and this is what underlies the differentiae. Again, in definitions the first constituent element, which is included in the 'what', is the genus, whose differentiae the qualities are said to be.[47]

In the above text genus is interpreted as something objectively existing, on the one hand, and as a term, on the other. It also becomes evident that generic terms are based on objective reality. Medieval logic distinguished between *genus naturale* and *genus logicum*. The latter, however, was supposed to be based on thought and not on objective reality. This view plays an important role in modern logic, too. Nevertheless, those who admit the objective basis of general terms would be inconsistent if they refused to accept the same view with regard to specific and generic terms as well.

Now let us see what is meant by generic and specific terms respectively.

> A 'genus' is what is predicated in the category of essence of a number of things exhibiting differences in kind. We should treat as predicates in the category of essence all such things as it would be appropriate to mention in reply to the question, 'what is the object before you?'; as, for example, in the case of man, if asked that question, it is appropriate to say 'He is an animal'. The question, 'Is one thing in the same genus as another or in a different one?' is also a 'generic' question; for a question of that kind as well falls under the same branch of inquiry as the genus: for having argued that 'animal' is the genus of man, and likewise also of ox, we shall have argued

---

[47] Aristotle, *Metaphysica*. 1024a29–1024b6.

that they are in the same genus; whereas if we show that it is the genus of the one but not of the other, we shall have argued that these things are not in the same genus.[48]

As this quotation has revealed, the Stagirite examined genus both from the point of view of intension and from that of extension. From the first point of view, genus implies the essence and inalienable nature of something. From the second, it is the total amount of objects reflected in it.

Porphyry regarded the relation of specific and generic terms as a basically extensive relation by claiming that genus includes species while species is included in genus but does not include it, for the genus is predicated of more things than the species.[49]

It was widely held in medieval logic that certain terms should be considered exclusively as species and others exclusively as genus.

> With regard to simple suppositions, one type is that of a general term *functioning as the subject,* for example, 'Man is a species'; another type is that of a general term *functioning as predicate* in an affirmative, for example, 'Every man is an animal'; the term 'animal' functioning as predicate has simple supposition, because it denotes only the nature of the genus; [50]

Kant, on the contrary, insisted on the relativity of specific and generic terms:

> The higher concept is called, with respect to its lower concept, *genus*; the lower concept, in regard to the higher, *species*.
> Just as higher and lower, so too are the *generic* and *specific* concepts different in logical subordination, not due to their nature but only in regard to their relation to one another (*termini a quo* or *ad quod*).[51]

He gave primacy to the aspect of intension, in contrast to Porphyry's approach.

---

[48] Aristotle. 1928. 102a32–102b3.
[49] Porphyrius, 15, 15.
[50] Peter of Spain, *Treatise on Suppositions*. English from Herman Shapiro *Medieval Philosophy* p. 297.
[51] Kant, *Schriften* p. 527.

The lower concept is not contained in the higher; for it contains *more* in itself than does the *higher*; but it is still contained *under* the latter, because the higher contains the ground of knowledge of the lower. [*op. cit.*, p. 529.]

Traditional logic overestimated both specific and generic terms. As a consequence the study of other kinds of terms was neglected.[52]

II

As to general terms, let us compare Aristotle's following two statements.

Whenever the subject, for which we must obtain the attributes that follow, is contained by something else, what follows or does not follow the highest term universally must not be selected in dealing with the subordinate term (for these attributes have been taken in dealing with the superior term; for what follows animal also follows man, and what does not belong to animal does not belong to man); but we must choose those attributes which are peculiar to each subject. For some things are peculiar to the species as distinct from the genus; for species being distinct there must be attributes peculiar to each. Nor must we take as things which the superior term follows, those things which the inferior term follows, e.g. take as subjects of the predicate 'animal' what are really subjects of the predicate 'man'. It is necessary indeed, if animal follows man, that it should follow all these also. But these belong more properly to the choice of what concerns man.[53]

This argumentation implies, on the one hand, that if the generic term (e.g. man) of some subject (e.g. Greek man) has been given, it is unnecessary to indicate other terms superior to the former (e.g. animal), since the intension of the term 'man' contains that of the term 'animal'. On the other hand, if the specific term of a subject has been given, it is unnecessary to add terms inferior to it since the properties of animal apply not to man only but also to Greek man.

And now let us see Aristotle's other thesis:

[52] Fogarasi, p. 133.
[53] Aristotle. 1928. 43b23–33.

...there is no necessity that all the attributes that belong to genus should belong also to the species; for 'animal' is flying and quadruped, but not so 'man', All the attributes, on the other hand, that belong to the species must of necessity belong also to the genus; for if 'man' is good, then animal also is good.[54]

That is to say, the intension of generic terms is wider than that of specific ones.

Porphyry found only the first of these Aristotelian theses to be valid. It is this view on which the division known as Porphyry's tree was based. The 'higher' the position of a term on this tree, the narrower is its intension:

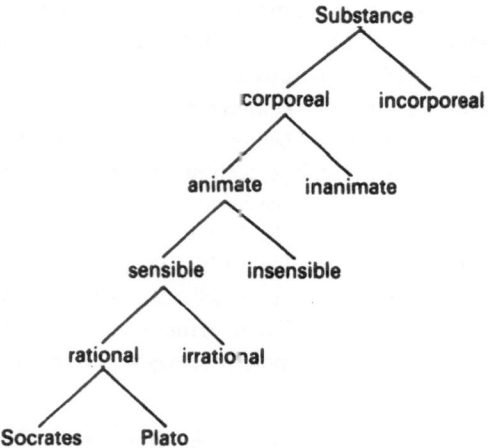

According to Bacon there are two ways to arrive at the general:

Both ways set out from the senses and particulars, and rest in the highest generalities, but the difference between them is infinite. For the one just glances at experiment and particulars in passing, the other dwells duly and orderly among them. The one, again, begins at once by establishing certain abstract and useless generalities, the other rises by gradual steps to that which is prior and better known in the order of nature.[55]

---

[54] *Op. cit.*, 111a25–28.
[55] Bacon, *Selection* p. 333 (*The New Organon* 1 (XXII)).

Bacon opposed the real general, which is based on an analysis of experiment, to the speculative general. At the same time, however, he failed to go beyond the concept expressed in Aristotle's first statement.

The views developed in traditional logic can be summed up as follows. The generalization (*generalisatio*) of terms is a process which leads from a given term to a wider one. By this operation the closest genus of the given term must be found. For example, by generalizing the term 'iron' we arrive at the 'metal'. In finding the closest genus, it may be instrumental to abstract from all those qualities in which the given term differs from the ones coordinated to it, that is, in this case from the terms 'copper', 'lead', etc. That is how we arrive at a more general term: 'metal'. If we want to continue generalizing, we must find the closest genus of this term, and so on.

The determination (*determinatio*) of terms as a logical operation is contrary and at the same time complementary to the former. This procedure aims at finding terms of narrower scope. If, for example, we want to limit the term 'intelligentsia' we arrive at such terms as 'physician', 'teacher', 'engineer', etc. That is, the determination of a term implies the finding of the closest specific term. The easiest way to do so is to attach specific properties to the given term, i.e. to increase its intension and thereby to decrease its extension.

As opposed to the formal approach, Hegel held the following view:

> The notion is generally associated in our minds with abstract generality, and on that account it is often described as a general conception. We speak, accordingly, of the notions of color, plant, animal, etc. They are supposed to be arrived at by neglecting the particular features which distinguish the different colors, plants, and animals from each other, and by retaining those common to them all. This is the aspect of the notion which is familiar to understanding; and feeling is in the right when it stigmatises such hollow and empty notions as mere phantoms and shadows. But the universal of the notion is not a mere sum of features common to several things, confronted by a particular which enjoys an existence of its own. It is, on the contrary, self-particularising or self-specifying, and with undimmed clearness finds itself at home in its antithesis. For the sake both of cognition and of our practical conduct, it is of the utmost importance that the real universal should not be confused with what is merely held in common.[56]

[56] Hegel, *System* pp. 358–359; Wallace tr. p. 292.

It was this approach which Marx kept in mind when developing e.g. the general term 'capital' in a concrete economy. To take another example, the term 'socialism' would be very poor indeed if it contained only those elements applicable in common. There is no need, however, to stick to the field of the social sciences. Even in a science as abstract as mathematics the generalization of the term 'number' has resulted in an expansion of not extension only, but also intension.

Let us now examine the relation of generalization to other operations. It was, first of all, the interrelation of abstraction and generalization which has aroused interest during the history of logic. The following view has prevailed:

> The cognition of reality at an abstract level is always a generalized cognition... By revealing the essence of reality, one arrives at such knowledge as reflects not only the essence of a thing, phenomenon, or relation, but also holds true of a whole series, and group, of things.[57]

Accordingly, abstraction is always accompanied by generalization. Let us look at this doctrine from both sides. (1) If, for example, I examine a student of history and disregard his, or her, knowledge of other subjects and the fact whether he, or she, is beautiful or ugly, I perform an abstraction. What need is here for generalization? (2) If only the operation revealing the properties several objects have in common (i.e. abstracting from all qualities which they do not have in common) can be described as generalization, abstraction is indispensable. If, however, the operation leading to a concrete general is also accepted as generalization, then abstraction has no part whatsoever in this respect.

For my part, I agree with those authors who, more or less explicitly, adhere to the following view:

> Closely connected with the process of classification is that of *abstraction*. To abstract is to separate the qualities common to all individuals of a group from the peculiarities of each individual. The notion 'triangle' is the result of abstraction in so far as we can reason concerning triangles, without any

---

[57] *Logika*. (Budapest 1956). p. 62.

regard to the particular size or shape of any one triangle. *All classification implies abstraction*, for in framing and defining the class I must separate the common qualities from the peculiarities. When I abstract, too, I form a general conception, or one which, generally speaking, embraces many objects. If, indeed, the quality abstracted is a peculiar property of the class, or one which belongs to the whole and not to any other objects, I may not increase the extent of the notion, so that Mr. Herbert Spencer is, perhaps, right in holding that *we can abstract without generalizing.*[58]

The operation opposite to generalization is determination. What was thought to be the contrary of abstraction in traditional logic?

> The operation of thought which detaches the peculiar attributes from each object compared and thus forms generic concepts is called *abstraction*; and the operation which attaches these attributes to the generic concepts again and thus represents the objects as species of their common genus is called *determination*. Abstraction and determination are *opposed* operations of thought.[59]

Accordingly, both abstraction and generalization have the same operation as their opposite. This opinion is based on the mistaken identification of the operations concerned. The operation contrary to abstraction is concretization.

Generalization and determination operate with the extension of terms, while abstraction and concretization operate with the intension of terms. We have the following analogy between these two pairs of operations:

	Extension	Intension
Increase	Generalization	Concretization
Decrease	Determination	Abstraction

I do not mean to say that only operations in the same line can go together. Both generalization and determination may be accompanied by both concretization and abstraction respectively. In this way, for example, the operation traditionally regarded as generalization proves to be generalizing abstraction.

[58] Jevons, p. 285.
[59] Drobisch, p. 21.

III

On the singular Aristotle wrote, among others, the following:

> In general those things the thought of whose essence is invisible, and
> cannot separate them either in time or in place or in definition, are most of
> all one, and of these especially those which are substances. For in general
> those things that do not admit of division are called one in so far as they do
> not admit of it; e.g. if two things are indistinguishable *qua* man, they are one
> kind of man; if *qua* animal, one kind of animal; if *qua* magnitude, one kind
> of magnitude.[60]

He illustrated singular terms, among others, with the following
examples: a certain man, a certain horse, Socrates, Aristomenes. He did
not question whether there was any difference between these terms, and if
so, what kind. Since it was the general terms which stood in the centre of
attention both with Aristotle and later philosophers, this question failed to
arouse special interest.

Mathematical logic has made the following distinction. Let us take the
following propositions to illustrate it:

(1)   The Danube is the biggest river of Hungary.
(2)   A man is sitting in the waiting-room.

The subject of the first proposition is regarded as an individual. The
second proposition, however, is interpreted as an existential statement
(there is such a man that is sitting in the waiting-room).

In my opinion there are indefinite (a man) and definite (Danube)
singular terms.

> All names are names of something, real or imaginary; but all things have
> not names appropriated to them individually. For some individual objects
> we require, and consequently have, separate distinguishing names; there is a
> name for every person, and for every remarkable place. Other objects, of
> which we have not occasion to speak so frequently, we do not designate by a
> name of their own; but when the necessity arises for naming them, we do so
> by putting together several words, each of which, by itself, might be and is

---

[60] Aristotle, *Metaphysica*. 1016b1–7.

used for an indefinite number of other objects; as when I say, *this stone*: 'this' and 'stone' being, each of them, names that may be used of many other objects besides the particular one meant, though the only object of which they can both be used at the given moment, consistently with their signification, may be the one of which I wish to speak.[61]

Opinions vary on the interpretation of singular terms. Sigwart held the following view:

> A singular term can never be so called merely because only one thing which corresponds to it happens to exist in empirical reality; nor would the logical nature of the term be affected if no object corresponding to it were given at all. Only terms whose attributes guarantee the uniqueness of an object corresponding to them can be called singular terms; thus the center of the earth is in this sense a singular term.[62]

If I understand it correctly only an object which is exclusively and necessarily one can correspond to a singular term, as has been demonstrated by the above example also. Accordingly, Napoleon e.g. is not a singular term since he is but one of the many men. In my opinion it is an arbitrary limitation which turns a special case into an absolute.

It is a common belief that singular terms, too, are the products of generalization.

> Formal logic places general and singular terms next to each other on the same level. But the term—when one considers its origin and function, the position it takes in the knowledge process—is in reality essentially the *general term*; its general character distinguishes it from sensations, from sensuality (*Anschauung*). The expression 'singular term' is very imprecise and can easily be misunderstood. More precisely, we ought to speak of terms which apply to single objects or to individuals. Sensations, notions, intuitions reflect the *individual* character of single, individual objects (persons). The term emphasizes the general even in single objects or persons. Nonetheless we can in a certain sense also speak of singular terms, namely, if we *transform* a notion which applied to a single thing, an individual, *into a concept*. Socrates is a term as the personified vehicle of a particular philosophical approach, of an intellectual conduct with general

[61] Mill, p. 27.
[62] Sigwart, p. 304.

8

significance. Julius Caesar is the embodiment of 'Caesarism'. Napoleon, too, is a term, but there are many French monarchs to whom no concept is attached. The so-called singular *term* is essentially the individualized use of general terms, their association with an individual or an individual event. Thus the terms 'first world war' or 'second world war' can only be called singular terms, because these terms are the individualized applications of the general term 'world war' to a unique historical chain of events. In truth we are dealing here with the unity of the general and the individual.[63]

I want to make the following remarks: (1) Upon the above considerations I find it mistaken to contrast the singular as reflected by experience to the general as expressed by terms (this view was put in medieval logic like this: *universale intelligitur, singulare sentitur*). (2) The criterion the above text has given for singular terms is not exact. Therefore, it would be difficult to decide on this basis whether e.g. Duns Scotus, the founder of Scotism, meets the implied requirements or not. To define the singular as a term which covers one certain object and property respectively is, of course, different. (3) Fogarasi argues that the above questions concern the unity of general and singular. By unity I mean the interdependence of the factors concerned, i.e. that they are conditional upon one another. It would be one-sided to classify only the individual as general.

IV

Now I shall turn my attention to the analysis of the concrete and the abstract, on the one hand, and of the singular and the general, on the other. Aristotle applied the ontological approach to this question:

Substance, in the truest and primary and most definite sense of the word, is that which is neither predicable of a subject nor present in a subject; for instance, the individual man or horse. But in the secondary sense those things are called substances within which, as species, the primary substances are included; also those which, as genera, include the species. For instance, the individual man is included in the species 'man', and the genus to which

[63] Fogarasi, pp. 133–134.

the species belongs is 'animal'; these therefore—that is to say, the species 'man' and and the genus 'animal'—are termed secondary substances.[64]

That is, primary substances correspond to singular objects while secondary ones correspond to general objects (species, genera).

Then he went on as follows:

> Everything except primary substances is either predicable of a primary substance or present in a primary substance. This becomes evident by reference to particular instances which occur. 'Animal' is predicated of the species 'man', therefore of the individual man, for if there were no individual man of whom it could be predicated; it could not be predicated of the species 'man' at all. Again, colour is present in body, therefore in individual bodies, for if there were no individual body in which it was present, it could not be present in body at all. Thus everything except primary substances is either predicated of primary substances, or is present in them...[65]

This passage treats not only objects but also their qualities. Their relations are characterized by the following: (1) a secondary substance can be predicated of a primary one as subject; (2) quality is included in a primary substance as subject.

I have an objection to this distinction. As to relation (1), the approach adopted here is extensive: a secondary substance can be predicated of a primary one but is not present in it (the extension of the term 'man' is not included in the extension of the term 'one individual man', but only the other way round). As to relation (2), the intensive aspect comes here to the foreground: quality is present in a primary substance (colour is one of the qualities of some object). If Aristotle had considered the aspect of intension also as in relation (1), he should have admitted the secondary substance to be present in the primary one since, contrary to Plato, he insisted that the general exists right in, and not outside, the individual things. As far as relation (2) is concerned, there is no reason why some quality could not be predicated of a primary substance as subject. Thus, according to what has been said above, Aristotle's limitation has proved to be unjustified.

[64] Aristotle. 1928. 2a12–18.
[65] Op. cit., 2a34–2b5.

Later on, Aristotle got closer to the right distinction.

> All substance appears to signify that which is individual. In the case of
> primary substance this is indisputably true, for the thing is a unit. In the
> case of secondary substances, when we speak, for instance, of 'man' or
> 'animal', our form of speech gives the impression that we are here also
> indicating that which is individual, but the impression is not strictly true;
> for a secondary substance is not an individual, but a class with a certain
> qualification; for it is not one and single as a primary substance is; the
> words 'man', 'animal' are predicable of more than one subject. Yet species
> and genus do not merely indicate quality, like the term 'white'; 'white'
> indicates quality and nothing further, but species and genus determine the
> quality with reference to a substance: they signify substance qualitatively
> differentiated.[66]

That is, secondary substances are concrete while qualities are abstract.

Three categories are mentioned in the above quotations: primary
substance (individual concrete object), secondary substance (general
concrete object), and quality. The third category has been left undifferentiated here. This defect, however, had been corrected by Aristotle
elsewhere by drawing a distinction between individual qualities (an
individual man's knowledge of languages) and general ones (knowledge of
languages).

Some relate both the abstract and the concrete to the general only.

> Those terms are *abstract* whose intension is made up of general qualities
> free from all specific and individual manifestations.
>
> Those terms are *concrete* whose intension is composed of general
> qualities with various specific characteristics.[67]

Überweg disagreed with this view:

> The *general* notion (as opposed to the singular notion) should not be
> confused with the *abstract* (as opposed to the concrete) notion. Both
> oppositions overlap one another. There are concrete and abstract singular
> notions and concrete and abstract general notions. The practice of some
> logicians, who identify *abstract* and *general,* cannot be condoned.[68]

---

[66] *Op. cit.*, 3b10–21.
[67] Marković, *Logika* p. 26.
[68] Überweg, p. 97.

Quine wrote the following:

> Besides the classification of terms into singular and general, there is a
> cross classification into *concrete* and *abstract*. Concrete terms are those
> which purport to refer to individuals, physical objects, events; abstract
> terms are those which purport to refer to abstract objects, e.g., to numbers,
> classes, attributes. Thus some singular terms, e.g., 'Socrates', 'Cerberus',
> 'earth', 'the author of *Waverly*', are concrete, while other singular terms,
> e.g., '7', '3 + 4', 'piety', are abstract. Again some general terms, e.g., 'man',
> 'house', 'red house', are concrete (since each man or house is a concrete
> individual), while others, e.g., 'prime number', 'zoological species', 'virtue',
> are abstract (since each number is itself an abstract object, if anything, and
> similarly for each species and each virtue).[69]

I share the view that it is inadmissible to identify concrete with singular
and abstract with general. Reading Hegel, Lenin wrote the following:

> A beautiful formula: 'Not merely an abstract universal, but a universal
> which comprises in itself the wealth of the particular, the individual, the
> single' (all the wealth of the particular and single!)!![70]

On the other hand, there are not only concrete singular terms which
reflect some object but also abstract singular terms which reflect some
quality, or property. All these can be grouped as follows:

	Singular	General
Concrete	Pablo Picasso	painter
Abstract	Picasso's artistic talent	artistic talent

V

Such formulas where it seems necessary to use zero often occur in
mathematical operations. There was, however, no way to express this in
traditional logic. Therefore Schröder introduced the figures 0 and 1 to

[69] Quine, p. 204.
[70] Lenin, *Philosophical Notebooks*. p. 99.

denote two special fields (*Gebiet*) in his algebra of logic. Field 0 is in a relation of inclusion (*Einordnung*) with all the fields of *a* and is contained by all fields of diversity, while all the fields of *a* are in a relation of inclusion with field 1 which contains all fields of diversity. In fact, he defined 0 as an empty field.[71]

If we want to understand further development, we should bear in mind the following: Russell, studying the philosophical problems of mathematics, was confronted with the issue of the relation of the existent and the nonexistent. To solve this problem, he applied and developed further Frege's idea expressed by the word 'denotation' (*Kennzeichnung*).

> By a 'denoting phrase' I mean a phrase such as any one of the following: a man, some man, any man, every man, all men, the present King of England, the present King of France, the centre of mass of the solar system at the first instant of the twentieth century, the revolution of the earth round the sun, the revolution of the sun round the earth. Thus a phrase is denoting solely in virtue of its *form*. We may distinguish three cases: (1) A phrase may be denoting, and yet not denote anything; e.g., 'the present King of France'. (2) A phrase may denote one definite object; e.g., 'the present King of England' denotes a certain man. (3) A phrase may denote ambiguously; e.g., 'a man' denotes not many men, but an ambiguous man.[72]

He has derived from this the following conclusion:

> My theory, briefly, is as follows. I take the notion of the *variable* as fundamental; I use '$C(x)$' to mean a proposition in which x is a constituent, where x, the variable, is essentially and wholly undetermined. Then we can consider the two notions '$C(x)$ is always true' and '$C(x)$ is sometimes true'. Then *everything* and *nothing* and *something* (which are the most primitive of denoting phrases) are to be interpreted as follows:
>
> C (everything) means '$C(x)$ is always true';
> C (nothing) means '«$C(x)$ is false» is always true';
> C (something) means 'It is false that «$C(x)$ is false» is always true'.[73]

---

[71] Schröder, pp. 188–189.
[72] Russell, 'On Denoting', p. 41.
[73] *Op. cit.*, p. 42.

Through denotation Russell has pointed out an interdependence between nothing and the non-existent.

> The interesting analogy between such 'empty' denotations and Meinong's 'nonexistent' and 'impossible' objects has not escaped Russell's attention. His attitude to them brings to light his typical view of the object of logic as well. In this opinion, classes are not objects with ideal or real existence but 'logical fictions', and a class is not the totality of its objects (if it were we could not speak of an empty class and the one-component class would be identical with its component) but merely a symbol denoting the extension of a propositional function.[74]

This analogy between the empty and the non-existent has turned into identity with many mathematical logicians.

> ... for some terms we would obviously search in vain for a class of things which exists in reality. In the world of Greek mythology the concept of Amazons plays an important role. However, no one maintains that there is a class of beings which corresponds to this term; to use another example, the concept of gods plays a large role in religions. Scientifically based atheism shows that no reality can be attributed to gods, and that they are only products of the imagination. Can we say that the concept of God is senseless? Obviously not! For we would then have to admit that it is possible with the help of senseless terms to construct true propositions, e.g., the proposition: there are no gods.
>
> In order to avoid this dilemma, we decide to denote a particular class, namely the class that contains no elements, as the null class and introduce the sign $\wedge$ for it. We shall then assign the null class to such terms as 'Amazons', 'gods', etc.[75]

There are, however, authors who do not agree with this identification:

> This being an unambiguous quality of the empty set, that H_r set [the real divisors of 5—Gy. T.] has no element, it does not mean to say that this set does not exist; it *exists* (it is its essence, quality, and determinant) by being *empty*, by having not a single element.[76]

---

[74] Pozsonyi, pp. 108–109.
[75] Klaus, p. 144.
[76] Halasy–Sólyom, p. 40.

All the efforts to identify existence with not-empty, and non-existence with empty, have at least two detects. On the one hand, they simplify the variations, and give the false impression that existence could be traced back to extension, on the other.

6. EXTENSION AND INTENSION OF TERMS

I

Up to the third century A.D., the extension and intension of terms were thought to constitute an undifferentiated unity. To my knowledge Porphyry was the first to break away from this view. He had founded what was to be called extensive logic. Both in the study of propositions and syllogistic, he attached primary importance to the extensive relations of terms to the detriment of intensive relations. Prantl has criticized him for this one-sided approach[77] while Bochenski has praised him for the same and regarded him as the initiator of the calculus of classes. [78] This extensive approach has been playing a dominant role in logic ever since.

Beginning from the seventeenth century, intensive relations were also taken into account. I do not mean to say that these relations had been totally ignored before. Aristotle was confronted with this problem in the *Topica*, and the schoolmen in their theory of supposition. It was, however, first in the Port-Royal logic (first edition: Paris 1662) that extension and intension had been explicitly compared:

> We distinguish the comprehension of a universal idea from the extension of a universal idea.
>
> The comprehension of an idea is the constituent parts which make up the idea, none of which can be removed without destroying the idea. For example, the idea of a triangle is made up of the idea of having three sides, the idea of having three angles, and the idea of having angles whose sum is equal to two right angles, and so on.
>
> The extension of an idea is the objects to which the word expressing the idea can be applied. The objects which belong to the extension of an idea are

[77] Prantl, p. 629.
[78] Bochenski, pp. 155–156.

called the inferiors of that idea, which with respect to them is called the superior. Thus, the general idea of triangle has in its extension triangles of all kinds whatsoever.[79]

As an illustration of the fact that this view has remained basically unchanged up until now, let me cite the following argumentation:

> The total amount of various objects generalized in a term is called the *extension of the term*, and the total of those essential properties on whose basis these objects were generalized is called the *intension of the term*...
>
> The essential properties reflected in the *intension of a term*, as it has already been stated, are those specifics and qualities of objects and phenomena on which the objective laws of reality are based. Therefore the intension of a term reflects the fundamental, and relatively most permanent, qualities and specifics of a certain group of objects and phenomena...
>
> The extension of a term, however, is not identical with the mere enumeration of the objects whose essential properties constitute the intension of the term concerned. If, e.g., one, or more, plants of those which belong to the term 'Hungary's flora' perish, it does not necessarily imply a change in the extension of this term. The extension of this term will change only if a whole species of plants becomes extinct, or a new species, which has not been in the country before, is acclimatized.[80]

It is primarily the total of species which is meant by the extension of terms in the above quotation. The underestimation of individual terms which have no species is therefore evident in this approach. In fact, there is some trouble even with the so-called lowest species (*infima species*) which directly breaks down into individual terms. Thus, it is reasonable for avoiding the problems inherent in limitation to start from the statement that the extension of a term includes every object reflected in the term, and in addition, also the species composed of these particular objects.

Intension, too, has been understood in a narrow sense. This is why I find the following criticism justified:

> If a term refers to a whole group of things, and not to one thing only, its intension comprises not only those properties which the group as a whole has in common but also the specific manifestations of these common

[79] Arnauld–Nicole, pp. 58–59, Eng. tr. p. 51.
[80] *Logika*. pp. 75–78.

properties in concrete particular cases. Whoever does not know these specific properties will be described as having a poor and abstract notion of this group of things. For example, whoever only knows that a planet is a celestial body revolving around the sun can rightly be said to have a very poor and abstract concept of planets. Only those have a socially concrete and rich notion of planets who know the data about the way the individual planets—Mercury, Venus, Earth, Mars, Jupiter, Saturn, Neptune, Uranus, and Pluto—revolve.[81]

It may, of course, be important in certain cases to select the very qualities they have in common. One must not think, however, that they qualify as the intension of the given term.

> The *intension* of a term is said to be the total of those properties of the objects on which the generalization of these very objects into the given term has been based ...
>
> In our view, to be expounded later, it is more exact to call the above total of properties the *basic intension* of the term while the intension proper would cover the total of those properties which can be logically deduced from those basic properties ...[82]

It has been a prevailing opinion since the appearance of the Port-Royal logic that the intension of a term is the total of the essential properties *in common,* and its extension is the total of the *species* belonging to this term. What these two definitions have in common is the emphasis they lay on the general in a rather one-sided way. Relying on what was said above, I think that also the not-general (to be explained later) has to be introduced into these definitions.

Traditional logic, though treating the intension of terms as well, has remained, nevertheless, an essentially extensive logic. The demand for an intensive logic was first expressed in Kant's transcendental logic which investigated the relation of the forms of thought to content. What Kant calls transcendental knowledge is oriented to the way we learn to know the objects, and not to the objects themselves.

Fichte tried to develop this view further by supplementing logic with what he called a study of science. While logic abstracts form from content,

[81] Marković, p. 23.
[82] Voishvillo, p. 162.

he thought, the study of science examines them in their unity. He regarded the study of science as a foundation for logic.

Hegel's logic represents the culmination of intensive logic in classical German philosophy and, to a certain extent, even in our times.

> The notion, in short, is what contains all the earlier categories of thought merged in it. It certainly is a form, but an infinite and creative form, which includes, but at the same time releases from itself, the fullness of all content. And so too the notion may, if it be wished, be styled abstract, if the name concrete is restricted to the concrete facts of sense or of immediate perception. For the notion is not palpable to the touch, and when we are engaged with it, hearing and seeing must quite fail us. And yet, as it was before remarked, the notion is a true concrete; for the reason that it involves Being and Essence, and the total wealth of these two spheres with them, merged in the unity of thought.[83]

This stand-point is, of course, unacceptable to the adherents of formal logic since most of them are of the opinion that the problems of intension should be treated in the special sciences and philosophy respectively. Since I want to disprove this view later on in detail let me here only refer to the old proverb: *naturam expellas furca, tamen usque recurret*.

II

It was with reference to the relation of the intension and the extension of terms that the rule of reciprocity had been stated. Accordingly, the greater the intension of a term the smaller its extension, and the smaller its intension the greater its extension.

Those logicians who accepted it as a universal rule did not strive to do more than to refine it, if at all. Let us see such an argument:

> The reader will now see clearly that a general law of great importance connects the quantity of extension and the quantity of intension, viz.—*As the intension of a term is increased the extension is decreased*. It must not be supposed, indeed, that there is any exact proportion between the degree in which one meaning is increased and the other decreased. Thus if we join the adjective *red* to metal we narrow the meaning much more than if we join the

---

[83] Hegel, *Encyclopädie* p. 354; Wallace tr. pp. 287–288.

adjective *white,* for there are at least twelve times as many white metals as red. Again, the term white man includes a considerable fraction of the meaning of the term *man* as regards extension, but the term *blind man* only a small fraction of the meaning. Thus it is obvious that in increasing the intension of a term we may decrease the extension in any degree.

In understanding this law we must carefully discriminate the cases where there is only an apparent increase of the intension of a term, from those where the increase is real. If I add the term *elementary* to *metal* I shall not really alter the extension of meaning, for all the metals are elements; and the elementary metals are neither more nor less numerous than the metals. But then the intension of the term is really unaltered at the same time; for the quality of an element is really found among the qualities of metal, and it is superfluous to specify it over again.[84]

Nevertheless, the above rule has been challenged more and more often. Fogarasi's position is the following:

If the reciprocity relation were generally valid, it would mean that thought must choose between terms rich in content but narrow in extension and terms with broad extension but impoverished content. This opposition corresponds to the classifying spirit of formal logic but not to dialectical thought, which seeks to unite richness of content with the broadest extension of terms.[85]

Thus, he has not rejected the rule of reciprocity as such, but failed to exactly define its range of validity.

Klaus held the following:

Scientific terms do not become poorer in extension through an increase in their content. The general term contains the more special term as a special case. In mathematics when we go from the equation $x^2 + y^2 = 1$ to the equation $ax^2 + by^2 = 1$, the extension of the term determined by this equation has certainly increased. However its content can in no way be said to have decreased, for the class of conic sections determined by the latter equation is more extensive and includes the former.

The basic error of the Port Royal law is that the concept formation undertaken by the activity of abstraction is done in such a way that more

[84] Jevons, pp. 40–41.
[85] Fogarasi, p. 130.

and more attributes are omitted from a concept and finally ever more general attributes remain. But the definiteness of our concepts cannot be said to get poorer with progressive abstraction. This conception overlooks the dialectical interrelation of the general and the particular. As a matter of fact a more thorough knowledge of the general supports an increased insight into the particular. This applies not only to mathematics. The concept of labour refers at first to concrete human labour. The physicist's concept of work is much more comprehensive. But it is in no way poorer than the former concept. It includes that concept as a special case and at the same time enables us to understand it better.

Thus the activity of abstraction consists not in the omission of attributes but rather, as we have seen, in making attributes variable, If one comprehends abstraction in this way, then the Port Royal law loses its significance, and it is hard to say exactly in what area it is strictly valid at all.[86]

That is, Klaus questions the rule itself and not its universal validity only.

The two Marxist authors just cited agree that the rule of reciprocity must not be thought to hold universally true of the relation of intension and extension. On the question whether the rule is valid at all and if so, according to what criteria, opinions vary. Let us proceed from the supposition that the rule holds true of all cases concerning the relation between genus and species. A generic term can be formed, as we have seen, also in a way that its intension covers the intension of a specific term. The rule obviously does not apply to such cases. It may be said valid, however, if a genus includes only the common properties of a species. Consequently, to reject this rule as such would be unjustified, but its scope of validity must be adequately restricted.

III

What are the relations of terms in propositions? Two typical stand-points have developed in this respect in *extensive logic*: the theory of subordination, and that of identity.

The logical trend, which can be called the *theory of extensive subordination (Subsumtionstheorie)* after B. Erdmann, has been dominant

---

[86] Klaus, pp. 160–161.

for a long time. Porphyry was the first to formulate the basic principle of this theory as a thesis: the predicate term can be at most of the same extension as the subject term but can never be narrower. Consequently, the quality of a proposition should be interpreted as follows: in affirmative propositions the extension of the subject is included in the extension of the predicate, while in negative propositions the extension of the subject is excluded from the extension of the predicate.

To avoid misunderstandings it should be noted that the extensive relations of terms had been dealt with before Porphyry as well. Aristotle wrote, for example, the following:

> That one term should be included in another as in a whole is the same as for the other to be predicated of all of the first. And we say that one term is predicated of all of another, whenever no instance of the subject can be found of which the other term cannot be asserted: 'to be predicated of none' must be understood in the same way.[87]

Apuleius, relying on this argumentation, declared that in propositions the subject is the minor term (*minor*) and the predicate is the major one (*maior*).

It was, however, Porphyry who had turned this view into a principle— and a prevailing opinion ever since.

> The most generally received notion of Predication decidedly is that it consists in referring something to a class, i.e., either placing an individual under a class, or placing one class under another class. Thus, the proposition, Man is mortal, asserts, according to this view of it, that the class man is included in the class mortal... If the proposition is negative, then instead of placing something in a class, it is said to exclude something from a class.[88]

The theory of extensive subordination was severely criticized by B. Erdmann.[89] Relying on his arguments, I find it necessary to say the following:

---

[87] Aristotle. 1928. 24b26–30.
[88] Mill, p. 93.
[89] Erdmann, pp. 248–249.

(1) The followers of the subordination theory are inevitably confronted with such affirmative propositions where the subject and the predicate are of the same extension. They regard them either as exceptions or extreme cases, or simply do not even mention them.

(2) However, particular affirmative propositions represent an even greater problem for them since it would be difficult to reveal in them, in general, a relation of subordination. One could think there is, indeed, subordination also in particular propositions, arguing that this given part of the subject of the proposition is wholly included in the extension of the predicate. This interpretation, however, does not conform to the definition of subordination, not even according to the adherents of this theory.

(3) The relation between the terms of a proposition can be not only subordination but also superordination, crossing, etc. Is it justified then to say in general that the predicate can never be narrower than the subject?

Another important trend within extensive logic is the *identity theory* (*Identitätstheorie des Umfangs*). Owing to the absolutism of the subordination theory, this view was but rarely represented until as late as the nineteenth century (Port-Royal logic, Ploucquet, etc.). Since then, however, for some time it stood in the centre of attention in close connection with quantification of the predicate.

The adherents of this school held that there is the possibility to demonstrate by quantification of the predicate that the extension of the subject and that of the predicate are identical in affirmative propositions, e.g., all iron is some metal. But is the extension of the terms still identical, e.g., in the proposition: all iron is all metal?

Both the subordination and the identity theories are one-sided. One can overcome their limitations by taking into consideration the above mentioned relations of the subject and the predicate.

IV

According to *intensive logic* which has developed from the seventeenth century onward, the intension of the predicate is included in the intension of the subject in affirmative propositions while it is totally excluded in negative ones. The proposition 'man is a living being', for example, expresses the relation that all the properties of living being are at the same time properties of man as well. In the proposition 'a stone is not a living being', however, the properties of living beings are excluded from among the properties of stone.

The relation of subordination and identity have also been considered to be of primary importance in intensive logic. According to the subordination theory the intension of the predicate is either included in, or excluded from, the intension of the subject.

But does the statement that the intension of the predicate is totally included within that of the subject hold true also of the proposition 'some students are sportsmen'? Obviously not, for the terms of this proposition cross one another. Here we witness the same unfounded generalization as in the theory of extensive subordination. An interrelation which holds true of a definite connection between terms has been extended to the relation of the subject and the predicate in general.

In Jevons' view identity is the basic relation of terms:

> An *affirmative* proposition is one which asserts a certain agreement between the subject and predicate, so that the qualities or attributes of the predicate belong to the subject. The proposition, 'gold is a yellow substance', states such an agreement of gold with other yellow substances, that we know it to have the colour yellow, as well as whatever qualities are implied in the name *substance*. A *negative* proposition, on the other hand, asserts a difference or discrepancy, so that some at least of the qualities of the predicate do not belong to the subject. 'Gold is not easily fusible' denies that the quality of being easily fused belongs to gold.[90]

Jevons distinguished between three kinds of identity:

(1)   Simple identity (e.g., the king of England—the king of India).
(2)   Partial identity (e.g., mammals—vertebral mammals).
(3)   Limited identity (e.g., gold in solid state—wrought gold).[91]

My only comment is that it would be difficult to define what this classification is based on, on the one hand, and it hardly needs proof that

[90] Jevons, 1872 p. 63.
[91] Jevons, 1874.

this view reveals very little about the manifold relation of subject and predicate, on the other.

Lotze interpreted propositions as stating a relation of identity between the intensions of two terms. He illustrated this view, among others, with the following example: the proposition 'some men are black' should be understood as 'some men, by whom only the black are meant, are black men'.

Erdmann rejected all views which considered subject and predicate to be totally identical and developed the theory of partial identity (*Einordnungstheorie*) by which he meant that the intension of the predicate is identical with a part of the intension of the subject.

I think our description of these views has provided a satisfactory survey of the theory of intensive identity. As a criticism of this view let me refer, *mutatis mutandis*, to what has been said about the theory of extensive identity.

The theory of intensive identity is just as narrow-minded and one-sided as that of intensive subordination. This is why neither of them could give a satisfactory solution to the problem of term relations. Nevertheless, it is due to these theories that intensive relations, so far neglected or underestimated, have been paid attention to.

Finally, let us see those concepts which take both intension and extension into consideration.

> The relation of the subject and the predicate mediated by the copula expresses *as intension* that the object of the proposition has a certain property... In intensive aspect the object of a categorical proposition always includes the property reflected in the predicate as a part, or moment, of its intension.
> From the extensive point of view a categorical proposition implies that the subject of the proposition is included in the extension of the predicate.[92]

The fact that the author cited above accepts both points of view should be described as a positive achievement. In other aspects, however, onesidedness prevails. It is manifest, first of all, in the adoption of the subordination theory which has already been criticized amply. Further on,

---

[92] *Logika*. (Budapest 1956) p. 149.

9

subordination is obviously restricted here to the rule of reciprocity. Let me note in this respect that in my opinion the intension of a more general term can include the intension of a less general one.

It is still to be made clear how to understand the intension and the extension of terms. According to the view prevailing in logic:

> The total amount of various objects generalized in a term is called the *extension of the term*, and the total of those essential properties on whose basis these objects were generalized is called the *intension of the term*.[93]

Is it really so that only objects belong to the extension of terms? Let us take the term of a quality, e.g., yellow. Its extension covers, among others, citrine, orange, and canary yellow. I think it need not be explained that they are not objects but qualities. In support of my statement that the extension of terms may just as well contain qualities, let me refer to the fact that the expression 'general property' has already been used in logic for a long time.

Now let us see the intension of terms. It is generally believed to be the totality of the properties of an object. In so far as this is true, the restriction of a term's intension to certain properties is not justified. In fact, intension can be correlated with the totality of properties, hence with the object. It is usually the basis for the intensive distinction between abstract and concrete terms. If, e.g., the extension of the term 'car' covers all cars, why is it not admissible that the intension of this very term contains all the properties of a car? Thus, the difference between the extension and the intension of terms consists in the fact that the former refers to the quantity of objects and properties and the latter to their quality—and not in the former including objects and the latter properties.

---

[93] *Op. cit.*, p. 75.

CHAPTER THREE

7. THE EXTENSION OF PROPOSITIONS

A

I

Aristotle in the 'Hermeneutics' divided propositions into two groups as regards their extension: universal and not-universal. An example of universal propositions is: all men are white. He ranked the following propositions among the not-universal: (a) man is white, (b) some men are white, (c) one man is white. For simplicity's sake I have cited here affirmative propositions only.[1]—In the *Prior Analytics*, however, the following classification can be found: universal, particular, and indefinite propositions.[2]

Having compared the two texts, I make the following remarks: (1) in the first case there are *two* subclasses, one of which covers universal propositions only, and the other all the rest. In the second case, there are *three* subclasses and not-universal propositions have been replaced by their species, namely particular and indefinite propositions. (2) The individual, or singular, proposition was mentioned by Aristotle, at least among the examples, in the first case while it is entirely left out of the second classification. (3) Universal propositions have a privileged position in both cases.

It is with Apuleius that the classification of propositions according to their quantity (extension) and quality respectively can first be found. The combination of these two aspects had produced four types of proposition (*A, E, I, O*—as denoted in the Middle Ages) which took a prominent part in traditional logic.

Boethius, the great systematizer of ancient logic, reckoned with all the versions which had occurred so far. Accordingly, he distinguished

---

[1] Aristotle. 1928. 17a38–18a8.
[2] *Op. cit.*, 24a16–22.

between the following propositions as regards their extension: universal (general), particular, indefinite, and individual propositions.

Mostly the same kinds which can be found in medieval logic in the parts on propositions (which were essentially commentaries on the 'Hermeneutics'). From the thirteenth century onward, however, mainly under the influence of Petrus Hispanus, the scope of logic books was enlarged by a chapter on the specifics of terms (*de terminorum proprietatibus*). The following can be read there concerning our subject: (1) according to whether the predicate refers to two or more things, dual and plural propositions are distinguished. (2) *Exclusive* propositions are also introduced, e.g., 'only some men are white'.

As to the role of the extension of propositions in inferences, Aristotle's approach was modified mainly in two points: (1) individual propositions were regarded as identical with general ones with the consideration in mind that in both cases the predicate refers to the extension of the subject as a whole. (2) Indefinite propositions were also identified with general ones.

The first to modify the medieval approach was Ramus who made a distinction between *general* and *special* propositions. He called a proposition special when the predicate did not refer to the whole extension of the subject; and he divided special propositions into particular and specific ones (by the latter, individual propositions are to be understood).

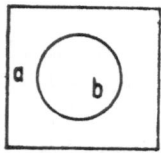

Fig. 19.

Kant distinguished between general, special (*besondere*) and singular propositions. (He used the words 'special' and 'particular' as synonyms.) He described a special proposition as reasonable only if the extension of the subject (a) is wider than that of the predicate (b).[3]

---

[3] Cf.: Kant, *Schriften* p. 532–534.

Hegel was interested, first of all, in the interrelations and transitions of the various kinds of proposition:

> The subject, receiving, as in the Singular judgment, a universal predicate, is carried out beyond its mere individual self. To say, 'This plant is wholesome', implies not only that this single plant is wholesome, but that some or several are so. We have thus the particular judgment (some plants are wholesome, some men are inventive, etc.). By means of particularity the immediate individual comes to lose its independence, and enters into an interconnection with something else. Man, as *this* man, is not this single man alone: he stands beside other men and becomes one in the crowd. Just by this means however he belongs to his universal, and is consequently raised.—The particular judgment is as much negative as positive. If only some bodies are elastic, it is evident that the rest are not elastic.
>
> On this fact again depends the advance to the third form of the Reflective judgment, viz. the judgment of allness (all men are mortal, all metals conduct electricity).[4]

Herbart argued as follows: If an individual proposition has a definite subject (e.g., Socrates), it can be treated as identical with a general proposition. If, however, its subject is indefinite (e.g., man) it belongs to particular propositions.[5]

Krug used the expressions particular, special, and plural propositions as synonyms.[6]—Sigwart, on the contrary, regarded plural propositions as an independent kind of proposition and described them as 'empirically general' propositions as opposed to universal ones which are 'necessarily general'.[7]

Überweg insisted that particular propositions were to be regarded as indefinite. Accordingly, a particular proposition means the following: at least one, but maybe all.[8]

Mathematical logic makes a distinction between general, existential, and individual propositions. Indefinite propositions are interpreted as general ones.

---

[4] Hegel, *Encyclopädie* pp. 376–377; Wallace tr. pp. 308–309.
[5] Herbart, p. 62.
[6] Krug, p. 157.
[7] Sigwart, p. 104.
[8] Überweg, p. 156.

For instance, if I formulate the sentence: 'The base determines the superstructure', it is not at first apparent where the variable elements are to be found here. The correct logical form of the statement reads: For every social order X it is the case that the base of X determines the superstructure of X.[9]

Existential propositions assert that there are things which have definite properties, or meet the demands of a definite relation. Existential propositions are understood as synonyms of particular ones.

II

Let us see now the relation of propositions with definite and indefinite extension respectively.

A premise then is a sentence affirming or denying one thing of another. This is either universal or particular or indefinite. By universal, I mean the statement that something belongs to all or none of something else; by particular, that it belongs to some or not to some or not to all; by indefinite, that it does or does not belong, without any mark to show whether it is universal or particular, e.g. 'contraries are subjects of the same science', or 'pleasure is not good'.[10]

Although Aristotle considered unquantified terms as general ones, he defined propositions with no indication to their extension as indefinite ones.

As a matter of fact, Aristotle found it possible in certain cases to replace an indefinite proposition by a particular one: "... the substitution of an indefinite for a particular affirmative will effect the same syllogism in all the figures."[11]

Theophrastus saw not only a certain agreement between indefinite and particular propositions but even called them identical.[12]—Apuleius identified indefinite propositions with particular ones, saying that in this way the requirement for the conclusion to always follow the weaker premise could be met.

[9] Klaus, p. 38.
[10] Aristotle. 1928. 24a16–22.
[11] Op. cit., 29a27–29.
[12] Cf.: Prantl, p. 356.

It was in modern logic that the view that indefinite propositions can be seen as general ones had become generally accepted.

> *Indefinite* or indesignate propositions are those which are devoid of any mark of quantity whatever, so that the form of words gives us no mode of judging whether the predicate is applicable to the whole or only part of the subject. *Metals are useful, Comets are subject to the law of gravitation*, are indefinite propositions. In reality, however, such propositions have no distinct place in logic at all, and the logician cannot properly treat them until the true and precise meaning is made apparent. The predicate must be true either of the whole or of part of the subject, so that the proposition, as it stands, is clearly incomplete; but if we attempt to remedy this and supply the marks of quantity, we overstep the proper boundaries of logic and assume ourselves to be acquainted with the subject matter or science of which the proposition treats. We may safely take the preceding examples to mean '*some metals* are useful' and '*all comets* are subject to the law of gravitation', but not on logical grounds. Hence we may strike out of logic altogether the class of indefinite propositions, on the understanding that they must be rendered definite before we treat them. I may observe, however, that in the following lessons I shall frequently use propositions in the indefinite form as examples, on the understanding that where no sign of quantity appears, the universal quantity is to be assumed. It is probable that wherever a term is used alone, it ought to be interpreted as meaning the whole of its class.[13]

In the first part of the above quotation, Jevons, following Aristotle, admits that indefinite propositions give no information about extension. Nevertheless, in the penultimate sentence he identifies indefinite propositions with general ones and, in doing so, he follows traditional logic. It is a rather inconsistent attitude, indeed.

That leads to the following question: if a proposition, lacking any indication of extension, is also to be regarded as general, what need is there for the words 'all' or 'every'? They would be entirely dispensable according to the above interpretation. Traditional logic usually formulates general propositions as: *all*, or *every*, *S* is *P*, and then states that the formula '*the S* is *P*' is equivalent to the former. If there were a real

---

[13] Jevons, 1872. p. 65.

equivalence, however, one could change the order and take '*S* is *P*' as the basic formula of a general proposition.

Further on, Jevons, assuming that the meaning of indefinite propositions is not clear, wanted to exclude this kind of proposition from logic. At the same time, however, he described it as a logical form and, on the basis of probability, identified it with general propositions. In my opinion, the indefinite proposition *is* a form of logic by being an assertion which affirms, or denies, something of something. I admit, however, that it is an incomplete proposition which has to be made more exact.

According to Jevons that can be performed by the special sciences only. It is, indeed, the task of special sciences to judge whether the assertion 'metals are useful' can be applied to all metals or to some only. What extension propositions may have at all is, however, a logical problem.

What considerations have led to the interpretation of indefinite propositions as general ones? The answer can be summed up as follows: (1) terms with no quantitative indication are considered to be general; (2) propositions with a general term as subject are regarded as general propositions; (3) consequently, propositions with an indefinite subject are general.

However, let us see the following examples: night is falling on the city; there is hydrogen on the planets; birds fly. The first sentence probably refers to a particular city. The second leaves open whether the statement refers to all planets or to some only. In the third the predicate can refer to some birds only because ostriches do not fly. Therefore I find the identification of indefinite propositions with general ones unjustified.

When do we use indefinite propositions? Mainly in the following cases: (1) we are uncertain of the extension and therefore cannot, or do not want to, take a stand, (2) the context itself suggest the extension of the proposition, so it would seem to be over-scrupulous to specifically indicate it. In the first case it would be ungrounded to regard the propositions as quantified. The trouble with the second is that subjective judgement plays a great part in defining whether something is self-evident or not.

As our historical survey has revealed, some have treated indefinite propositions as particular ones, others as general. Relying on what has been said, I find both interpretations arbitrary and mistaken. Quantified propositions should be distinguished from indefinite ones.

III

Aristotle in the 'Hermeneutics' took the position that 'all', or 'every', quantifies propositions and not terms: "... the word 'every' does not make the subject a universal, but rather gives the proposition a universal character."[14] In the *Prior Analytics*, however, he wrote the following: "If *A* is predicated of all *B*, and *B* of all *C*, *A* must be predicated of all *C* ..."[15] In this statement 'all' serves to quantify terms.

Both interpretations are justified. A quantifier may refer to a term (*de re*) just as much as to a proposition as a whole (*de dicto*). What is meant in the first case are the objects reflected in the term, while in the second the relation expressed in the proposition, e.g.:

(1)   *De re*: all-*S* is *P*
(2)   *De dicto*: all (*S* is *P*)

The latter can be formulated as follows: it holds true of all cases (situations) that *S* is *P*. For simplicity's sake the brackets will be omitted later on. If there is no hyphen the quantifier is to be understood *de dicto*. Let me still note of *de dicto* that there the quantifier may also stand after, and not only in front of, the terms since the two propositions are equivalent, e.g., 'all apples are fruit', and, 'apple is a fruit in all cases'.

The *de re* and *de dicto* versions are in many cases equivalent to one another, e.g., 'all-men are mortal'—'all men are mortal'. It may also account for the underestimation of the distinction between these two kinds of interpretation. Nevertheless, there are cases where this equivalence presumably does not hold true any more, e.g., 'all-my-buttons are plastic'—'my buttons are plastic in all cases'. These assertions differ even in form. The first is an unquantified proposition with a quantified subject and an unquantified predicate. The latter is a quantified proposition with unquantified terms.

Let us see first the unquantified propositions of which mainly the following versions have been considered in traditional logic:

A: all-*S* is *P* I: some-*S* is *P*
E: no-*S* is *P* O: some-*S* is not *P*

---

[14] Aristotle. 1928. 17b13.
[15] *Op. cit.*, 25b38–39.

In order to assess them I take Figure 12 as a basis. Proposition A is f in case (1) for not all of neS are neP (e.g., 'all Greek gods are Egyptian gods'). Case (2) is also f in the same respect (e.g., 'all angels are archangels'). Diagram (3), however, reveals that neS is a part of neP (e.g., all water fairies are fairies). In case (4) neS is identical with neP (all devils are devils).

In cases (5)–(8) we get a false assertion because S and P are alien, e.g.,

(5)   All devils are angels.
(6)   All elves are monads.
(7)   All men are snake-haired.
(8)   All white is black.

Cases (9)–(10) are t because there is no such eS which is not eP, e.g.,

(9)   All apples are apples.
(10)   All irons are metal.

Finally, the result of diagrams (11)–(12) is f because there is such an eS which is not eP, e.g.,

(11)   All flowers are tulips.
(12)   All firemen are tall.

If this analysis is continued the result will be the following table of evaluation:

TABLE 6

	(1)	(2)	(3)	(4)	(5)	(6)	(7)	(8)	(9)	(10)	(11)	(12)
A	f	f	t	t	f	f	f	f	t	t	f	f
E	f	f	f	f	t	t	t	t	f	f	f	f
I	t	t	t	t	f	f	f	f	t	t	t	t
O	t	t	f	f	t	t	t	t	f	f	t	t

What would be the results according to Figure 10? As the reader probably remembers the original provision, shade-lines denote the nonexistent and dots the existent. Let us try to assess proposition A on the basis of diagram (1). No unambiguous result will be obtained since

(a)   All devils are devils (t)
(b)   All devils are angels (f)

The above assertions were called predicative and categorical propositions respectively in traditional logic. The expression 'predicative proposition' can be found first with Apuleius. Accordingly, Martianus Capella spoke about the predicative syllogism. Boethius retained the term 'predicative' with regard to propositions but wrote about the categorical syllogism as regards inferences. Later on the term 'categorical' has become generally accepted also with reference to propositions.

The expression 'categorical proposition' has been interpreted in different ways. Let us see some. "Propositions in which the assertion is not dependent on a condition, are said, in the language of logicians, to be *categorical*."[16] Let us take the following proposition as an example: some clovers have four leaves. Is it justified to consider this proposition as categorical? The fact that the existence of some relation is not subject to certain conditions still does not make it categorical. But even if it were categorical, to imply it would not be sufficient—it has to be explicitly stated, e.g., some clovers categorically have four leaves. Now, the indefensibility of the above interpretation has become even more striking.

According to Eisler a categorical proposition is generally an affirmative or a negative proposition (S is P, S is not P).[17] This definition is incomplete for the propositions discussed under the name 'categorical' refer not only to the quality, but also to the quantity of the proposition.

Finally, let us see another approach: "The most important kind of propositions is the class of *categorical propositions* in which we express whether one object or another does, or does not, possess some property or properties (qualities, attributes, relations)."[18] Let us take the formula 'all-S is P' as an example. It is not postulated or indicated here whether or not the subject refers to an object and the predicate to a property. It is all but implied by the author. Explicitly formulated: all-S objects are of P property. This formula, however, immediately reminds us of several other alternatives for the relation of the subject and the predicate in this respect: object—object, property—object, property—property. Consequently the definition cited above has resulted from a narrow-minded attributive approach.

[16] Mill, p. 83.
[17] Eisler, p. 622.
[18] *Logika*. (Budapest 1956) pp. 147–148.

B

I

So far those propositions have been studied whose subject is quantified and whose predicate is an indefinite term. Is it possible to quantify the predicate as well?

> If, then, a man states a positive and a negative proposition of universal character with regard to a universal, these two propositions are 'contrary'. By the expression 'a proposition of universal character with regard to a universal', such propositions as 'every man is white', 'no man is white', are meant. When, on the other hand, the positive and negative propositions, though they have regard to a universal, are yet not of universal character, they will be contrary, albeit the meaning intended is sometimes contrary. As instances of propositions made with regard to a universal, but not of universal character, we may take the propositions 'man is white', 'man is not white'. 'Man' is a universal, but the proposition is not made as universal character; for the word 'every' does not make the subject a universal, but rather gives the proposition a universal character. If, however, both predicate and subject are distributed, the proposition thus constituted is contrary to truth; no affirmation will, under such circumstances, be true. The proposition 'every man is every animal' is an example of this type.[19]

Like that, the quantification of the predicate is false according to Aristotle. His opinion was accepted as valid, with a few exceptions, up to the very end of the nineteenth century. One of the exceptions was, first of all, Theophrastus who thought that the predicate should also be defined quantitatively.[20] The schoolmen touched upon this question indirectly only when dealing with exclusive propositions (*exponibilia*). By saying, for example, that only some snakes have a poison-fang, we have unambiguously defined the extension of the predicate in this proposition. This recognition, however, was but a by-product of the study of exclusive propositions.

The *Outline of a New System of Logic* wherein Bentham declared the quantification of the predicate to be necessary was published in 1827. For

---

[19] Aristotle. 1928. 17b4–17.
[20] Cf.: Prantl, p. 356, and Bochenski, p. 115.

almost twenty years nobody reacted to this notion of Bentham. It was not before 1848 that Hamilton took notice of it in his *New Analytic of Logical Forms* (Edinburgh) and pointed out: even if the extension of the predicate is usually not expressed linguistically its analysis is just as important for logic as that of the subject. This analysis is especially necessary when the predicate becomes subject by conversion of the proposition. Hamilton had elaborated this operation, and therefore even today he is thought to be the first to deal with this question.

According to Hamilton, having performed the possible combinations, we arrive at the following eight propositions:

(1)   All *S* is all *P*.
(2)   All *S* is some *P*.
(3)   No *S* is any *P*.
(4)   No *S* is some *P*.
(5)   Some *S* is all *P*.
(6)   Some *S* is some *P*.
(7)   Some *S* is no *P*.
(8)   Some *S* is not some *P*.

Let us examine this conception with special emphasis on the arguments used in it. Terms can be used with various extensions in categorical propositions. There are two possibilities in this respect. Terms can be taken in their whole extension which implies that the given proposition refers to all the objects reflected in the term. When the proposition refers to one part of the objects only, the term is used in its partial extension only. A term taken in its whole extension is called distributed (*d*), and if it is taken in its partial extension only, we speak of it as undistributed (*ud*).

Let us now examine the categorical propositions one after the other. The subject of proposition *A* is *d* since we state something of all *S*, e.g., all fluids are flexible.

At the same time the proposition concerned does not give any information about the whole extension of the predicate. With regard to our example, not only fluids can be flexible but also rubber, steel, etc. Thus, the predicate in universal affirmative propositions is *ud*.

In proposition *E* the negation belongs to all *S*, so the subject is *d*, e.g., no metal is transparent.

In this proposition the subject is excluded from the whole extension of the predicate. If, however, the predicate is present in its whole extension in a proposition, it is *d*.

In propositon *I*, it is only a part of the terms used as subject of which something is asserted, therefore the subject is *ud*, e.g., some birds migrate to warmer countries in autumn.

A particular affirmative proposition reveals something only of those *P* objects which are identical with certain *S* objects. Since this proposition does not contain any knowledge of the other *P* objects, its predicate is *ud*.

Finally, the subject in proposition *0* is *ud*, e.g., certain people cannot speak foreign languages.

In order to define the extension of the predicate we assume that in this case all *P* objects are known to be excluded from the extension of *S* objects. Consequently the predicate of proposition *0* is *d*.

Now let us sum up what has been said so far. The extension of the proposition is the main thing to be considered when defining the extension of the subject. In universal propositions the subject is used in its whole extension and is therefore *d*. In particular propositions, however, it is but a part of the terms used as subject of which something is stated, therefore the subject is *ud*.

In order to define the extension of the predicate it should be considered whether the proposition concerned is affirmative or negative. In affirmative propositions our knowledge of the extension of the predicate is incomplete, indefinite, therefore the predicate is *ud*. In negative propositions we have definite knowledge of the extension of the predicate as we know exactly what is not included in it since the subject is excluded from the whole extension of the predicate. Therefore the predicate of negative propositions is *d*.

TABLE 7

	S	P
A	d	ud
E	d	d
I	ud	ud
0	ud	d

It is in relying on the above findings that the law of distribution has been formulated: the subject is d in universal propositions and is ud in particular ones—the predicate is d in negative propositions and is ud in affirmative ones.

II

It has already been mentioned that according to Aristotle it is mistaken to speak of the universality of the predicate. He even discussed it at length:

> But that which follows one must not suppose to follow as a whole, e.g. that every animal follows man or every science music, but only that it follows, without qualification, as indeed we state it in a proposition: for the other statement is useless and impossible, e.g. that every man is every animal or justice is all good. But that which something follows receives the mark 'every'.[21]

Aristotle has proven the impossibility of the operation concerned neither here nor elsewhere.

It is often used as an argument against the quantification of the predicate that one does not say in everyday speech that 'every apple is every fruit'. That does not mean, however, that one never expresses the extension of the predicate, and even less, that everyday language is unable to do so. Let me illustrate it with the following examples: 'Newton was a genius'; 'this film is worth its weight in gold'; 'the orator is greeted by many people'.

What can be the reason for the relatively small number of such examples? One could see already with the quantification of the subject that everyday language often uses the indefinite instead of the definite. We say rather that men are mortal than all men are mortal. It is even more so in the case of the predicate.—Everyday language does not strive for precision and often uses short forms. Americans, for example, when ordering a milk-shake, usually just say: 'please, give me a shake'. If, however, they were actually shaken contrary to the common meaning of the expression they would be stunned, indeed.

But even if no appropriate example could be found in everyday language, would that be sufficient reason to reject the quantification of the

[21] Aristotle. 1928. 43b17–22.

predicate? This problem can easily be solved with the help of *termini technici*, or notation.

One can often hear people say: 'it was so beautiful no words could describe it'. Why should one, after all, doubt the existence of something we cannot appropriately formulate in words? The shortcomings of the language must not be blamed on logic. Linguistic forms should conform to logical forms, and not the other way round.

Another counter-argument runs like this: an assertion with a quantified predicate is senseless. In my opinion whether or not an assertion is senseless is not a logical but rather a psychological criterion. The subjective factor has a great part in defining what one finds senseless. This reference to senselessness seems to be a convenient excuse for avoiding the solution of the problem. As to the example objected to, I think it has, indeed, sense in the following formulation: the term 'every apple' is contained in the extension of the term 'every fruit'.

The extension of the predicate has been taken into account even by Aristotle, whether he liked it or not, when converting universal affirmative propositions: "... the terms of the affirmative must be convertible, not however universally, but in part, e.g. if every pleasure is good, some good must be pleasure..."[22] Traditional logic has established the following rule with regard to conversion: a term which has not been distributed in the premises must not be distributed in the conclusion, either.

It is a rule of the categorical syllogism that the extreme terms in the conclusion must have an extension exactly the same as, or less than, in the premises. If one of the extreme terms is not used in its whole extension in the corresponding premise, no complete knowledge of it can be obtained from the conclusion, either. The conclusion cannot contain knowledge of a wider scope than is already there in the premises if it follows from these premises, indeed.

Let us examine the following inference from this point of view:

> All soldiers must be disciplined.
> This man is not a soldier.
> _____
> Hence this man does not have to be disciplined.

[22] *Op. cit.*, 25a8–10.

Both terms are distributed in the conclusion, but the major term was not distributed in the major premise. So in this case the above rule has been broken. A fallacy of this kind is called an illicit process of the minor and the major term respectively.

All historians of logic agree that the theory of the quantification of the predicate has had an important part in the development of the De Morgan–Boole calculus, and consequently, in the evolution of mathematical logic.

> De Morgan proceeded from the quantification of predicates as introduced by Hamilton (this method had already been used by Leibniz who tried to exactly define the relations of terms, and it boils down to the exact definition of the extensional relations of the subject and the predicate: *omne A est omne B, omne A est aliquot B*, etc.). Then, he systematized the various forms of proportion he had arrived at and tried to develop an arithmetical calculus for denoting their interrelations and their being deduced from one another.[23]

Jevons wrote the following:

> It would not in the least be possible to give in an elementary work a notion of the system of indirect inference first discovered by the late Dr. Boole, the Professor of Mathematics at the Queen's College, Cork. This system was founded as mentioned in the last lesson upon the Quantification of the Predicate...[24]

However, that trend of mathematical logic which has been initiated by Frege and is still dominant does not find this question important any more.

As opposed to this view, I share the following opinion:

> ... all propositions directly expressing some knowledge of the objects which belong to the subject contain (though do not express) some knowledge of the objects which constitute the extension of the predicate. We know, e.g., that no ancient Greek man knew about the existence of the American continent. This proposition directly expresses our knowledge of ancient Greeks. It also implies, however, some knowledge of those who are

---

[23] Pozsonyi, p. 102.
[24] Jevons, (London 1872) p. 191.—Cf. Bochenski, p. 306.

included in the extension of the predicate term, namely those who knew
about the existence of the American continent. We learn from this
proposition also that none of those knowing of the existence of the
American continent was an ancient Greek.

Thus, when something is affirmed (or denied) of the objects which belong
to the subject something is indirectly affirmed (or denied) of the predicate as
well.[25]

I do not want to overestimate the quantification of the predicate but I
find its underestimation just as mistaken. By ignoring the extension of the
predicate, we shall have a one-sided and incomplete notion of the
propositions and the conclusions which follow from them. I agree with
Menne in the following:

Eine der wichtigsten logischen Entdeckungen der Neuzeit vor Beginn der
Algebra der Logik und der Logistik war die Quantifikation des
Prädikates.[26]

When the law of distribution is discussed, categorical propositions are
usually characterized by a *single* definite term relation:

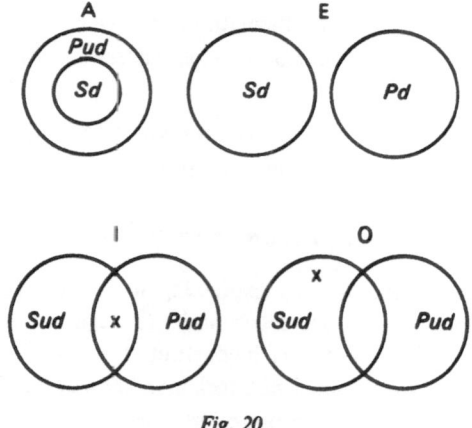

Fig. 20.

[25] *Logika.* (Budapest 1956) p. 55.
[26] Menne, p. 64.

The interrelations denoted above are interpreted as follows:

A: all S is identical with some P
E: all S differs from all P
I: some S is identical with some P
O: some S differs from all P

At first sight all possibilities of combination seem to be adequately covered. Let us, however, take into account the term relations (4)–(8) in Figure 10. Accordingly, this is how distribution will look like in the case of true propositions:

TABLE 8

	(4)	(5)	(6)	(7)	(8)
A	—	d–d	d–ud	—	—
E	d–d	—	—	—	—
I	—	ud–ud	ud–ud	ud–ud	ud–ud
O	ud–ud	—	—	ud–ud	ud–ud

The propositions concerned, if interpreted *de re*, define the extension of their subject. The predicate of proposition E is unambiguously d. But all we are justified to state of the other propositions in general is that their predicate is ud. The reason why proposition O has been differently interpreted is that: (1) the Eulerian diagram of this proposition has been misinterpreted, and (2) this is how distribution could be used for selecting the true moods of categorical syllogisms.

The traditional approach proceeded from the assumption that there is but one quantifier in a simple proposition. Some think it refers to the subject only while others believe it refers to the propositions as a whole. Some even assumed that the same quantifier refers both to the subject and the proposition as a whole. The adherents of distribution admitted as justified also a second quantifier which refers to the predicate. In my view, to start with, the quantification of the proposition and that of the terms should be distinguished. The latter is justified in the case of the predicate just as much as of the subject. Every logical element should be quantified separately.

10*

The law of distribution has resulted from the efforts to reconcile the quantification of the predicate with syllogistics and to support it with another argument. Although this 'law' has been disproved, it has, nevertheless, contributed to the further development of quantification. The consistent application of this operation is a step forward in revealing the structure of the proposition and also in advancing from the abstract to the more concrete.

III

What is the relation between the quantification of terms and the assertion, or predication, of existence? The prevailing opinion is:

> ... not only an existential but any kind of simple proposition implies the knowledge of whether or not the object of the proposition concerned exists in reality.[27]

Still, some maintain a contrary view:

> A proposition of the form *A is B* never has the force to include and imply the proposition '*A* exists' merely because the subject and predicate are joined by 'is'.[28]

Brentano, assuming the existence of the subject to be stated in categorical propositions, upheld the view that all categorical propositions can be converted into existential ones without the slightest modification of their meaning by replacing the copula by 'is' or 'is not' used in existential propositions.[29] His examples are the following:

Categorical proposition	Existential proposition
Some man is ill.	There is an ill man.
No stone is animate.	There is no animate stone.
All men are mortal.	There is no immortal man.
Some man is not a scholar.	There is an unscholarly man.

Brentano, of course, has selected his examples in a way that they should conform to his thesis. Therefore, the subject of his categorical pro-

---

[27] Tavanets, p. 34.

[28] Sigwart, pp. 94–96.

[29] Brentano, *Psychologie* p. 56.

positions is an existential term in each case. There would be, indeed, no problem if the propositions at issue contained but existential terms. Let us take, however, the following proposition: 'some mermaid is ill'. Now, is the existential proposition 'there is an ill mermaid' really true?

In addition, the above examples allow the combination of the subject and the predicate of these categorical propositions into one term in the existential ones. Let us, however, see the following proposition: 'some apple is fruit'. How could one transform it into an existential proposition? On the basis of the above evidence I think that categorical propositions generally must not be converted into existential ones.

Menne has pointed out by the means of mathematical logic[30] that it is in principle impossible to adequately transcribe the propositions discussed above with the help of two-valued truth functions. A many-valued system is, however, out of the question since the law of excluded middle, which has a primary importance in traditional logic, is not maintained in it.

According to Menne this problem can be solved only by the calculus of classes. Proceeding from the possible relations between classes, he selected those seven where no empty class occurs. A definite combination of these relations by disjunction makes the adequate transcription of the propositions concerned possible. Since the notation is complicated and the introduction is lengthy, let me refer to the source for details.[31] Another reason for this is that I think, while agreeing with the basic idea, that the problem can be solved in a simpler way as well.

Let us see, first of all, the logistic interpretation of particular propositions. Accordingly, the proposition 'some S is P' corresponds to the formula 'there is such x thing that Sx and Px.' Tavanets has written the following on this issue:

> Particular propositions, as regards the character of their predicate and subject, stand out in sharp contrast to the corresponding existential propositions. In the particular affirmative proposition 'Some metals are heavier than water', for example, the subject of the proposition is the term 'metals' and the predicate is the term 'objects heavier than water'. In the corresponding existential proposition, namely 'There are metals heavier than water', the subject is, however, the term 'metals heavier than water'

---

[30] Menne, pp. 41–45.
[31] *Op. cit.*, pp. 45–56.

and the predicate is the term 'objects existing in reality'. The meaning of the two propositions is also different. In the existential proposition all that we affirm is that a certain class of objects ('metals heavier than water') exists in reality. In the particular proposition, however, in addition to the real existence of a certain other class of objects ('metals'), it is also directly asserted that a certain property ('to be heavier than water') belongs to some objects of this class of objects.

It is clearly senseless to identify, or to replace by one another, propositions which are so much different. It would but result in a decrease of the actual richness of the kinds of propositions.[32]

I agree with Tavanets in so far as the formula 'some S is P' must not be identified with the formula 'there is an S which is P'. On what, however, has he based his conception? On the assumption that the former formula contains more than the latter by predicating not only the existence of the subject but also that a certain property belongs to some objects.

Let us compare this with the approach of mathematical logic. According to the latter these two formulas are equivalent for: (1) 'some S' means that there is at least one S; (2) if there is an S which is P, then there is at least one, so 'some S is P'. Let us see the following explanation: "There is need for existential propositions as well—and because existence occurs primarily in particular propositions when definite individuals are pointed out they are always treated simply as existential propositions."[33]

This correlation is unacceptable to me for, on the one hand, I am not willing to admit, e.g., the existence of dark elves in the proposition 'Some dark elves are elves'. On the other, if S exists it does not necessarily follow that there must be at least one. I have tried to prove in the preceding that existential terms may be empty.

Let us now draw into the analysis also propositions A and E.

Let us suppose that in the proposition 'all S is P'—for example, 'we consume all the fish that have been caught' or 'all outstanding students in our group get an increased scholarship'—class S may prove to be empty. Then however, there is no reason not to suppose that the same class may be empty also in particular propositions: 'some S is P', that is, 'at least some, but maybe all, outstanding students in our group get an increased

[32] Op. cit., p. 83.
[33] Op. cit., p. 11.

scholarship'. In our view the emptiness of the subject of a particular proposition does not contradict the Aristotelian interpretation of categorical propositions more than the emptiness of the subject of universal propositions. The illusion that the subject of a particular proposition cannot be empty results from the way this proposition is formulated—with the help of the existential quantifier in the predicate calculus of mathematical logic.[34]

The formulations which are in the same line in the following system are considered to be equipollent in mathematical logic:

(I)	(II)
All S is P	There is no S which is not P
No S is P	There is no S which is P
Some S is P	There is some S which is P
Some S is not P	There is some S which is not P

Let us have a closer look at the differences. In column (I) the operators refer to the extension of the subject while in column (II) to its existence. Consequently in case (I) quantifiers, and in case (II) existential operators, are involved. The difference between these propositions which belong to group (I) and (II) respectively is fully illustrated by a comparison of Tables 5 and 4.

In one of the works already cited, after a list of examples, the following can be read:

> In this manner we assert universal, particular, and singular propositions. In formal logic their differentiating character is their quantity, which is expressed in the so-called *quantity operators* (*quantifiers*). They tell us whether the whole class S is contained in P or whether only some of its elements.[35]

On the next page, however, the following is written:

> Formal logic expresses the quantitative relationships in the following manner: For a term whose entire extension is treated in a proposition, it

---

[34] Uemov, 1959.
[35] Händel–Kneist, p. 98.

uses the so-called *universal operator* (∀), and to denote the existence of elements within a class, about which a particular statement is being made, it uses the so-called *existential operator* (∃).[36]

The problem it raises is the following: as the first part of the quotation reveals, the quantifier refers to the extension of the subject and can equally be interpreted as universal or particular. In the second part, however, we find that the universal quantifier refers to the whole extension of the subject while the so-called existential operator refers to the existence of the elements contained within the subject. In my view the first distinction is the correct one while the second mixes up the aspects of extension and existence.

On this basis the following distinction can be made:

(I)		(II)	
(1)	universal quantifier	(1)	existential operator
(2)	particular quantifier	(2)	nonexistential operator

Mathematical logic has combined (I/2) with (II/1) without sufficient reason ('some' with 'there is such which'). As to (I/1), the following has happened: an equivalence has been stated between 'all' and 'there is no such which', on the one hand, and the universal quantifier has been said to refer both to empty and not-empty terms, on the other. In my opinion this is inconsistent.

Let me sum up what has been said so far. Traditional syllogisms treated the universal and the particular only. These were correlated with existential terms only, since non-existential terms were regarded as non-existent.—Mathematical logic has reckoned with the non-existent, too, but only with regard to the universal. The particular has been correlated with the existent only. As I see it the universal and the particular can equally refer to existential and non-existential terms.

---

[36] *Op. cit.*, pp. 39–40.

C

I

In logical literature many kinds of definition have been given of universal propositions. Let me quote two typical approaches. One of them is the following:

> By the quantity of a proposition we always mean the extension covered by the subject term since it is always the subject which tells how many entities the predicate refers to. So we have the following propositions: (a) There are *universal propositions* whose subject is a general term, e.g., 'man is a rational being', 'animal is a sensible being' . . . [37]

I have the following objections: (1) There are propositions whose subject is a general term and still, the proposition itself is not universal, e.g., 'in some cases all pupils gathered at the meeting of the literary circle'. In my opinion the universal character of the proposition is independent of the universal character of the subject. (2) Not only simple but also compound propositions can be universal, e.g., 'in every case when it is raining the pavement becomes wet'. To assert of a compound proposition that its subject is a general term is senseless.

According to the other approach:

> If the proposition affirms the predicate to belong to the whole of the subject, it is an universal proposition, as in the example 'all metals are elements', which affirms that the quality of being undecomposable or being simple in nature is true of all metals.[38]

It has often been pointed out that this definition is not satisfactory since it can refer to singular propositions as well, e.g., 'Mendeleev was a chemist'. This false definition has served as a basis for the view that singular propositions can be treated as universal ones in the syllogism. Let me note, among others, that this definition also ignores compound propositions.

---

[37] Huszár, p. 106.
[38] Jevons, 1872. pp. 63–64.

The definition of universal propositions depends primarily on what is meant by universal. According to Husserl, for example, what we conceive in a general proposition is not this or that particular case, not the collective notion as a whole, and not even its specific form which, by itself, cannot be separated from it. What is meant here is the ideal form which, in the arithmetical sense, is inevitably uniform, be it objectified in any act, and which has nothing to do with the individual singularity of temporary and ephemeral real things.[39]

This is essentially Plato's theory of ideas being applied to propositions in a way that the objects corresponding to general terms have just as independent an existence as those reflected in singular terms.

Mill's position was just the opposite. He thought:

> ... a general truth is but an aggregate of particular truths; a comprehensive expression, by which an indefinite number of individual facts are affirmed or denied at once.[40]

The same has been formulated by Russell like this: A universal proposition is the sum of singular ones and its truth depends upon the truth of the elementary propositions used as examples. When, however, the number of instances is infinite, in the majority of cases it is practically impossible to decide whether or not a universal proposition is true. Accordingly, in the strict sense of the word, universal propositions are not theses which are either true or false.

Husserl was wrong to maintain that the universal has nothing to do with the singular for the universal is composed of singulars. Positivists, on the other hand, are mistaken in their view that the universal is but a merely numerical summary of the singulars.

Let us see now the interpretation of universal propositions in mathematical logic. Quine wrote as follows:

> The existential quantifier has a companion-piece in the *universal quantifier* '(x)', which corresponds to the words 'each thing x (in the universe) is such that'. Application of '(x)' to the expression:
>
> (10) x is identical with x

[39] Husserl, p. 148.
[40] Mill, p. 186.

in the fashion:

(11) $(x)(x$ is identical with $x)$

is called *universal quantification* of (10); and the result (11) is also spoken of as the universal quantification of (10). To say that (11) is true is to say that, no matter what object in the universe be imagined named by 'x' in (10), (10) becomes true. Thus (11) goes into words fairly literally as:

(12)   Each thing is such that it is identical with itself,

or more briefly:

Everything is identical with itself.

Similarly the quantification:

(13) $(x)(x$ is a man $\supset x$ is mortal)

goes into words fairly literally as:

Each thing is such that if it is a man then it is mortal,

or more briefly:

All men are mortal.[41]

The main concern of the above approach is not so much the universality of the propositions but rather the universal character of the things the predicates refer to. I think a universal proposition asserts whether a certain relation is valid or not valid in every case. This definition applies both to simple and compound propositions, e.g.

(1)   In every case man is mortal.
(2)   In every case if light passes through a prism it will be refracted.

The particular proposition is usually, and traditionally, defined as a proposition whose predicate belongs but to a part of the extension of the subject. This definition leaves it open what is meant by the expression 'a part'. As the historical survey has revealed, some meant by it 'at least one but maybe all', others 'only some'.

[41] Quine, p. 85.

These two interpretions exclude one another. Therefore we must either decide in favour of one of them, or admit that we have here two different kinds of proposition. In my view the latter is the right solution, so later I shall tell the particular and the exclusive particular propositions apart. A particular proposition, as I see it, asserts that some relation holds good in some cases and does not hold good in others.

Relying on the above, let us have a closer look at the following versions:

(1)   Universal affirmative: in every case S is P
(2)   Universal negative: in no case S is P
(3)   Particular affirmative: in some cases S is P
(4)   Particular negative: in some cases S is not P

Let us evaluate them on the basis of Figure 12. The interpretation of the diagrams requires the following modification: 'in every case' should be implied at the beginning of every term relation (e.g., in every case eS is inferior to eP). Why is this modification necessary? Let us start with the original interpretation of diagram 10: eS is inferior to eP. As regards this relation, the *de re* proposition 'all-S is P' is true, irrespective of the fact whether the subordination holds good in every case, or in some cases, or in just one case. This fact, however, is not irrelevant when the proposition 'all S is P' is to be evaluated. If the subordination is valid in every case it will be t, but if it is in some cases only then it will be f.

As a result of this modified interpretation we obtain the values of Table 6 for the propositions concerned.

II

Now I shall include singular propositions as well in this analysis. Aristotle assessed this issue as follows:

> Some things are universal, others individual. By the term 'universal' I mean that which is of such a nature as to be predicated of many subjects, by 'individual' that which is not thus predicated. Thus 'man' is a universal, 'Callias' an individual.
>
> Our propositions necessarily sometimes concern a universal subject, sometimes an individual.

If, then, a man states a positive and a negative proposition of universal character with regard to a universal, these two propositions are 'contrary'. By the expression 'a proposition of universal character with regard to a universal', such propositions as 'every man is white', 'no man is white' are meant. When, on the other hand, the positive and negative propositions, though they have regard to a universal, are yet not of universal character, they will not be contrary, albeit the meaning intended is sometimes contrary. As instances of propositions made with regard to a universal, but not of universal character, we may take the propositions 'man is white', 'man is not white', 'Man' is a universal, but the proposition is not made as of universal character; for the word 'every' does not make the subject universal, but rather give the proposition a universal character.[42]

The attitude reflected in the above text suggests the following possibilities of combination:

(1)   Universal assertion with universal subject
(2)   Not universal assertion with universal subject
(3)   Universal assertion with individual subject
(4)   Not universal assertion with individual subject

In the text, however, only the first two cases are set forth. An important statement has been made about them, namely that the universal character of the subject is irrespective of whether the proposition is universal or not universal. In the proposition 'Every man is white', for example, the subject is universal not because the proposition is universal but because man is a general term.

Is it possible for the subject only, or also for the predicate, to be universal and individual respectively? Aristotle has given the following answer:

Of things themselves some are predicable of a subject, and are never present in a subject. Thus 'man' is predicable of the individual man, and is never present in a subject. By being 'present in a subject' I do not mean present as parts are present in a whole, but being incapable of existence apart from the said subject. Some things, again, are present in a subject, but are never predicable of a subject. For instance, a certain point of

---

[42] Aristotle. 1928. 17a38–17b14.

grammatical knowledge is present in the mind, but is not predicable of any subject; or, again, a certain whiteness may be present in the body (for colour requires a material basis), yet it is never predicable of anything. Other things, again, are both predicable of a subject and present in a subject. Thus while knowledge is present in the human mind, it is predicable of grammar.

There is, lastly, a class of things which are neither present in a subject nor predicable of a subject, such as the individual man or the individual horse. But, to speak more generally, that which is individual and has the character of a unit is never predicable of a subject. Yet in some cases there is nothing to prevent such being present in a subject. Thus a certain point of grammatical knowledge is present in a subject.[43]

Since one single contradictory instance is sufficient to disprove a mistaken generalization, let me refer to the following example: this man is Socrates. The predicate of this proposition is a singular term. Why was Aristotle so reluctant to accept propositions of this kind as justified? Because he considered only those propositions in which the subject is 'present' in the predicate, i.e., the former is subordinated to the latter. It is a well-known fact, however, that apart from subordination there can be many different relations between the terms of a proposition. Thus Aristotle's approach is based on an arbitrary limitation.

Assuming that the terms of a proposition can be both general and singular terms, we arrive at the following variations:

	Subject	Predicate
(I)	General term —	General term (e.g., All irons are metal)
(II)	Singular term —	General term (e.g., Austria is a republic)
(III)	General term —	Singular term (e.g., A famous man is Socrates)
(IV)	Singular term —	Singular term (e.g., Budapest is the capital of Hungary)

Singular propositions are completely left out of the Aristotelean syllogistics. In the course of the history of logic, however, the approach already mentioned has developed:

---

[43] *Op. cit.*, 1a20–1b9.

> As far as logical form is concerned, singular judgements are equivalent in use to general judgements. For in both the predicate holds for the subject without exception. In the singular statement, e.g., *Caius is mortal*, there can no more be an exception than in the general statement, *All men are mortal*. For there is only one Caius.[44]

In the *Critique of Pure Reason*, however, Kant had already modified his opinion:

> Logicians are justified in saying that, in the employment of judgements in syllogisms, singular judgments can be treated like those that are universal. For, since they have no extension at all, the predicate cannot relate to part only of that which is contained in the concept of the subject, and be excluded from the rest. The predicate is valid of that concept, without any such exception, just as if it were a general concept and had an extension to the whole of which the predicate applied. If, on the other hand, we compare a singular with a universal judgment, merely as knowledge, in respect of quantity, the singular stands to the universal as unity to infinity, and is therefore in itself essentially different from the universal. If, therefore, we estimate a singular judgment (*judicium singulare*), not only according to its own inner validity, but as knowledge in general, according to its quantity in comparison with other knowledge, it is certainly different from general judgments (*judicia communia*), and in a complete table of the moments of thought in general deserves a separate place—though not, indeed, in a logic limited to the use of judgments in reference to each other.[45]

Inference had been understood exclusively as deductive inference for a long time. Since the beginning of modern times, however, inductive inferences have attracted attention. With regard to them, singular propositions could not be treated as universal any more. Therefore the above approach has become more and more restricted to the field of categorical syllogisms.

In addition, the following definition was widely accepted in traditional logic:

> The *singular* proposition is one which has a singular term for its subject, as in
>
> Socrates was very wise.
> London is a vast city.[46]

[44] Kant, *Schriften*. p. 532.
[45] Kant, *Kritik*. p. 142; Kemp Smith, p. 107.
[46] Jevons, 1872. p. 64.

It goes almost without saying that this definition refers to simple propositions only since a compound one has no subject. It must also be kept in mind that the singular character of the proposition is independent of the singularity of its subject. Singular propositions assert that some relation does or does not hold good in this case.

From the point of view adopted before let me distinguish between *de re* and *de dicto* propositions. Let us first examine *de re* propositions with an indefinite singular term: (1) one-S is P, (2) one-S is not P. With the following consideration in mind a detailed analysis of such propositions can be spared: 'some-S' means 'at least one, but maybe all'. If therefore a proposition whose subject is a particular term is true it must be true also with an indefinite singular term. Thus, the values of the propositions concerned coincide with the values of 'some-S is P' and 'some-S is not P' respectively.

Propositions with definite singular terms differ from the above: (1) this-S is P, (2) this-S is not P. Let us take Figure 12 as a basis. The values of the propositions at issue in cases (3)–(10) agree with those of 'some-S is P' and 'some-S is not P' respectively. But what happens in case of subordination (2 and 11) and crossing (1 and 12)? Let us see, for example, diagram 12 which has revealed that the assertion 'all books are interesting' is surely f while the assertion 'some books are interesting' is surely t. The proposition 'this (very) book is interesting', however, is generally neither surely t, nor surely f, and consequently can be both. In order to decide whether it is t or f further information is required.

The indefinite and the definite singular must be distinguished also in the *de dicto* interpretation. Indefinite singular propositions are: (1) in one case S is P, (2) in one case S is not P. Definite singular propositions are: (1) in this case S is P, (2) in this case S is not P. The values of these propositions always agree with those of *de re* propositions with the slight modification that the specification is 'in every case'.

III

Two kinds of exclusive quantified propositions are usually distinguished: (a) the exclusive singular proposition, e.g., 'only this man could be the culprit', (b) the exclusive particular proposition, e.g., 'only some birds fly'.

What makes an exclusive singular proposition differ from a singular one is that the latter admits while the former excludes universal propositions. This exclusiveness is stressed by the following interpretation: this S is P and not all S is P. The values of the proposition discussed (1) and those of the proposition 'only this S is not P' (2) are:

TABLE 9

	(1)	(2)	(3)	(4)	(5)	(6)	(7)	(8)	(9)	(10)	(11)	(12)
(1)	f/t	f/t	f	f	f	f	f	f	f	f	f/t	f/t
(2)	t/f	t/f	f	f	f	f	f	f	f	f	t/f	t/f

The proposition 'only some S is P' is usually interpreted in logical literature as follows: some S is P and some S is not P. Let us see an argumentation on this issue:

> Particular propositions of the type 'some S is P' (or: 'some S is not P') have two meanings: there are *indefinite* and *definite* particular propositions.
>
> In the indefinite particular proposition 'some S is P' the word 'some' implies that at least some (maybe only one) S is P, but maybe even all S is P. It can be put like this: "there are S's that are P's". How many S's are P's is not specified by the proposition.
>
> 'Some S is P' may also mean, however, that *only* some S is P and there are S's that are not P's. In this sense we have here an exclusive particular proposition which asserts (or denies) something of a *definite* group of objects.[47]

In everyday life, propositions of the type 'some S is P' are used in the second meaning in most, or at least many, cases. They are understood as 'some but not all S is P' (it is the 2 *o* 3 version in the table of quantifiers). Logic, however, uses these propositions in their first meaning: some but at least one S is P (2 *o* 3 *o* 4).

With regard to the propositions 'only some S is P' everyday usage requires the exclusion of 'all' only and leaves open whether or not 'this' is also excluded. Consequently, even this proposition can be interpreted in two different ways: (a) some but not all (2 *o* 3); (b) some but not this (2).

[47] *Logika*. (Budapest 1956) p. 153.

The latter (2) differs from the former (1) by excluding not only 'all' but also 'this'.

I suggest we accept the second interpretation of the formula 'only some S is P'. On the one hand, it conforms more to the linguistic expression implying that all other quantifiers are excluded (some but no other). On the other, it will be seen later that this approach has been adopted in the field of modalities from the very beginning. Therefore it makes the comparison of quantifiers and modalities easier.

I shall evaluate the following propositions in this spirit:

(1)   Only some S is P
(2)   Only some S is not P

TABLE 10

	(1)	(2)	(3)	(4)	(5)	(6)	(7)	(8)	(9)	(10)	(11)	(12)
(1)	t/f	t/f	f	f	f	f	f	f	f	f	t/f	t/f
(2)	f/t	f/t	f	f	f	f	f	f	f	f	f/t	f/t

'Some' can equally mean 'mostly' and 'few'. First of all, the question arises whether the word 'mostly' should be understood as meaning 'may be all', or only a limited number. Aristotle used it in the latter sense.

> So occurrences are universal (for they are, or come-to-be what they are, always and in every case); others again are not always what they are but only as a general rule: for instance, not every man can grow a beard, but it is the general rule. In the case of such connexions the middle term too must be a general rule. For if A is predicated universally of B and B of C, A too must be predicated always and in every instance of C, since to hold in every instance and always is of the nature of the universal. But we have assumed a connexion which is a general rule; consequently the middle term B must also be a general rule. So connexions which embody a general rule — i.e. which exist or come to be as a general rule — will also derive from immediate basic premisses.[48]

The application of these kinds of propositions in logic will be treated when inferences are discussed.

[48] Aristotle. 1928. 96a8–20.

Exceptive propositions were interpreted as follows by Petrus Hispanus.[49]

All *A*, with the exception of *B*, is *C*; All *A* which is something other than *B* is *C*, otherwise *B* is *A*, otherwise *B* is not *C*.

No *A*, with the exception of *B*, is *C*; No *A* which is something other than *B* is *C*, otherwise *B* is *A*, and all *B* is *C*.

Jevons has given the following characterization:

> *Exceptive* propositions affirm a predicate of all the subject with the exception of certain defined cases, to which, as is implied, the predicate does not belong. Thus, 'all the planets, except Venus and Mercury, are beyond the earth's orbit', is a proposition evidently equivalent to two, viz. that Venus and Mercury are not beyond the earth's orbit, but that the rest are. If the exceptions are not actually specified by name an exceptive proposition must often be treated as a particular one. For if I say 'all the planets in our system except one agree with Bode's law', and do not give the name of that one exception, the reader cannot, on the ground of the proposition, assert of any planet positively that it does agree with Bode's law.[50]

The above statements are differentiated versions of the exclusive particular proposition. Apart from them, there are, of course, still many possible variations.

IV

As I have discussed before, 'only some' has been interpreted so far as 'some but not all' (2 *o* 3). Since I have suggested another interpretation for 'only some' a new name should be introduced to denote 2 *o* 3. (As a matter of fact, 'some but not all' could also be used but it is too lengthy.) So my suggestion is to use the word 'certain'. The values of 'certain *S* is *P*' are the following:

TABLE 11

(1)	(2)	(3)	(4)	(5)	(6)	(7)	(8)	(9)	(10)	(11)	(12)
t	*t*	*f*	*f*	*f*	*f*	*f*	*f*	*f*	*f*	*t*	*t*

---

[49] Cf.: Prantl, p. 69.
[50] Jevons, 1872. p. 68.

11*

The values of 'certain S is not P', being equivalent to the formula 'certain S is P', need not be enlisted. This equivalence is obvious on the ground of the above interpretation.

The only quantifier left is represented by 2 *o* 4 and should be interpreted as 'some but not only this S is P'. Further on, the following short form will be used: 'several S is P'. Its values are:

TABLE 12

(1)	(2)	(3)	(4)	(5)	(6)	(7)	(8)	(9)	(10)	(11)	(12)
t|*f*	*t*|*f*	*t*	*t*	*f*	*f*	*f*	*f*	*t*	*t*	*t*|*f*	*t*|*f*

Now let us see the summary of the quantified affirmative propositions and their negations:

0	S is P and not S is P.
1	No S is P (some, but not only some, S is not P).
2	Only some S is P (some, but not this, S is P).
3	Only this S is P (this, but not all, S is P).
4	All S is P (this, but not only this, S is P).
1 *o* 2	Not this S is P (some S is P or no S is P).
1 *o* 3	Not several S is P (only this S is P or no S is P).
1 *o* 4	Not certain S is P (all S is P or no S is P).
2 *o* 3	Certain S is P (some, but not all, S is P).
2 *o* 4	Several S is P (some, but not only this, S is P).
3 *o* 4	This S is P (only this or all S is P).
1 *o* 2 *o* 3	Not all S is P (some S is not P).
1 *o* 2 *o* 4	Not only this S is P (several S is P or no S is P).
1 *o* 3 *o* 4	Not only some S is P (this S is P or no S is P).
2 *o* 3 *o* 4	Some S is P (at least one, but maybe all, S is P).
1 *o* 2 *o* 3 *o* 4	S is P or not S is P.

Their values are indicated in the following table (see Table 13).

Versions (16)–(19) are negations of versions (1)–(8). The negative propositions and their negations (e.g., only this S is not P—not only this S

is not *P*) can be traced back to versions (1)–(8), too, by equivalence and
negation. The relations indicated in Table 13 will be analysed in detail
within the treatment of immediate inferences.

TABLE 13

	(1)	(2)	(3)	(4)	(5)	(6)	(7)	(8)	(9)	(10)	(11)	(12)
(1)	*f*	*f*	*f*	*f*	*f*	*f*	*f*	*f*	*f*	*f*	*f*	*f*
(2)	*f*	*f*	*f*	*f*	*t*	*t*	*t*	*t*	*f*	*f*	*f*	*f*
(3)	*t/f*	*t/f*	*f*	*f*	*f*	*f*	*f*	*f*	*f*	*f*	*t/f*	*t/f*
(4)	*f/t*	*f/t*	*f*	*f*	*f*	*f*	*f*	*f*	*f*	*f*	*f/t*	*f/t*
(5)	*f*	*f*	*t*	*t*	*f*	*f*	*f*	*f*	*t*	*t*	*f*	*f*
(6)	*t/f*	*t/f*	*f*	*f*	*t*	*t*	*t*	*t*	*f*	*f*	*t/f*	*t/f*
(7)	*f/t*	*f/t*	*f*	*f*	*t*	*t*	*t*	*t*	*f*	*f*	*f/t*	*f/t*
(8)	*f*	*f*	*t*	*t*	*t*	*t*	*t*	*t*	*t*	*t*	*f*	*f*
(9)	*t*	*t*	*f*	*f*	*f*	*f*	*f*	*f*	*f*	*f*	*t*	*t*
(10)	*t/f*	*t/f*	*t*	*t*	*f*	*f*	*f*	*f*	*t*	*t*	*t/f*	*t/f*
(11)	*f/t*	*f/t*	*t*	*t*	*f*	*f*	*f*	*f*	*t*	*t*	*f/t*	*f/t*
(12)	*t*	*t*	*f*	*f*	*t*	*t*	*t*	*t*	*f*	*f*	*t*	*t*
(13)	*t/f*	*t/f*	*t*	*t*	*t*	*t*	*t*	*t*	*t*	*t*	*t/f*	*t/f*
(14)	*f/t*	*f/t*	*t*	*t*	*t*	*t*	*t*	*t*	*t*	*t*	*f/t*	*f/t*
(15)	*t*	*t*	*t*	*t*	*f*	*f*	*f*	*f*	*t*	*t*	*t*	*t*
(16)	*t*	*t*	*t*	*t*	*t*	*t*	*t*	*t*	*t*	*t*	*t*	*t*

8. IMMEDIATE INFERENCES

A

I

Aristotle was the first to analyse such procedures as, for example,
conversion, subordination, etc. However, he failed to look at these
procedures as independent operations but treated them only as instruments for syllogism and proof.

In the second century Galen came forward with the claim that these
operations had to be studied relatively independently of syllogisms. He

laid the foundations for the theory of the equipollence of differently shaped propositions (*aequipollentia*). His contemporary, Apuleius, already differentiated from one another the procedures based on the logical square (which he used for the first time) and on equipollence respectively.

It was Boethius who, at the beginning of the 6th century, systematized the statements of his forerunners and at the same time introduced a new terminology still in use. He distinguished between three groups. He put inferences based on the logical square into the first. They are the inferences regarding subalterns (*subalternans*), contrary (*contrarius*), contradictory (*contradictorius*), and subcontrary (*subcontrarius*).—The second group consists of inferences based on the equipollence of differently shaped propositions (Boethius called them '*consentientia*' but later the term '*aequipollentia*' has become widely used).—The third group covers the various kinds of conversion (*conversio*). These are the following: simple conversion (*conversio simplex*), conversion by limitation, or *per accidens* (*conversio per accidens*), and contraposition (*conversio per contrapositionem*).[51]

As the procedures concerned were being treated relatively independently of the syllogism, they were regarded as operations related to propositions rather than inferences and were discussed in the chapter concerning propositions. This tendency was supported by the Aristotelean tradition as well which identified inference with syllogism.

This identification had already been objected to by certain Arab logicians, namely Avicenna and Averroës. It was, however, not before the 13th century that Duns Scotus drew a definite distinction between syllogisms and consequentia. He included into the latter, apart from syllogism, conversion, equipollent inference, and enthymeme as well. In his view what the latter three ways of inference have in common is that the conclusion follows from one premise only. Later Occam made it even more obvious that, for example, conversion is also an inference by always using the word 'hence' (*ergo, igitur*) before a converted proposition.

From this time on, the opinions of logicians differed as to whether or not these procedures were to be seen as inferences. Wolff introduced the term

[51] Cf.: Prantl, pp. 692–698.

immediate inference (*consequentia immediata*) as opposed to ratiocinative inference (*ratiocinatio*) and was the first to discuss these immediate inferences right after syllogisms.

Kant forced immediate inferences into the bed of Procrustes, i.e., into his table of categories.

(1)   Quantity: subaltern inference (*per iudicia subalternata*).

(2)   Quality: opposition (*per iudicia opposita*)

    a) inference concerning the contradictory (*per iudicia contradictoriae opposita*)

    b) inference concerning the contrary (*per iudicia contrariae opposita*)

    c) inference concerning the subcontrary (*per iudicia subcontrariae opposita*).

(3)   Relation: conversion (*per iudicia conversa*)

    a) modified conversion (*conversio per accidens*) .

    b) simple conversion (*conversio simpliciter talis*).

(4)   Method: contraposition (*per iudicia contraposita*).[52]

The alterations can be summed up as follows:

(1)   He divided inferences based on the logical square into two groups (according to quantity and quality).

(2)   He picked out contraposition from conversions and described it as an independent kind of inference.

(3)   He did not consider equipollence as an immediate inference and left it out, arguing that it results but in a mere change of words. It is, however, not improbable that it was left out only because no adequate place had been found for it in this system.

Überweg was contented with merely listing the various kinds of inferences.[53]

Drobisch has chosen the following classification: (1) equipollence, (2) subordination, (3) opposition, (4) conversion and contraposition.[54]

---

[52] Kant, *Schriften* pp. 545–551.
[53] Überweg, pp. 206–207.
[54] Drobisch, pp. 72–92.

And finally, let us see the system of a modern author: (1) subordination, (2) opposition, (3) conversion and contraposition.[55]

These are but a small sample of the various opinions. It might, however, suffice in order to see the reason for this chaos.

Let me underline two main causes: (1) the basis of classification is problematic (some have not even tried to define it while others have chosen more or less arbitrary definitions), (2) all classifications cover but the narrow field of immediate inferences selected upon practical considerations.

These two reasons are closely interlinked. On the one hand, if the material to be systematized is so far from complete as in this case it will be rather difficult to find an essential property necessary for any classification. On the other, if the basis of classification is missing or misleading it is difficult to reveal operations still unknown. The vicious circle typical of the present situation has resulted from the working together of these two reasons.

II

Some authors describe immediate inferences based on the logical square as immediate inferences relying on the characteristics of the relations between propositions. This definition is supposed to imply that the existing logical relations of propositions make it possible to arrive at conclusions. Traditional logic used to perform four kinds of operation on the basis of the logical square: inferences concerning the subaltern, the contrary, the subcontrary, and the contradictory.

The results of inferences based on the logical square were usually summed up in the following table (see Table 14).

The instances denoted by a line in this table are usually not discussed because of their triviality. An overall survey of inferences, however, requires the examination of theses cases as well. The operation based on the identity of two propositions will be called inference concerning the identical.

---

[55] Klaus, pp. 200–202.

TABLE 14

	A	E	I	O
If A is t, then	—	f	t	f
If A is f, then	—	impossible to decide	impossible to decide	t
If E is t, then	f	—	f	t
If E is f, then	impossible to decide	—	t	impossible to decide
If I is t, then	impossible to decide	f	—	impossible to decide
If I is f, then	f	t	—	t
If O is t, then	f	impossible to decide	impossible to decide	—
If O is f, then	t	f	t	—

It was a widely held view in traditional logic that the inferences under discussion should be classified according to the changes in extension and quality respectively.[56]

(1)   Inference concerning the subalterns: extension has changed.

(2)   Inference concerning the contrary: quality has changed.

(3)   Inference concerning the subcontrary: quality has changed.

(4)   Inference concerning the contradictory: extension and quality have changed.

The relations of propositions which served as a basis for the immediate inferences discussed above were described in traditional logic in this way:

(1)   Equipollent (*aequipollens*) propositions express the same relations in different forms, e.g., 'not all men are mortal'—'some men are not mortal'. The extreme cases of equipollent propositions are the identical ones which are even formulated in the same way, e.g., 'all men are mortal'—'all men are mortal'. All identical propositions are equipollent but not all equipollent ones are identical.

[56] Kant, *Schriften* pp. 546–548.

(2)   Subaltern propositions are those where the connection expressed in one of them contains the connection stated by the other, e.g., 'all crimes are a threat to society'—'some crimes are a threat to society'.

(3)   Subcontrary propositions are compatible, e.g., 'some books are interesting'—'some books are not interesting'.

(4)   Contradictory propositions typically deny the whole content of one another, e.g., 'all workshops of the factory have been transformed'—'some workshops of the factory have not been transformed'.

(5)   Contrary propositions express two extreme relations, e.g., 'all books are interesting'—'no book is interesting'.

The above relations were usually divided into two groups: cases (1)–(3) were described as a relation of congruence and cases (4)–(5) as that of incongruence. Some, however, followed Aristotle[57] by isolating cases (3)–(5) and calling them contrary propositions. The negations of these relations (e.g., not-contradictory propositions) had not been treated.

Propositions related to one another as in cases (1)–(5) were called comparable and the rest incomparable. Comparable propositions were characterized by the fact that some relation can be detected between them, e.g., 'every effect has a cause'—'there is no effect without a cause'. Propositions with no relation to one another are incomparable, e.g., 'everybody is expected to be a law-abiding citizen'—'no sportsman can achieve outstanding results without regular training'.

Traditional logic described and tried to systematize the relations of propositions which had occurred in the course of study, but failed to raise the questions as to which are the possible relations of propositions.

Let us take two propositions, p and q, where q can be t or f or indefinite according to the given value of p. If both values of p are reckoned with q, we can have $3^2 = 9$ values. They are summed up in the following table:

[57] Aristotle. 1928. 63b23–30.

TABLE 15

p	q (1)	(2)	(3)	(4)	(5)	(6)	(7)	(8)	(9)
(1) t	t	t	t	t/f	t/f	t/f	f	f	f
(2) f	t	t/f	f	t	t/f	f	t	t/f	f

They are to be understood as follows:

(a) only true   (d) subcontrary   (g) contradictory
(b) subalternate   (e) independent   (h) contrary
(c) identical   (f) subalternant   (i) only false

Six out of the nine instances had obviously been treated in traditional logic. As far as incomparable propositions are understood as assertions with no interdependence, the other three (1), (5), (9) were also implied, though not differentiated or discussed in exact terms.

What relations are there between the arbitrarily quantified propositions? When the extension of terms has been discussed, a complete survey has been given on the quantifiers under the given conditions. Let me list them again in order to make my next argumentation easier to follow:

(1) denied universe   (9) some but not all
(2) nothing (not even one)   (10) some but not only this
(3) some but not this   (11) this
(4) only this   (12) not all
(5) all   (13) not only this
(6) not this   (14) this or nothing
(7) only this or nothing   (15) some
(8) nothing or all   (16) universe

We have also had an overall picture of the relations of propositions (versions (a)–(i)). The possible interrelations of propositions of type '*S* is *P*' which can be quantified in 16 ways are illustrated in the following table (see Table 16).

TABLE 16

	(1)	(2)	(3)	(4)	(5)	(6)	(7)	(8)	(9)	(10)	(11)	(12)	(13)	(14)	(15)	(16)
(1)	c	e	e	e	e	e	e	e	e	e	e	e	e	e	e	g
(2)	i	c	h	h	h	b	b	b	h	h	h	b	b	h	g	a
(3)	i	h	c	h	h	b	h	h	b	b	h	b	b	g	b	a
(4)	i	h	h	c	h	h	b	h	b	h	b	b	g	b	b	a
(5)	i	h	h	h	c	h	h	b	h	b	b	g	b	b	b	a
(6)	i	f	f	h	h	c	e	e	e	e	g	b	b	d	d	a
(7)	i	f	h	f	h	e	c	e	e	g	e	b	d	b	d	a
(8)	i	f	h	h	f	e	e	c	g	e	e	d	b	b	d	a
(9)	i	h	f	f	h	e	e	g	c	e	e	b	d	d	b	a
(10)	i	h	f	h	f	e	g	e	e	c	e	d	b	d	b	a
(11)	i	h	h	f	f	g	e	e	e	e	c	d	d	b	b	a
(12)	i	f	f	f	g	f	f	d	f	d	d	c	d	d	d	a
(13)	i	f	f	g	f	f	d	f	d	f	d	d	c	d	d	a
(14)	i	f	g	f	f	d	f	f	d	d	f	d	d	c	d	a
(15)	i	g	f	f	f	d	d	d	f	f	f	d	d	d	c	a
(16)	g	e	e	e	e	e	e	e	e	e	e	e	e	e	e	c

This table reveals the interrelations of quantified propositions and at the same time gives an overall view of immediate inferences based on the interrelations of propositions.

III

Aristotle summed up the rules of conversion of categorical propositions as follows:

> It is necessary then that in universal attribution the terms of the negative premiss should be convertible, e.g. if no pleasure is good, then no good will be pleasure; the terms of the affirmative must be convertible, not however universally, but in part, e.g. if every pleasure is good, some good must be pleasure; the particular affirmative must convert in part (for if some pleasure is good, then some good will be pleasure); but the particular negative need not convert, for if some animal is not man, it does not follow that some man is not animal.[58]

[58] Aristotle. 25a5–13.

Boethius classified the various instances of conversion by drawing a distinction between simple conversion (*conversio simplex*) and conversion by limitation, or *per accidens* (*conversio per accidens*). Accordingly, universal negative propositions and particular affirmative ones can be simply converted. Universal affirmative propositions can be converted by limitation except for the cases noted by Apuleius and to be discussed later among the other instances of converting universal affirmative propositions. Particular negative propositions, however, must not be converted at all.[59]

Now we have to take a great leap in time since the theory of conversion stayed essentially unchanged, with the exception of certain corrections, until the 19th century. Then, however, Trendelenburg questioned the traditional approach as a whole. Having analysed the versions of conversion one by one, he concluded that, with the exception of universal negative propositions, the whole theory of conversion was dubious.[60] To fully understand the significance of this statement, we must first study the specific kinds of conversion.

Let me start with the conversion of the universal negative proposition the validity of which is beyond dispute in the literature of logic: no S is P, hence no P is S. There even was a view which accepted the conversion of a false proposition as valid inference. Albert von Sachsen illustrated it with the following example: no man is mortal, consequently no mortal is man.[61] This view, however, soon became forgotten. Traditional logic conventionally dealt with the conversion of true propositions only. If, however, in the case of inferences based on the logical square which are also immediate inferences, one may draw conclusion even from false propositions, it must be allowed, at least in principle, for conversion as well.

What will be the difference when the proposition concerned is interpreted *de re* and *de dicto* respectively? The *de re* interpretation leads to the following formula: no-S is P, consequently no-P is S. As this formula has revealed, here even the extension of the terms has been altered in addition to conversion. This complex operation will be discussed later. If

[59] Cf.: Prantl, p. 698.
[60] Trendelenburg, p. 335.
[61] Cf.: Prantl, p. 75.

only the positions of the terms were exchanged the following conclusion
would be arrived at: all-*S* is not *P*, consequently *P* is not all-*S*. In this case
the operation has resulted in a quantified predicate which rarely occurs in
everyday language. Now it is easy to understand why Aristotle, avoiding
these complications, had the *de dicto* interpretation in mind in the
quotation early in this chapter: ". . . in universal attribution the terms of
the negative premiss should be convertible." In my terminology: in no
case *S* is *P*, consequently in no case *P* is *S*.

A particular affirmative proposition can be simply converted: in some
cases *S* is *P*, consequently in some cases *P* is *S*. — Naturally, universal
negative and particular affirmative propositions can be simply converted,
even if they contain non-existential terms.

Particular negative propositions, as Aristotle has written, must not be
converted. There were, however, some logicians who could not put up
with this restriction. Jevons, for example, thought that a new operation
which he called 'conversion by negation' should be introduced in this case:
first the original negative proposition is transformed into an affirmative
one which is later simply converted. He took the following proposition as
an example: 'some existing things are not material substances'. He
transformed it into an affirmative proposition: 'some existing things are
immaterial substances', which, in turn, can be simply converted: 'some
immaterial substances are existing things'.[62]

Yet, Jevons has made a double mistake here. (1) He has failed to
recognize that this operation is not new at all, but rather a version of the
contraposition of the predicate: some *S* is not *P* — some not-*P* is *S*. So
what he called conversion by negation is not a special case of conversion,
but an immediate inference different from (though including) conversion.
(2) On the page preceding the one where he suggests this operation the
following rule has been formulated: "the quality of the proposition
(affirmative or negative) must be preserved". (This rule has not been
explicitly formulated, but only implicitly required, by most logicians with
regard to conversion.) However, by performing a so-called conversion by
negation, he has turned a negative proposition into an affirmative one.
That is, Jevons has contradicted himself.

---

[62] Jevons, 1872. p. 83.

IV

As it is true that apart from immediate inferences conversion has been treated most often in the history of logic, it holds equally true that the most intensive interest has been shown in the conversion of universal affirmative propositions. That is quite understandable, considering how difficult it could be to put up with the Aristotelean statement: such propositions can be converted by limitation only.

Apuleius applied the method of looking for cases when a universal affirmative proposition can be converted without limitation. He had found two such instances. One of them is any definition where the extension and intension of the subject coincide with the extension and intension of the predicate. The other case is when the predicate contains a property (*proprium*) which belongs to the given subject only. For example: 'all men live in society' — 'all social beings are men'.

It would be unfair to Aristotle if we failed to note that he was also aware of these relations. It is, however, Apuleius who deserves credit for analysing them *from the point of view of convertibility*. He was the first to realize, even if within limits only, that during conversion it is reasonable to take into consideration the relations of the terms as well.

Kant has made, among others, the following remark on this subject:

> Some general affirmative judgements, too, can indeed be simply converted. But the reason for this lies not in their *form* but in the particular constitution of their *matter*.[63]

I have the following objections: on the one hand, there is such a universal affirmative proposition whose convertibility is directly based on its form: all S is S. In the proposition 'all S is P', in turn, the possibility of conversion is indirectly implied in its form. Supposing, though not allowing, the existence of some kind of dividing line between form and content, Kant was not right in finding the ground for the simple conversion of universal affirmative propositions in the intension of these propositions. How should this ground be understood anyway? As Apuleius has realized already, it is but the relation of the terms of the

63 Kant, *Schriften* p. 550.

proposition. Traditional logic, also Kant himself, has treated this interrelation as a formal problem under the title 'The Relations of Terms'.

If this is so, why should one consider the term relations within a proposition as a material (intensive) factor? It would be justified only if in the case of universal affirmative propositions we should decide *one by one* whether or not they can be simply converted. As soon, however, as it is stated that every (true) definition can be simply converted this operation will be raised to a general level, and therefore can be performed by means of formulas.

Later on some logicians (for some time I also) went even further than Kant and, with the supposed goal of guaranteeing the consistency of formal logic, declared that the simple conversion of universal affirmative propositions cannot be allowed even as an exception.

Their argumentation runs like this: when the subject and the predicate of a proposition have identical extensions there seems to be no need to limit the extension of the predicate during conversion. What has been ignored here, however, is that the universal affirmative proposition as a form of thought gives only the information that the extension of S term is contained in that of P term and therefore a part of P objects coincides with the objects which belong to S term. It would be a special task to decide if all, P, or only some P, coincide with the extension of S term. But as soon as this question is solved conversion becomes unnecessary since it is needed only to give a clear notion of the objects expressed in the predicate of the proposition. In such a case conversion would become an empty transposition of terms.

To disprove this argument universal affirmative propositions should be compared to particular affirmative ones. It is formally known of the latter that its predicate is not indicated in its whole extension. To perform the operation of conversion here is still thought to be sensible. Does the conversion of 'some S is P' into 'some P is S' convey more than the conversion of 'all S is identical with P' into 'all P is S'? It is either just a useless transposition of terms in both cases, or both operations have their function in logic.

Some mathematical logicians have acknowledged this consideration as justified:

Furthermore, 'If *SaP*, then *PaS*' is true, assuming that subject and predicate have the same extension.[64]

Furthermore, those rejecting the simple conversion of universal affirmative propositions reason as follows: If all we know of a proposition is that it is a universal affirmative one, it can be stated of the relation of its terms that the extension of the subject is contained in that of the predicate. For example: 'every sport is physical training'. This can be represented by circles:

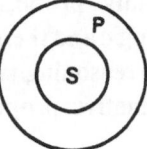

Fig. 21.

Whoever simply converts such a proposition will end up with a false proposition: 'all physical training is sport'.

Yet, even some of the well-known representatives of traditional logic[65] have admitted that, if the subject and the predicate are identical terms, a universal affirmative proposition can rightly be described as follows:

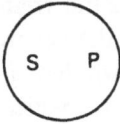

Fig. 22.

And the simple conversion of this universal affirmative proposition is to be regarded as a valid inference.

Accordingly, as the advocates of this view have pointed out, universal affirmative propositions can have two kinds of conversion:

(1) All *P* is *S* (with some additional knowledge simply converted).

(2) Some *P* is *S* (converted by limitation).

[64] Klaus, p. 202.
[65] Überweg, p. 157.

12

This position, however, is not at all unproblematic. If accepted, it would create a totally exceptional situation, i.e., the same immediate inference could be performed on a given proposition in two ways (simply and by limitation respectively). There is no such duality in any other immediate inferences. Some A proposition, for example, does not allow for two kinds of obversion or two kinds of subordination, etc.

Is duality justified in this case, and if not, how could it be avoided? In the history of logic all who opposed this approach argued that the simple conversion of universal affirmative propositions implies a deviation from the general method of formal logic and therefore should be rejected. I have tried to prove before that this reasoning is mistaken and that the simple conversion of universal affirmative propositions is justified and well-founded.

Now is it not the conversion by limitation that gives the key to the solution of this problem? As far as I know Trendelenburg has been to this day the only logician who had doubts about this way of conversion. As he has formulated it, conversion *per accidens* is an expedient which provides too little in essential cases and therefore can be said to be false.[66]

Which are, then, the essential cases? Trendelenburg has referred to the Pythagorean proposition as an example: in all right-angled triangles the square on the hypotenuse equals the sum of the squares on the other sides. By conversion *per accidens*: some triangles where the square on the hypotenuse equals the sum of the squares on the other sides are right-angled. It is, however, well known that *all* such triangles are right-angled.

Trendelenburg has correctly recognized the limitations of the formal approach in this case, but having pointed them out, failed to overcome them. In order to do so we should proceed from the fact that conversion by limitation is not an inference with one variable only. (What I call immediate inference with one variable is where one factor, e.g. quality, or extension, of the convertend has been modified during the given process.) As a result of this process not only the terms will be transposed, but also the extension of the proposition will be altered. That is, what we do here is not only conversion but also subordination.

---

[66] Cf.: Trendelenburg, p. 335.

With all this in mind, we see now even Aristotle's view cited above in a new light. As we have witnessed, the Stagirite began the discussion of conversion with the analysis of universal negative propositions. This was, however, a deviation from the order he usually followed. His basic method was to treat the universal propositions before the particular ones, and the affirmative propositions before the negative ones. Therefore he usually began the discussion of categorical propositions with the universal affirmative proposition. Why did he depart from this order in this case? I think because the conversion of universal affirmative propositions was problematic for him.

He could not assert, on the one hand, that the universal affirmative proposition can be converted just as much as the universal negative or the particular affirmative ones since it would have led to a logical fallacy. He could not declare either that it must not be converted at all, similar to the particular negative proposition, because the conversion of universal affirmative propositions was an indispensable operation for him to trace back the modes of categorical syllogism to those of the first figure. That is why he had chosen the solution to present the conversion of the universal affirmative proposition as if it would stand in the same line with the conversion of universal negative and particular affirmative propositions.

Is it not because the conversion of the universal negative proposition creates no problem, as opposed to the universal affirmative one, that the former has been discussed first, contrary to the usual order? And what accounts for placing the analysis of the universal affirmative proposition between the universal negative and the particular affirmative ones, though the latter two are converted in the same way, contrary to the former? In order to answer these questions let me first analyse immediate inferences with one and several variables in the next chapter.

V

Before we have already met inferences where conversion is combined with another operation, I have not denied their validity, but noted the fact that they are not immediate inferences with one variable.

First I shall analyse the operation when, parallel to conversion, even the extension of the convertend is changed. If this operation is performed on a universal affirmative proposition it will be a procedure which Aristotle treated at the same level with simple conversion. E.g., 'all chairs are pieces of furniture' — 'some pieces of furniture are chairs'. Since here subordination is combined with conversion I shall use from now on the expression 'conversion into subalternate' instead of the traditional 'conversion *per accidens*'.

What happens if there are non-existential terms in the convertend? This question was raised first by Herbart. He uses the following conversion as an example: 'the rage of the Homeric gods is fearful' — 'some fearful phenomenon is the rage of the Homeric gods'. In his view this inference is valid only if the existence of an ideal reality is admitted. In my opinion this idealistic specification is unnecessary because the above conversion is valid even if the terms refer to nonexistent things.

The conversion into subalternate can be performed on universal affirmative propositions just as much as on universal negative ones. Boethius was the first to expound this consideration. E.g., 'no heat-conductor is glass' — 'some glass is not a heat-conductor'. The proposition, of course, could be converted also like this: 'no glass is a heat-conductor'. The conversion into subalternate in this case leads to what is called weakened conclusion as when, e.g., Barbari is used instead of Barbara.

With regard to particular affirmative propositions, Trendelenburg argued as follows: let us take an example, namely, 'some parallelograms are squares' — 'some squares are parallelograms'. Since all squares are parallelograms, in this and the corresponding cases it would be justified to conclude to a universal affirmative proposition. Trendelenburg called this operation 'conversion by expansion' (*erweiternde Conversion*) as the counterpart of the conversion by limitation.

Here we are confronted with the following problem: for the conversion into subalternate one still could find a place within the limits of traditional logic. If, however, the conversion of particular affirmative propositions should go together with the alteration of the extension of the convertend, these limits prove to be too narrow, indeed.

The particular affirmative proposition is usually represented in this way in traditional logic:

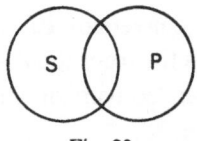

Fig. 23.

In this case terms which cross one another are used, e.g., 'some ploughs are made of wood'. If this proposition werè to be transformed into a universal affirmative one in the course of conversion into subalternant, the result would be an obviously false proposition: 'everything made of wood is a plough'.

Such particular affirmative propositions where identical terms are used can, however, also occur, e.g., 'some parallelograms are geometrical figures whose opposite sides are parallel and equal'. By conversion into subalternant one may conclude from this particular affirmative proposition to a universal affirmative one: 'all geometrical figures whose opposite sides are parallel and equal are parallelograms'.

If all we know of a proposition is that it is a particular affirmative one the above operation does not necessarily lead to a true conclusion. In this case, however, we have the additional knowledge that there are identical terms in the proposition concerned. Therefore no logical fallacy is incurred if this proposition is converted with the goal to arrive at a universal affirmative one.

Nowhere in logical literature have I met the idea that a particular negative proposition could be converted into subalternant. I think this operation must not be performed in general, but on the ground of adequate term relations. If, e.g., the terms are mutually exclusive then the universal negative converse follows from a particular negative proposition.

Conversion can be combined not only with inference concerning subalterns but with inferences concerning the contrary, the subcontrary, and the contradictory as well. These operations have probably not been treated because they can be used with the help of additional knowledge only. E.g., all S is subordinated to P, consequently some P is not S ('all foxes are animals, consequently some animals are not foxes').

Finally, it is obvious that conversion can be combined with inference concerning the identical. It is the traditional simple conversion itself that will be called from now on 'conversion into identical' to make our terminology uniform.

So we have summed up the combinations of conversion with the inferences based on the logical square. Supposing that the convertend is *t*, we get the following values as a result of conversion:

TABLE 17

Conversion		*A*	*E*	*I*	*O*
into					
	identical	*t/f*	*t*	*t*	*t/f*
	subalternate	*t*	*t*	—	—
	subalternant	—	—	*t/f*	*t/f*
	subcontrary	—	—	*t/f*	*t/f*
	contrary	*f*	*f*	—	—
	contradictory	*t/f*	*f*	*f*	*t/f*

TABLE 18

	(1)	(2)	(3)	(4)	(5)	(6)	(7)	(8)	(9)	(10)	(11)	(12)	(13)	(14)	(15)	(16)
(1)	*c*	*e*	*e*	*e*	*e*	*e*	*e*	*e*	*e*	*e*	*e*	*e*	*e*	*e*	*e*	*g*
(2)	*i*	*c*	*h*	*h*	*h*	*b*	*b*	*b*	*h*	*h*	*h*	*b*	*b*	*b*	*g*	*a*
(3)	*i*	*h*	*e*	*h*	*e*	*e*	*h*	*e*	*e*	*b*	*e*	*e*	*b*	*e*	*b*	*a*
(4)	*i*	*h*	*h*	*e*	*e*	*h*	*e*	*e*	*e*	*e*	*b*	*e*	*e*	*b*	*b*	*a*
(5)	*i*	*h*	*e*	*e*	*e*	*e*	*e*	*e*	*e*	*e*	*e*	*e*	*e*	*e*	*b*	*a*
(6)	*i*	*f*	*e*	*h*	*e*	*e*	*e*	*e*	*e*	*e*	*e*	*e*	*b*	*e*	*d*	*a*
(7)	*i*	*f*	*h*	*e*	*e*	*e*	*e*	*e*	*e*	*e*	*e*	*e*	*e*	*b*	*d*	*a*
(8)	*i*	*f*	*e*	*e*	*e*	*e*	*e*	*e*	*e*	*e*	*e*	*e*	*e*	*e*	*d*	*a*
(9)	*i*	*h*	*e*	*e*	*e*	*e*	*e*	*e*	*e*	*e*	*e*	*e*	*e*	*e*	*b*	*a*
(10)	*i*	*h*	*f*	*e*	*e*	*e*	*e*	*e*	*e*	*e*	*e*	*e*	*e*	*d*	*b*	*a*
(11)	*i*	*h*	*e*	*f*	*e*	*e*	*e*	*e*	*e*	*e*	*e*	*e*	*d*	*e*	*b*	*a*
(12)	*i*	*f*	*e*	*e*	*e*	*e*	*e*	*e*	*e*	*e*	*e*	*e*	*e*	*e*	*d*	*a*
(13)	*i*	*f*	*f*	*e*	*e*	*f*	*e*	*e*	*e*	*e*	*d*	*e*	*e*	*d*	*d*	*a*
(14)	*i*	*f*	*e*	*f*	*e*	*e*	*f*	*e*	*e*	*d*	*e*	*e*	*d*	*e*	*d*	*a*
(15)	*i*	*g*	*f*	*f*	*f*	*d*	*d*	*d*	*f*	*f*	*f*	*d*	*d*	*d*	*c*	*a*
(16)	*g*	*e*	*e*	*e*	*e*	*e*	*e*	*e*	*e*	*e*	*e*	*e*	*e*	*e*	*e*	*c*

Now let us go beyond the scope of *AEIO* propositions by examining quantified propositions in general. Their detailed analysis can be spared upon the following consideration. Table 6 contains the values of the quantified propositions. The values of the converses differ from these values only in the case of subalterns. All we have to do in order to evaluate the converses is to exchange the values of the second column with those of the third one, and the values of the tenth column with those of the eleventh in Table 6. By the help of these two tables, then, the relations between convertends and converses, on the one hand, and the immediate inferences based on them, on the other, can be revealed.

The vertical numbers in the above table represent the quantified propositions, and the horizontal numbers their converses.

The letters denote the relations of the propositions discussed above.

B

I

The immediate inferences based on propositions with negative terms have been treated by Aristotle mainly in the 'Hermeneutics'. Let me cite his following statement for its comprehensive character:

> Since the contrary of the proposition 'every animal is just' is 'no animal is just', it is plain that these two propositions will never both be true at the same time or with reference to the same subject. Sometimes, however, the contradictories of these contraries will both be true, as in the instance before us: the propositions 'not every animal is just' and 'some animals are just' are both true. Further, the proposition 'no man is just' follows from the proposition 'every man is not-just' and the proposition 'not every man is not-just', which is the opposite of 'every man is not-just', follows from the proposition 'some men are just'; for if this be true, there must be some just men... The propositions 'everything that is not man is just', and the contradictory of this, are not equivalent to any of the other propositions; on the other hand, the proposition 'everything that is not man is not just', is equivalent to the proposition 'nothing that is not man is just'.[67]

---

[67] Aristotle. 1928. 20a16–24, 37–40.

The above quotation contains the following kinds of proposition: universal and particular, affirmative and negative propositions with positive and negative terms. Let us sum up their combinations:

TABLE 19

(1) All S is P	(5) No S is P
(2) All S is not-P	(6) No S is not-P
(3) All not-S is P	(7) No not-S is P
(4) All not-S is not-P	(8) No not-S is not-P
(9) Some S is P	(13) Some S is not P
(10) Some S is not-P	(14) Some S is not not-P
(11) Some not-S is P	(15) Some not-S is not P
(12) Some not-S is not-P	(16) Some not-S is not not-P

With regard to these propositions a denotation problem should be solved. Traditional logic denoted all universal affirmative propositions as A, irrespective of the differences in their inner structure. It could do so because the specifics of propositions with negative terms were of no interest for it. Since we are now interested in these very propositions the traditional denotation is not satisfactory any more. Further on the following signs will be used:

$$A_1: \text{all } S \text{ is } P$$
$$A_2: \text{all } S \text{ is not-}P$$
$$A_3: \text{all not-}S \text{ is } P$$
$$A_4: \text{all not-}S \text{ is not-}P$$

I shall do the same also with propositions $(E, I, \text{and } O)$.

In the history of logic propositions with negative terms were avoided as much as possible in inferences. It was partly because in this way, as Prantl has rightly remarked, certain difficulties could be removed in advance and partly because their discussion was considered to be unnecessary.

Obversion, for example, which is an immediate inference whereby the quality of the obvertend is altered and its predicate is replaced by a contradictory term, was regarded as an exception.

According to the basic kinds of categorical propositions, the following modes of obversion are distinguished:

All S is P	$\rightarrow$	No S is not-P
No S is P	$\rightarrow$	All S is not-P
Some S is P	$\rightarrow$	Some S is not not-P
Some S is not P	$\rightarrow$	Some S is not-P

Here only premises with positive terms have been used. All we have to do in order to go beyond this narrow scope is to turn to Table 19 which contains the modes with negative terms as well. Let us take these premises and obvert them. Regardless of the quality of the terms we shall arrive at equivalence of the premise and the conclusion in every instance.

So far only categorical propositions have been concerned. Let us now extend our study to quantified propositions in general. Let us take their 16 kinds as a basis and obvert them. We shall see that the relation of the premise and the conclusion has remained unchanged. Thus, whatever quantified proposition is obverted it will always result in equivalence.

De Morgan was the first among mathematical logicians to deal systematically with propositions with negative terms, and the immediate inferences based on them. He regarded the negation of terms as an independent operation whereas until then it had been studied in relation to obversion and contraposition only. This operation was later called inversion by Keynes. Among the modern authors it is Menne who has studied this issue in detail.[68]

II

Now let us see how to define the values of the modes enlisted in Table 19. It would be rather hard work if we had to examine them one by one. There are, however, certain important facts and means which help to make the solution of this problem more simple.

First of all, we have already obtained the values of A, E, I, and O propositions as a result of the analysis of categorical propositions (Table 6).

Let us define now the values of A_2 proposition. As we know, the obversion of the proposition 'all S is not-P' results in E_1 proposition, i.e., 'no S is P'. Since obversion does not change the meaning of a proposition

[68] Menne, pp. 69–72.

the values of A_2 proposition are equivalent to those of E proposition to be found in Table 3. By obversion we arrive at equivalence also between E_2–A, I_2–O, and O_2–I propositions. Consequently, the values of all propositions with index 2 are already known.

The values of A_3 proposition cannot be traced back to the former ones, so we must turn directly to the relations of terms. Let me repeat that A_3 denotes the proposition 'all not-S is P'. However, the circles used so far are not sufficient to represent the extension of not-S, therefore the Venn-diagrams denoting the universe as well have to be applied.

Let us define the values of A_3 proposition with the help of these diagrams. Not-S contains all elements outside the extension of S. If there are elements, existent or nonexistent, in the universe, then there must be

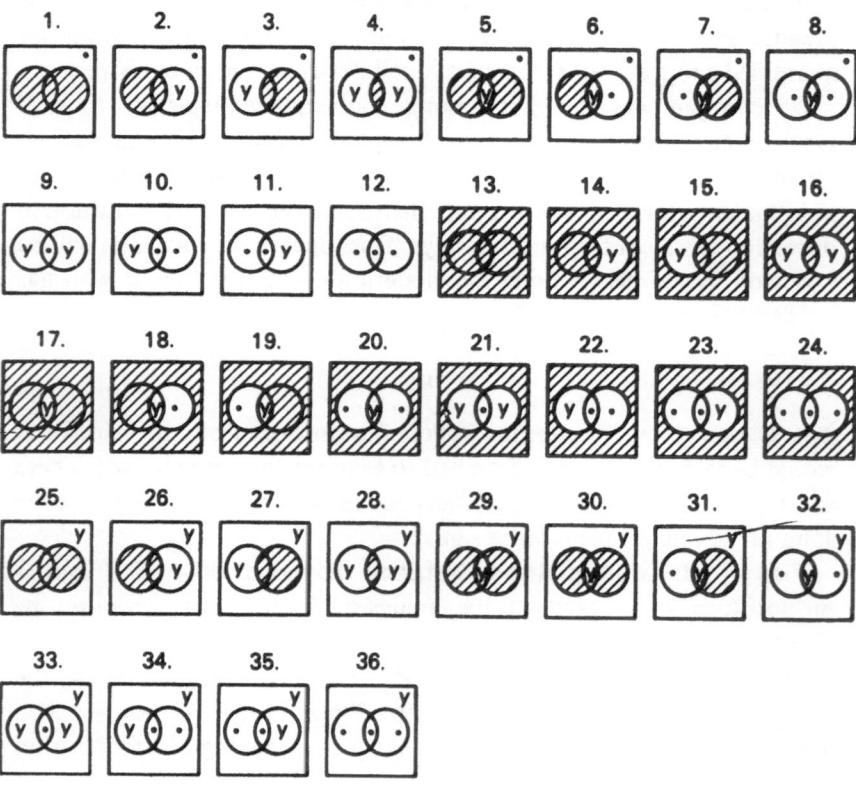

Fig. 24.

not-P as well. Consequently the proposition 'all not-S is P' is f in cases (1)–(24).

The result may seem to be t in cases (25)–(36) for the universe under discussion has no elements. Here, however, another aspect has also to be reckoned with. Let us take, e.g., case (33) where P belongs to the extension of S, being identical with it. 'All not-S is P' leads to contradiction in this instance for it is equivalent to 'all not-S is S', e.g., all not-metal is metal. In general it can be stated of all instances where P belongs into the extension of S that the proposition 'all not-S is P' is f.

The values of A_4 proposition can be obtained as follows: if P belongs to the extension of S, then the proposition at issue is t because not-S can be but not-P. If, however, it does not belong there, the value is f.

E, I, O propositions can be traced back to A propositions by an inference concerning the identical, on the one hand, and the contradictory, on the other, therefore their detailed analysis is unnecessary. All the values according to diagrams (13)–(24) coincide successively with the values based on diagrams (1)–(12). Relying on the above findings, the following table can be constructed (see Table 20).

Having compared the values of A propositions in Table 20, one can make the following observations. The values of A, A_2, and A_4 propositions in diagrams (1)–(12) may be both t and f. On the contrary, the values of A_3 proposition in the same diagrams are but f. By taking all this into account, we go far beyond the scope of traditional logic. Consequently, now we have got an explanation for: (a) why Aristotle was contented with merely stating that the meaning of the proposition 'all not-S is P' differs from that of the other propositions, (b) why these propositions have not been used as premises in inferences, (c) why Kant failed to include these propositions into his system.

Table 20 can be made much more simple by the methods applied so far. The values of propositions I and O are, as we know, the negations of propositions E and A respectively. The values of propositions E coincide with those of propositions A (e.g., E with A_2). Accordingly, it is sufficient to deal with propositions A.

The evaluation helps us to define all the relations between the propositions discussed and the immediate inferences based on them. The

TABLE 20

		(1)	(2)	(3)	(4)	(5)	(6)	(7)	(8)	(9)	(10)	(11)	(12)
(1)	A	f	f	t	t	f	f	f	f	t	t	f	f
(2)	A_2	f	f	f	f	t	t	t	t	f	f	f	f
(3)	A_3	f	f	f	f	f	f	f	f	f	f	f	f
(4)	A_4	f	t	f	t	f	f	f	f	t	f	t	f
(5)	E	f	f	f	f	t	t	t	t	f	f	f	f
(6)	E_2	f	f	t	t	f	f	f	f	t	t	f	f
(7)	E_3	f	t	f	t	f	f	f	f	t	f	t	f
(8)	E_4	f	f	f	f	f	f	f	f	f	f	f	f
(9)	I	t	t	t	t	f	f	f	f	t	t	t	t
(10)	I_2	t	t	f	f	t	t	t	t	f	f	t	t
(11)	I_3	t	f	t	f	t	t	t	t	f	t	f	t
(12)	I_4	t	t	t	t	t	t	t	t	t	t	t	t
(13)	O	t	t	f	f	t	t	t	t	f	f	t	t
(14)	O_2	t	t	t	t	f	f	f	f	t	t	t	t
(15)	O_3	t	t	t	t	t	t	t	t	t	t	t	t
(16)	O_4	t	f	t	f	t	t	t	t	f	t	f	t

		(25)	(26)	(27)	(28)	(29)	(30)	(31)	(32)	(33)	(34)	(35)	(36)
(1)	A	f	f	t	t	f	f	f	f	t	t	f	f
(2)	A_2	f	f	f	f	t	t	t	t	f	f	f	f
(3)	A_3	t	f	t	f	t	t	t	t	f	t	f	t
(4)	A_4	f	t	f	t	f	f	f	f	t	f	t	f
(5)	E	f	f	f	f	t	t	t	t	f	f	f	f
(6)	E_2	f	f	t	t	f	f	f	f	t	t	f	f
(7)	E_3	f	t	f	t	f	f	f	f	t	f	t	f
(8)	E_4	t	f	t	f	t	t	t	t	f	t	f	t
(9)	I	t	t	t	t	f	f	f	f	t	t	t	t
(10)	I_2	t	t	f	f	t	t	t	t	f	f	t	t
(11)	I_3	t	f	t	f	t	t	t	t	f	t	f	t
(12)	I_4	f	t	f	t	f	f	f	f	t	f	t	f
(13)	O	t	t	f	f	t	t	t	t	f	f	t	t
(14)	O_2	t	t	t	t	f	f	f	f	t	t	t	t
(15)	O_3	f	t	t	t	f	f	f	f	t	f	t	f
(16)	O_4	t	f	t	f	t	t	t	t	f	t	f	t

TABLE 21

	(1)	(2)	(3)	(4)	(5)	(6)	(7)	(8)	(9)	(10)	(11)	(12)	(13)	(14)	(15)	(16)
(1)	c	h	e	e	h	c	e	e	b	g	e	e	g	b	e	e
(2)	h	c	e	h	c	h	h	e	g	b	b	e	b	g	e	b
(3)	e	e	c	h	e	e	h	c	e	e	b	g	e	e	g	b
(4)	e	h	h	c	h	e	c	h	b	e	g	b	e	b	b	g
(5)	h	c	e	h	c	h	h	e	g	b	b	e	b	g	e	b
(6)	c	h	e	e	h	c	e	e	b	g	e	e	g	b	e	e
(7)	e	h	h	c	h	e	c	h	b	e	g	b	e	b	b	g
(8)	e	e	c	h	e	e	h	c	e	e	b	g	e	e	g	b
(9)	f	g	e	f	g	f	f	e	c	d	d	e	d	c	e	d
(10)	g	f	e	e	f	g	e	e	d	c	e	e	c	d	e	e
(11)	e	f	f	g	f	e	g	f	d	e	c	d	e	d	d	c
(12)	e	e	g	f	e	e	f	g	e	e	d	c	e	e	c	d
(13)	g	f	e	e	f	g	e	e	d	c	e	e	c	d	e	e
(14)	f	g	e	f	g	f	f	e	c	d	d	e	d	c	e	d
(15)	e	e	g	f	e	e	f	g	e	e	d	c	e	e	c	d
(16)	e	f	f	g	f	e	g	f	d	e	c	d	e	d	d	c

numbers in Table 21 refer successively to the propositions discussed above and the letters to the corresponding relations of these propositions.

All the combinations of the obversions of categorical propositions can be found in Table 21. At the same time it reveals also that obversion is a special mode of inversion.

So far only the inversions of categorical propositions have been dealt with. There is no reason why we should not perform this operation on any quantified proposition. Due to the relatively great number (64^2) of instances, however, I shall omit their detailed analysis here.

III

Concerning the operation of contraposition Aristotle held the following:

> Seeing that the modes of opposition are four in number, you should look for arguments among the contradictories of your terms, converting the order of their sequence, both when demolishing and when establishing a view, and you should secure them by means of induction — such arguments

e.g. as that 'If man be an animal, what is not an animal is not a man': and
likewise also in other instances of contradictories. For in those cases the
sequence is converse: for 'animal' follows upon 'man', but 'not-animal'
does not follow upon 'not-man', but conversely 'not-man' upon 'not-
animal'. In all cases, therefore, a postulate of this sort should be made, e.g.
that 'If the honourable is pleasant, what is not pleasant is not honourable,
while if the latter be untrue, so is the former'. Likewise, also, 'If what is not
pleasant be not honourable, then what is honourable is pleasant'.[69]

Accordingly, in the process of contraposition the order of sequence of
the terms of the original proposition is converted and they are replaced by
contradictory terms.

Let us have a closer look at one of the examples: the honourable is
pleasant. As seen above, Aristotle has interpreted it as follows: all
honourable is pleasant (all-S is P). The contraposition of this proposition
produces a quantified predicate in the conclusion which, as we know, he
considered senseless.

Let us try to interpret this proposition *de dicto*: all honourable is
pleasant (in every case S is P). The contraposition of this proposition is a
correct inference. Aristotle, however, has made a distinction between
indefinite and quantified propositions — and rightly so. Consequently, for
him 'the honourable is pleasant' was not equivalent to 'all honourable is
pleasant'.

This distinction allows us to quantify an indefinite proposition both as
particular and universal respectively. A particular affirmative proposition,
however, must not be contraposed. From 'some S is P' it does not follow
that 'some not-P is not-S'. It can be checked as follows: let us convert the
conclusion, i.e., some not-S is not-P. As Table 20 reveals, the conclusion
does not follow from the given premise. Accordingly, the unquantified
proposition (S is P) must not be contraposed. Aristotle would have acted
correctly only if he had proceeded from the proposition 'all honourable is
pleasant'.

Galen applied contraposition already to quantified propositions: all S is
P; all not-P is not-S. Apuleius added to them the contraposition of
particular negative propositions: some S is not P; some not-P is not not-S.

---

[69] Aristotle. 1928. 113b15–25.

Boethius examined the categorical propositions one after the other and concluded:

All S is $P(t)$	$\rightarrow$ All not-P is not-$S(t)$
No S is P	must not be contraposed
Some S is P	must not be contraposed
Some S is not $P(t)$	$\rightarrow$ Some not-P is not not-$S(t)$

It was Boethius who first raised the question: What happens if in the course of the conversion of the terms only one of them is replaced by its contradictory (while the quality and the extension of the proposition remains unchanged)? Let us try it first on the predicate:

All S is P	must not be contraposed
No S is $P(t)$	$\rightarrow$ No not-P is $S(f)$
Some S is $P(f)$	$\rightarrow$ Some not-P is $S(t)$
Some S is not P	must not be contraposed

With regard to the subject we get the following results:

All S is $P(t)$	$\rightarrow$ All P is not-$S(f)$
No S is $P(t)$	$\rightarrow$ No P is not-$S(t)$
Some S is $P(f)$	$\rightarrow$ Some P is not-$S(t)$
Some S is not $P(f)$	$\rightarrow$ Some P is not not-$S(t)$

It was a widely held view in traditional logic that contraposition is merely obversion and conversion applied together and can be performed in two ways, namely by contraposition of the subject and the predicate respectively.

The contraposition of the predicate implies that first the proposition concerned is obverted, then this obverse is converted. The following modes are accepted as correct:

All S is P	$\rightarrow$ No not-P is S
No S is P	$\rightarrow$ Some not-P is S
Some S is P	must not be contraposed
Some S is not P	$\rightarrow$ Some not-P is S

When the subject is contraposed, the given proposition is first converted, then the converse is obverted. This operation leads to the following traditional modes:

All *S* is *P*	→ Some *P* is not not-*S*
No *S* is *P*	→ All *P* is not-*S*
Some *S* is *P*	→ Some *P* is not not-*S*
Some *S* is not *P*	must not be contraposed

These traditional modes represented a step forward as compared to the ancient view because they had reckoned also with the potential inferences resulting from an alteration in the quality and the extension of the proposition. To regard as contraposition only those operations which involve obversion as well was, however, a step backward since in this way all the modes where the quality of the propositions remains unchanged had been left out. This restriction disappears when contraposition is interpreted as an operation where conversion goes together with a change of the quality of at least one term.

IV

Obversion, conversion, and contraposition are usually designated by the collective name of equipollent inference in traditional logic and the following table is used to compare their modes:

TABLE 22

	Obversion	Conversion	Contraposition	
			predicate	subject
A	*E*	*I*	*E*	*O*
E	*A*	*E*	*I*	*A*
I	*O*	*I*	—	*O*
O	*I*	—	*I*	—

It follows from the above that this system covers but a very narrow field of correct inferences. It was upon practical considerations and with many limitations that traditional logic had selected certain kinds of immediate inference for study.

Some of the limitations are manifest, for example, in the following view:

> In the so-called immediate inferences the subject and the predicate are the
> same, and it is only the extension and the quality of the propositions that
> changes. An example: 'all capitalists are exploiters, consequently so are
> certain capitalists as well'. The second proposition 'follows' from the first
> because all capitalists as S contain some capitalists as S. In the same way: if
> all metals are good conductors, then so are the individual metals as well.[70]

It is not true that only the extension and the quality of propositions are
changed in immediate inferences. The extension of terms, for example,
may also change. Is the subject really the same in these two propositions:
all S is P; all not-S is P? Yes, it is in so far as both subjects have S involved.
It is obvious, however, that S does not refer to the same objects as not-S.

This selection upon practical considerations whose positive aspects I do
not deny has led to inconsistencies, unnecessary repetitions, and the
mixing up of operations of different kinds. In order to overcome these
deficiencies one has to adopt a combinative method. As many modes have
to be considered as possible in order to arrive at as few necessary modes as
only possible.

It can be stated after an analysis of traditional inferences that they
modify the following factors of a categorical proposition:

(1)   the quality of the proposition
(2)   the extension of the proposition
(3)   the quality of the terms
(4)   the extension of the terms
(5)   the sequence of the terms.

I shall classify the individual kinds of immediate inferences according to
how many of the above factors of the proposition have been modified in
the course of these inferences:

0 factor

(1)   (Inference concerning the identical).

---

[70] Fogarasi, p. 232.

13

1 factor

(2) The quality of the proposition
(3) The extension of the proposition
(4) The quality of the terms
(5) The extension of the terms
(6) The sequence of the terms

2 factors

(7) The quality of the proposition, the extension of the proposition.
(8) The quality of the proposition, the quality of the terms
(9) The quality of the proposition, the extension of the terms.
(10) The quality of the proposition, the sequence of the terms.
(11) The extension of the proposition, the quality of the terms.
(12) The extension of the proposition, the extension of the terms.
(13) The extension of the proposition, the sequence of the terms.
(14) The quality of the terms, the extension of the terms.
(15) The quality of the terms, the sequence of the terms.
(16) The extension of the terms, the sequence of the terms.

3 factors

(17) The quality of the proposition, the extension of the proposition, the quality of the terms.
(18) The quality of the proposition, the extension of the proposition, the extension of the terms.
(19) The quality of the proposition, the extension of the proposition, the sequence of the terms.
(20) The quality of the proposition, the quality of the terms, the extension of the terms.
(21) The quality of the proposition, the quality of the terms, the sequence of the terms.
(22) The quality of the proposition, the extension of the terms, the sequence of the terms.

(23)   The extension of the proposition, the quality of the terms, the extension of the terms.

(24)   The extension of the proposition, the quality of the terms, the sequence of the terms.

(25)   The extension of the proposition, the extension of the terms, the sequence of the terms.

(26)   The quality of the terms, the extension of the terms, the sequence of the terms.

4 factors

(27)   The quality of the proposition, the extension of the proposition, the quality of the terms, the extension of the terms.

(28)   The quality of the proposition, the extension of the proposition, the quality of the terms, the sequence of the terms.

(29)   The quality of the proposition, the extension of the proposition, the extension of the terms, the sequence of the terms.

(30)   The quality of the proposition, the quality of the terms, the extension of the terms, the sequence of the terms.

(31)   The extension of the proposition, the quality of the terms, the extension of the terms, the sequence of the terms.

5 factors

(32)   The quality of the proposition, the extension of the proposition, the quality of the terms, the extension of the terms, the sequence of the terms.

The above summary covers immediate inferences concerning not only the categorical but also all kinds of quantified propositions.

The following consideration makes the survey of immediate inferences easier. As is well known, categorical syllogisms are usually systematized according to their figures. Let us define by analogy the figures of immediate inferences. The restriction that the conclusion should always be a proposition 'S is P', similarly to the specification used for categorical syllogism, leads to two figures:

13*

(I) $\underline{S \text{ is } P}$ (II) $\underline{P \text{ is } S}$

          $S \text{ is } P$ $S \text{ is } P$

By disregarding the above restriction, we get two more figures:

(III) $\underline{S \text{ is } P}$ (IV) $\underline{P \text{ is } S}$

          $P \text{ is } S$ $P \text{ is } S$

Modes have already been mentioned quite often also with regard to immediate inferences. Let us see them as they are usually designated in traditional logic:

$$AA \quad EA \quad IA \quad OA$$
$$AE \quad EE \quad IE \quad OE$$
$$AI \quad EI \quad II \quad OI$$
$$AO \quad EO \quad IO \quad OO$$

The four figures allow for 64 modes altogether. Since any term can be substituted for S and P, figure (III) can be traced back to figure (II), and figure (IV) to figure (I). Consequently, only 32 modes have to be examined which has just been done. When discussing the inferences based on the relations of propositions, I have treated figure (I) and in the analysis of conversion also figure (III). As soon as negative terms are also reckoned with obversion and contraposition will easily fit into these figures, too.

With all the quantified propositions examined, the number of modes will increase. Still, two figures are just as sufficient as before.

Why did traditional logic fail to treat the figures of immediate inferences? Aristotle had distinguished the figures of the categorical syllogism according to the position of the middle term. Since there is no middle term in immediate inferences it seemed unnecessary to speak of figures there. Further on I shall try to prove the insufficiency on the ground of the Aristotelean classification. Any distinction of the figures requires the consideration of the position of the other terms just as much as that of the middle one. With this in mind, it is both possible and necessary to speak of figures also with regard to immediate inferences.

9. THE CATEGORICAL SYLLOGISM

I

Philosophers have paid attention first to the terms from among the basic forms of thought. That is natural enough for, in order to find his way in the extremely complicated world, man has to generalize occurrences and therefore to develop terms, or notions. Everyday terms, however, could not be used for scientific purposes due to their ambiguity and obscurity. They had to be defined with exactitude which meant at first the subordination of the given term to a more general one. This way of definition called for induction whereby more and more general terms could be arrived at, building up to categories.

Plato who had thoroughly analysed terms and their relations regarded categories not only as final results but also as starting points for revealing by division the whole range of terms inferior to them. Thus, he completed induction by a logical process of a deductive kind.

This process served as a direct source for Aristotle's syllogistic. Aristotle realized that necessary connections could be revealed among terms by using the relation of specific and generic terms (inferior and superior terms) as a basis. The principle of the first figure was built on this very relation:

> Whenever three terms are so related to one another that the last is contained in the middle as in a whole, and the middle is either contained in, or excluded from, the first as in or from a whole, the extremes must be related by a perfect syllogism. I call that term middle which is itself contained in another and contains another in itself: in position also this comes in the middle. By extremes I mean both that term which is itself contained in another and that in which another is contained.[71]

The last term should be understood as the minor term (*terminus minor*) and the first term as the major one (*terminus maior*) while the name middle term has remained unchanged. In figure (I) Aristotle introduced the following signs for denoting the terms: major term (*A*), middle term (*B*),

[71] Aristotle. 1928. 25b31–38.

minor term (*C*). These terms became denoted by letters *P*, *M*, *S* successively in traditional logic.

Following the above principle, Aristotle examined what conclusions follow from premises composed of universal propositions. He had found a correct inference in two instances:

(1)   If *A* is the predicate of all *B*, and *B* is the predicate of all *C*, then it necessarily follows that A is the predicate of all *C*.

(2)   If *A* is the predicate of no *B*, and *B* is the predicate of all *C*, then it necessarily follows that *A* is the predicate of no *C*.

What happens if *A* is the predicate of all *B* and *B* is the predicate of no *C*? There is no syllogistic relation between the extreme terms in this case.

> ... for nothing necessary follows from the terms being so related; for it is possible that the first should belong either to all or to none of the last, so that neither a particular nor a universal conclusion is necessary. But if there is no necessary consequence, there cannot be a syllogism by means of these premisses.[72]

Upon the same consideration it is also impossible to draw a true conclusion when *A* is the predicate of no *B* and *B* is the predicate of no *C*.

Accordingly, such combinations of propositions as *AA* and *AE*, to use a designation developed later, provide the necessary ground for drawing a true conclusion. On the contrary, no syllogism can be formed from the premises *AE* and *EE* respectively. No more combinations are possible for the relations of universal propositions.

Then Aristotle started to analyse such premises where one of the theses is universal and the other is particular. Such an interrelation of the premises leads to true conclusions, again, in two cases:

(3)   If *A* refers to all *B* and *B* refers to some *C*, then it necessarily follows that *A* refers to some *C*.

(4)   If *A* refers to no *B* and *B* refers to some *C*, then it necessarily follows that *A* does not refer to some *C*.

Aristotle had proved one by one that, if the theses concerned are related to one another as above, no syllogism can be developed in any other case.

[72] Aristotle. 26a4–7.

That is, a true conclusion can be drawn only from the theses *AI* and *EI* respectively, and not from the rest, i.e., *AO*, *EO*, *IA*, *IE*, *OA*, and *OE*.

Finally, both theses can be particular. No valid conclusion can be drawn from them be they affirmative or negative. Such theses are the following: *II*, *IO*, *OI*, and *OO*.

Summing up the analysis of figure (I), Aristotle pointed out the following:

> It is evident also that all the syllogisms in this figure are perfect (for they are all completed by means of the premises originally taken) and that all conclusions are proved by this figure, viz. universal and particular, affirmative and negative.[73]

That is, in figure (I) each one of the propositions *A*, *E*, *I*, *O* appears as a conclusion. And 4 instances out of the 16 possible combinations of the premises allows for the development of a syllogism.

Everything seems to be in perfect order. Let us have a closer look, however, at case 3 where the minor premise is a particular proposition. If *B* refers to some *C* it follows that certain *C*'s are contained in *B*, but there may be such a *C* as well which is not contained. The principle of figure (*I*), however, requires the last term (*C*) to be contained in the middle (*B*) wholly, or not at all.

> That one term should be included in another as in a whole is the same as for the other to be predicated of all of the first. And we say that one term is predicated of all of another, whenever no instance of the subject can be found of which the other term cannot be asserted: 'to be predicated of none' must be understood in the same way.[74]

Consequently, either case (3) (and similarly, case (4)) is problematic, or the above principle does not hold true of these two instances.

Überweg has already pointed out that the Aristotelean principle of figure (I) can be applied in its strict sense to Barbara only.[75] That is, that very mode has a middle term "which is itself contained in another and contains another in itself".

---

[73] Aristotle. 1928. 26b29–33.
[74] *Op. cit.*, 24b26–30.
[75] Überweg, p. 332.

It has taken several decades to recognize that even Überweg's definition
has to be specified. As Łukasiewicz has pointed out, the principle of figure
I applies to Barbara only if it is confined to true premises. As we know,
Aristotle used variables in his syllogisms. For example:

> If all B is A,
> and all C is B,
> then all C is A.

Any term can be substituted for these variables. Łukasiewicz has chosen
the following example:

> If all crows are birds,
> and all animals are crows,
> then all animals are birds.

This syllogism is correct, although one of the premises is f. Thus, the
principle of figure (I) cannot be applied in general even to Barbara.[76]

Günther Patzig has come upon further problems.[77] He took as a
starting point Aristotle's interpretation of the extreme terms: one of them
is contained in another and the other contains another. The extreme terms
in case (1) meet the demands of this definition since C is contained in B as
in a whole and A contains B as in a whole. The above definition holds true
of case (2) only under the condition that B is excluded from A as from a
whole. With regard to cases (3) and (4), however, this definition has to be
radically changed for the minor term is neither contained in nor excluded
from B as in and from a whole respectively.

Aristotle was also aware of this problem and made the following
modification in this definition of the extreme terms with regard to
instances (3)–(4): "I call that term the major in which the middle is
contained and that term the minor which comes under the middle."[78]

But how is the expression 'comes under the middle' to be understood?
Let us substitute, for instance, for C the concrete term 'watch' and for B
the term 'wrist-watch'. Then the minor premise in case (3) will be the

---

[76] Cf.: Łukasiewicz, p. 29.
[77] Patzig, p. 106.
[78] Aristotle. 26a22–23.

following true particular proposition: 'some watches are wrist-watches'. Has it any sense to say that the term 'watch' comes under the term 'wrist-watch'?

Why are we confronted with such nonsense? Because Aristotle interpreted the relation of two terms *exclusively* as the relation of specific and generic terms. Such an interpretation fits case (1) very well, and that is why this instance has perfectly conformed to Aristotle's notion. Inspired by this success, he tried to prove that this very term relation serves as a basis for necessary inference in the other three cases as well. As we have seen, this attempt was foredoomed to failure. Therefore Aristotle was forced to make concessions and to modify his original notion. With regard to cases (3)–(4) he had to reinterpret the definition of the extreme terms and to replace the expression 'to be contained' which refers to the relation of specific and generic terms by the rather indefinite formulation 'to come under'.

Aristotle's basic idea should be recognized as positive by all means. In his interpretation, the syllogism is not a purely formal relation of the terms. He defined the function of the middle term so as to give the real cause for the connection between the extreme ones. He believed the syllogism could not provide scientific knowledge if this demand were not met. The fact that the terms were related in certain cases even without this condition being fulfilled was of little interest for him. It is due to the low level of objective knowledge that from a logical point of view he could not satisfactorily realize his basic conception in his syllogistic. Even so he has had an immense influence on posterity.

II

Theophrastus was mainly interested in the formal relations of terms. As a result he held that figure (I) is that combination of premises where the middle term plays the role of the subject in the major premise and that of the predicate in the minor one. In this way he could reconcile the principle of figure (I) with the correct modes, and the problem Aristotle had encountered had been solved formally. In exchange, however, the function of the middle term as the expression of the real cause for the connection between the extreme terms had to be sacrificed.

By reinterpreting the function of the middle term, Theophrastus could arrive at nine correct inferences in figure (I). Thus, he complemented the four Aristotelian modes by the following ones:

(5)   If A refers to all B, and B refers to all C, then C refers to some A.

(6)   If A refers to no B, and B refers to all C, then C refers to no A.

(7)   If A refers to all B, and B refers to some C, then C refers to some A.

(8)   If A refers to all B, and B refers to no C, then C does not refer to some A.

(9)   If A refers to some B, and B refers to no C, then C does not refer to some A.

To make orientation easier let me transcribe them in the traditional form:

(5)	(6)	(7)	(8)	(9)
$M\,a\,P$	$M\,e\,P$	$M\,a\,P$	$M\,a\,P$	$M\,i\,P$
$S\,a\,M$	$S\,a\,M$	$S\,i\,M$	$S\,e\,M$	$S\,e\,M$
$P\,i\,S$	$P\,e\,S$	$P\,i\,S$	$P\,o\,S$	$P\,o\,S$

The first three are identical with the theses constructing the premises of the first three Aristotelian modes. What then makes Theophrastus's syllogisms specific? The fact that *the order of the sequence of the terms in the conclusions is reversed* as compared to the Aristotelian syllogisms (instead of 'A is C'—'C is A').

Aristotle also knew these syllogisms.

> ... if one is affirmative, the other negative, and if the negative is stated universally, a syllogism always results relating the minor to the major term, e.g. if A belongs to all or some B, and B belongs to no C: for if the premisses are converted it is necessary that C does not belong to some A.[79]

This passage concerns syllogisms (8) and (9).

---

[79] *Op. cit.*, 29a21–25.

With regard to syllogisms (5)–(7) he wrote the following:

> Since some syllogisms are universal, others particular, all the universal syllogisms give more than one result, and of particular syllogisms the affirmative yield more than one, the negative yield only the stated conclusion. For all propositions are convertible save only the particular negative: and the conclusion states one definite thing about another definite thing. Consequently all syllogisms save the particular negative yield more than one conclusion, e.g. if A has been proved to belong to all or to some B, then B must belong to some A: and if A has been proved to belong to no B, then B belongs to no A. This is a different conclusion from the former.[80]

Thus, the innovation of Theophrastus consists not in the discovery of these syllogisms, but in their inclusion into figure (I).

One of the late peripatetics, Ariston (about 50 B.C.), adopted even a combinatory method to explore as many formal relations as possible. He succeeded in finding three more correct modes:

(10)   If A belongs to all B, and B belongs to all C, then A belongs to some C.

(11)   If A belongs to no B, and B belongs to all C, then A does not belong to some C.

(12)   If A belongs to no B, and B belongs to all C, then C does not belong to some A.

Modes (10) and (11) are the weakened or, to put it differently, subaltern version of modes (1) and (2) respectively. It means that there is a particular proposition instead of a universal one in the conclusion. An example of mode (10) is: 'if all metals are heat-conductors, and all irons are metal, then some irons are heat-conductors'. Mode (12) is a weakened version of mode (6) where the terms are transported in the conclusion. The above modes of figure (I) are called as follows:

(1)	Barbara	(7)	Dabitis
(2)	Celarent	(8)	Fapesmo
(3)	Darii	(9)	Frisesomorum
(4)	Ferio	(10)	Barbari
(5)	Baralipton	(11)	Celaront
(6)	Celantes	(12)	Celantop.

[80] *Op. cit.*, 53a3–13.

This innovation was introduced not for convenience's sake only. While Aristotle built syllogisms, first of all, on term relations, this way of designation put the stress on the relations between propositions. Terms were mainly used in examples only.

The Aristotelean and the Theophrastean modes of figure (I) were distinguished from one another as early as by Boethius. He called the former direct (*directi*) and the latter imperfect (*imperfecti*). From the time of Petrus Hispanus on, the term 'indirect modes' (*indirecti modi*) has been used instead of 'imperfect modes'.

Modern logic did not strive to increase the number of modes. On the contrary, it tried to prove that several modes are arbitrary and therefore superfluous. The majority of logicians returned to the original four modes introduced by Aristotle.

Their interest was concentrated on the principal ground of correct modes. The axiom of the categorial syllogism served here as a starting point: all that we know of all the objects expressed by a certain term in a generalized form can be asserted of any concrete object, or any group of objects, which are included in the extension of the given term.

It follows from this connection that one of the premises of categorical syllogisms in figure (I) must be a universal proposition, i.e., a proposition which refers to all objects expressed by a certain term in a generalized form. And the other premise must be an affirmative proposition for the fact that certain objects are included in the extension of the given term can be expressed but in an affirmative form. This rule holds true of the propositions used in syllogisms. Yet, methods for checking the validity of syllogisms on the ground of term relations have also been developed.

III

Aristotle defined the principle of figure (II) as follows:

> Whenever the same thing belongs to all of one subject, and to none of another, or to all of each subject or to none of either, I call such a figure the second; by middle term in it I mean that which is predicated of both subjects, by extremes the terms of which this is said, by major extreme that which lies near the middle, by minor that which is further away from the

middle. The middle term stands outside the extremes, and is first in position.[81]

Let us analyse the first sentence of this quotation. It concerns the relation between the middle and the extreme terms. Let us illustrate the combinations described above:

(1)	(2)	(3)
to all of each subject	to all of one subject to none of another	to none of either

When applied to modes, this classification leads to a division in group (2). That is to say, it is not indifferent to *which* of the extreme terms the middle one belongs, as to a whole or not at all respectively. Accordingly, there are four, and not only three, combinations here (let me illustrate them by signs introduced at a later date):

(1)	(2)	(3)	(4)
PaM	PaM	PeM	PeM
SaM	SeM	SaM	SeM

These four combinations can be found with Aristotle as well.[82] Why, then, did he distinguish but three combinations in his definition of figure (II)? At first sight one could think that he had combined two instances for simplicity's sake. His argumentation in the *Organon*,[83] however, reveals that he, indeed. reckoned with three kinds of relation only between the middle and extreme terms:

(1) $S\ M\ P$
(2) $P\ S\ M$
(3) $M\ P\ S$

In reality there is also a fourth instance at least: $P\ M\ S$. The definition of figure (II), however, was based on the fetishism of these 'three kinds of

[81] *Op. cit.*, 26b34–27al.
[82] *Op. cit.*, 27a3–25.
[83] *Op. cit.*, 41a5–20.

relation'. So, it is not the case that Aristotle simply contracted two combination possibilities into one.

The fact that in the above quotation the middle terms has been defined with regard to its position in the premises is also conspicuous. This approach must not be thought to be accidental for the Stagirite wrote the following about the distinction of various figures:

> If then the middle term is a predicate and a subject of predication, or it is a predicate, and something else is denied of it, we shall have the first figure; if it both is a predicate and is denied of something, the middle figure: if other things are predicated of it, or one is denied, the other predicated, the last figure.[84]

Logicians have usually accepted figure (II) as justified. Kant, on the contrary, regarded only figure (I) as legitimate and the others in themselves as unjustified, being acceptable only in so far as they can be traced back to the first figure. With regard to figure (II), he quoted the following example:

> No spirit is divisible.
> All material is divisible.
> Consequently no material is spirit.

He found this argument possible only because it can be traced back to figure (I) by a conversion of the major premise

> No spirit is divisible,
> and therefore nothing divisible is spirit.
> All material is divisible.
> Consequently no material is spirit.[85]

Aristotle described the specifics of figure (III) as follows:

> But if one term belongs to all, and another to none, of a third, or if both belong to all, or to none, of it, I call such a figure the third; by middle term in

---

[84] *Op. cit.*, 47b1–5.
[85] Kant, *Die falsche Spitzfindigkeit der vier syllogistischen Figuren. (1762). — Frühschriften.* p. 13.

it I mean that of which both the predicates are predicated, by extremes I mean the predicates, by the major extreme that which is further from the middle, by the minor that which is nearer to it. The middle term stands outside the extremes, and is last in position.[86]

By its essence, this argumentation is analogous to the quotation concerning figure (II). Our criticism of the latter — *mutatis mutandis* — holds true of this one as well.

Laurentius Valla, contrary to the opinion of the majority, discarded this figure as useless and superfluous.[87] He argued that no living man would ever reason in this way, therefore this figure is unnatural. Considering that perfect induction is often presented in the form of figure (III), the assumption that such reasoning does not occur in human thinking can hardly be accepted. Furthermore, the basic question for logic is not whether we usually reason this way in everyday thinking, but whether or not these inferences are valid, or correct.

Bakradze proceeded from the following example:

> Sodium is metal.
> Sodium is lighter than water.
> _____
> Some metals are lighter than water.

Then he commented it as follows:

> The third figure basically does not offer any conclusion. In one of the premises one property of the object is predicated ('metal'), in the other another property ('lighter than water'), and the conclusion asserts that the object concerned possesses both properties. I use the expression 'the object concerned' because the conclusion obviously refers to the same object. In fact, the proposition functioning as conclusion all but combines the given premises: sodium is metal and lighter than water. Instead of a genuine conclusion which would be drawn this way:
>
> M is P
> P is M
> _____
> S is P

[86] Aristotle. 1928. 28a10–15.
[87] Valla, *Dialecticae disputationes.* (1543).

we get the following:

$$M \text{ is } P$$
$$M \text{ is } S$$
$$\overline{\phantom{M \text{ is } S \text{ and } P}}$$
$$M \text{ is } S \text{ and } P$$

which is not a conclusion at all.[88]

This view has been disproved by Tavanets already who has rightly pointed out that in the conclusion of the given inference it is not of sodium but of metal that something is predicated. And if the conclusion predicates something of a thing of which the premises have neither affirmed nor denied anything, then this conclusion, no doubt, offers some new knowledge.[89]

IV

Aristotle has given the following explanation why he finds but three figures possible:

> ... no syllogism can establish the attribution of one thing to another, unless some middle term is taken, which is somehow related to each by way of predication. For the syllogism in general is made out of premises, and a syllogism referring to *this* out of premises with the same reference, and a syllogism relating *this* to *that* proceeds through premises which relate this to that. But it is impossible to take a premiss in reference to B, if we neither affirm nor deny anything of it; or again to take a premiss relating A to B, if we take nothing common, but affirm or deny peculiar attributes of each. So we must take something midway between the two, which will connect the predications, if we are to have a syllogism relating this to that. If then we must take something common in relation to both, and this is possible in three ways (either by predicating A of C, and C of B, or C of both, or both of C), and these are the figures of which we have spoken, it is clear that every syllogism must be made in one or other of these figures.[90]

Under the given conditions not only three combinations are possible. Out of three terms (here A and B are the extreme terms and C the middle

[88] Bakradze, *Logika* p. 316.
[89] Cf.: Tavanets, Chapter III. Sect. 2.
[90] Aristotle. 1928. 41a3–18.

one) four such formulas can be constructed where, apart from C, one of the two others is also present (A is C, B is C, C is A, and C is B). Furthermore, if these formulas are combined in a way that the first thesis contains C and A, and the second C and B, then not three but four pairs can be formed: (1) A is C, C is B; (2) C is A, C is B; (3) A is C, B is C; (4) C is A, B is C.[91]

Trendelenburg has defended Aristotle's operation in the following argumentation: if the three terms necessary for a syllogism are examined, with their *relation of subordination* in mind, three figures can be formed for the middle term takes either a middle (figure (I)), or an upper (figure (II)), or a lower (figure (III)) position with regard to the other terms in subordination. Those who reckon with four figures use another criterion for classification, namely the various positions of the middle term in the premises. Aristotle, however, has built his system on the *inner* relation of the three terms used in inferences. Later the *external* aspect, namely whether the middle term is the subject or the predicate in the premises, has become prevailing.[92]

Überweg, on the contrary, was of the opinion that the relation of subordination can be used as a basis in figure (I) only, but even there not for all modes. In mode Darii, for example, the minor premise is a particular proposition with the implication that the extension of the subject may, and may not, be included in the extension of the predicate. — As to figure (II), the relation of the terms can be described as subordination only if an analogy between figures (I) and (II) is assumed.

Consequently, Trendelenburg was mistaken in his view that Aristotle had taken as a basis nothing but the subordination of the terms and that the function of the middle term in the premises became a main concern only later. As we have seen, this latter aspect is already manifest with Aristotle as well.[93]

Patzig has rightly noted that Aristotle was confronted with the following dilemma: if he had defined figure (IV) by the relation of the terms it would have coincided with the definition of figure (I). If he had

---

[91] Cf.: Patzig, p. 117.
[92] Trendelenburg, p. 342.
[93] Cf.: Überweg, pp. 258–266.

chosen the position of the middle term as a basis he would not have arrived
at an unambiguous definition. All this is to say that figure (IV) could not
have been reconciled to the system developed for the first three figures.[94]

The credit for complementing the Aristotelean figures by figure (IV) had
been attributed to Galen for centuries. This belief can be traced back to
one of the works of Averroës where he named Galen as the author.[95] M.
Minas published a fragment of an unknown Greek commentator in 1844
where, again, Galen was credited for having developed figure (IV). Prantl
reviewed a work by Johannes Italus (11th c. A. D.) in 1861 where the
latter, among others, had criticized the so-called Galenian figure.[96]

Überweg found all these sources insufficient upon the consideration
that there is no trace of the figure concerned in the still existing works of
Galen. From Galen to Boethius, that is for four hundred years, no logician
had even mentioned this figure, and instead, following Theophrastus, all
of them discussed the corresponding modes within figure (I).[97]

M. Wallies published a fragment by another anonymous author in 1899
where the following can be found: there are two kinds of categorical
syllogism, namely the simple and the compound one. The former has three
figures, the latter four. Aristotle had the simple syllogism in mind when he
spoke of three figures. Galen, however, referred to the compound one
when he distinguished four figures.

Let us see an example of the compound syllogism:

> All men are mortal.
> All Greek are men.
> Socrates is Greek.
> _____
> Socrates is mortal.

A compound syllogism consists of two simple syllogism. If the three
figures of simple syllogism are joined in pairs, in principle the following
nine combinations will be the result:

> I–I, I–II, I–III, II–II, II–I, II–III, III–III,
> III–I, III–II.

---

[94] Patzig, pp. 117–136.
[95] Cf.: Prantl, p. 571.
[96] *Op. cit.*, Vol. II., p. 302.
[97] Überweg, p. 267.

Yet, only four out of the nine can be described as independent syllogistic figure: I–I, I–II, I–III, II–III. Two combinations (II–II, III–III) must be discarded because they contain premises with two negative and two particular propositions respectively. Three combinations (II–I, III–I, III–II) are simply superfluous since they differ from the accepted figures only in the sequence of their premises.

Łukasiewicz became interested in this fragment and analysed it as follows: first, he established all the possible combinations of the four terms (*A, B, C, D*) used in compound syllogism (in the following propositions the first term is the subject, and the second is the predicate):

	Minor Premise	Middle Premise	Major Premise	Conclusion
(1)	*A* is *B*	*B* is *C*	*C* is *D*	*A* is *D*
(2)	*A* is *B*	*B* is *C*	*D* is *C*	*A* is *D*
(3)	*A* is *B*	*C* is *B*	*C* is *D*	*A* is *D*
(4)	*A* is *B*	*C* is *B*	*D* is *C*	*A* is *D*
(5)	*B* is *A*	*B* is *C*	*C* is *D*	*A* is *D*
(6)	*B* is *A*	*B* is *C*	*D* is *C*	*A* is *D*
(7)	*B* is *A*	*C* is *B*	*C* is *D*	*A* is *D*
(8)	*B* is *A*	*C* is *B*	*D* is *C*	*A* is *D*

The following pairs of figures correspond to these combinations: I–I, I–II, II–III, II–I, III–I, III–II, I–III, and I–I. By disregarding all those combinations which differ from one another in the sequence of the premises only, we arrive at the same four syllogistic figures as before when they were joined in pairs.

Relying upon these findings, Łukasiewicz has made the following statements. It is due to some misunderstanding that Galen has been regarded as the inventor of figure (IV). It had probably been discovered by a so far unknown author who must have lived not before the 6th century A.D. This author presumably did not agree with Aristotle and the Peripatetics, therefore the four figures found with Galen came in handy for him.[98]

---

[98] Łukasiewicz, pp. 38–42.

14*

The fourth figure based on the simple syllogism was first deduced from the extensive relations of the terms. According to the present state of research into the history of logic, Albalag (13th c.) was the first to use the function of the terms in the syllogistic propositions as an argument for necessity of the fourth figure.

It was also in the 13th century that the names for denoting the modes of categorical syllogism had been invented. The modes of figure (IV) were called as follows: Bamalip, Calemes, Dimatis, Fesapo, and Fresison. Their formulas are successively the following:

(1)	(2)	(3)	(4)	(5)
$P \, a \, M$	$P \, a \, M$	$P \, i \, M$	$P \, e \, M$	$P \, e \, M$
$M \, a \, S$	$M \, e \, S$	$M \, a \, S$	$M \, a \, S$	$M \, i \, S$
$S \, i \, P$	$S \, e \, P$	$S \, i \, P$	$S \, o \, P$	$S \, o \, P$

Among the modern logicians Leibniz and Lambert should be mentioned. The former proved figure (IV) to be justified in his work *De arte combinatoria* while the latter demonstrated at especially great length in his work *Neues Organon* that four figures are of equal rank and characterized them as follows: the function of figure (I) is to demonstrate the properties of things (*dictum de omni et nullo*), figure (II) points out the differences between things (*dictum de diverso*), figure (III) gives the examples and exceptions (*dictum de exemplo*) and figure (IV) the species of the genus (*dictum de reciproco*). — Mathematical logicians, if they comment on this subject at all, in general accept figure (IV) as justified.

V

Now let us turn our attention to authors who have rejected figure (IV) in modern times. Kant judged the inferences of figure (IV) to be unnatural and therefore unnecessary in his essay 'Die falsche Spitzfindigkeit der vier syllogistischen Figuren'. He illustrated his thesis, among others, with the following example:

> No fool is a scholar;
> Consequently *no scholar is a fool.*

> Some scholars are pious people;
> Consequently *some pious people are scholars*;
> Hence: some pious people are not fools.

This example has been chosen with the obvious intent to discredit figure (IV) as both its content and form demonstrate. Its content is so silly that it is not worth talking about. Its form offers no reason whatever for the inclusion of the converses of each premises in the conclusion.

For Kant, however, it was still not enough, so he declared that no correct affirmative conclusion can be drawn from figure (IV). He started from the following example:

> Every individual spirit is simple;
> Every simple thing is undecaying (unverweslich),
> Hence: some undecaying things are spirit.

It is obvious, as he wrote, that this conclusion does not follow from the above premises. For having examined the middle term, one has no ground to state that some undecaying things are spirit just because they are simple, since what is simple is not necessarily spirit.[99]

As to the above argument, I have but the following remarks. Kant pointed out as an objection that here the middle term provides no justification for the major term. This is true. But is it really a condition for the validity of an inference? Even in Kant's former example illustrating the negative conclusion ('some pious people are not fools') the middle term 'scholar' failed to play this role, and nevertheless, he accepted the inference at issue as valid.

I have still to comment on Kant's other statement as well, namely that figure (IV) is unnatural. What is the criterion for some figure to be natural? As is well-known, Aristotle interpreted the relation of the subject and the predicate primarily as a connection of specific and generic terms. Still, in those modes of figure (IV) where the conclusion is an affirmative proposition the subject is a generic term and the predicate is a specific term. That is, we find here a relation of superordination instead of the subordination declared to be natural. Let us take Dimatis as an example:

[99] Kant, *Frühschriften*. Vol. II. pp. 14–15.

Some Greeks are men.
All men are mortal.
_____
Some mortals are Greek.

Is the conclusion a true proposition? No doubt, it is. Then why is it bad that its terms are in a relation of superordination? The fetishism of subordination compelled Kant to develop his mistaken theses. This is, among others, why he called figure (IV) unnatural and its affirmative modes false.

Hegel took the following stand: traditional logic attached great importance to the difference which of the two premises is the major and which the minor one respectively.

> This is the ground of the ordinary *fourth figure* of the syllogism, a figure unknown to Aristotle and which in any case is concerned with a wholly empty and pointless distinction. In it the immediate position of the terms is the *reverse* of their position in the first figure. Since subject and predicate of the negative conclusion in the formal treatment of the judgment do not have the definite relationship of subject and predicate, but either can take the place of the other, it is indifferent which term is taken as subject and which as predicate; therefore equally indifferent which premiss is taken as major and which as minor. This indifference, aided as it is by the determination of particularity (especially when it is observed that this can be taken in the comprehensive sense) makes this fourth figure a sheer futility.[100]

First let us see the following reasoning: according to the formal approach the terms of a negative proposition can be converted. If the conclusion of an inference in figure (IV) is a negative proposition the conversion of its terms results in a proposition which is identical with the conclusion of the corresponding mode of figure (I). The difference between figures (I) and (IV) can be seen, among others, in the sequence of their premises. If identical conclusions are arrived at by conversion in these two figures, then the sequence of the premises is, indeed, indifferent. Consequently, figure (IV) proves to be superfluous.

Let us illustrate the above argumentation with concrete examples. First, we should take an inference which belongs to figure (IV) and leads to a negative proposition in its conclusion:

[100] Hegel, *Wissenschaft* pp. 138–139; A. V. Miller tr., p. 679.

> All squares are parallelograms.
> No parallelograms is a circle.
> _____
> No circle is a square.

This example corresponds to the mode called Calemes. Let us convert the terms of the conclusion and the order of sequence of the premises:

> No parallelogram is a circle.
> All squares are parallelograms.
> _____
> No square is a circle.

We have arrived at the mode Celarent of figure (I). Relying on these findings, one could believe that figure (IV) has been proved to be unnecessary. In reality, however, there are still other modes in this figure, for example, Fesapo and Fresison. Let us see an example of Fesapo:

> No parallelogram is a circle.
> All circles are geometrical figures.
> _____
> Some geometrical figures are not parallelograms.

The conclusion of this inference, being proposition O, must not be converted according to formal logic. By reversing the sequence of its premises, we get such a combination of theses *(AE)* from which no valid conclusion can be drawn within figure (I). The same applies to mode Fresison. Thus, by converting the terms of the conclusion and changing the sequence of the premises, one cannot identify the above modes with any mode of figure (I).

What is true of the relation Calemes-Celarent Hegel extended to the relation of figures (IV) and (I) in general. It is, as we have seen, indefensible. Consequently, the above reasoning fails to disprove the necessity of figure (IV).

Moreover, the following assertion can be read in the above quotation:

> In it [i.e. in figure (IV) Gy. T.] the immediate position of the terms is the *reverse* of their position in the first figure...

It is true in so far as the figures at issue are dealt with in a totally abstract way.

FIGURE (IV)   *FIGURE (I)*

P is M M is P

M is P S is M
_____   _____

S is P S is P

These formulas, however, are useless in themselves. It is not from the premises formulated in this way that the conclusion follows. These formulas guarantee a valid inference only in case of definite modes. Such is, for example, the following mode:

$$P\ a\ M$$
$$M\ e\ S$$
$$\overline{S\ e\ P}$$

If the terms of the premises in this mode of figure (IV) are changed according to Hegel's idea and the sequence of the premises is ignored as something indifferent, then the result will be the following formula:

$$M\ i\ P$$
$$S\ e\ M$$
$$\overline{S\ e\ P}$$

However, there is no such valid mode in figure (I).Thus, the connection that the terms of the premises in figure (IV) are transposed as compared to figure (I) does not allow for the replacement of figure (IV) by figure (I).

Fogarasi wrote the following about figure (IV):

> All whales are mammals.
> All mammals are warm blooded.
> _____
> Therefore: Some warm blooded animals are whales.
>
> This is the fourth, the so-called Galenic figure. For this figure we can only cite artificial, sophistical examples, which have no practical value.[101]

Under what conditions could one accept as justified the assertion that only fictive examples can be presented in this figure? As I see it, only if the

_____

[101] Fogarasi, p. 230.

artificial nature of concretization followed from the structure of figure
(IV). To my knowledge this has not yet been proved convincingly, and I
doubt whether it will ever be, for the difference between figure (IV) and the
other figures is not so great as to justify the assertion criticized above. To
decide whether only fictive or even natural examples can be quoted to
illustrate the figure concerned is not a logical problem. The main concern
of logic is whether or not this figure allows for valid inferences.

Finally, let me mention the efforts to find a compromise between the
advocates of three and four figures respectively. Such a view was
represented, for example, by Überweg who decided in favour of the
following classification:

(I.1)	(I.2)	(II)	(III)
MP	*PM*	*PM*	*MP*
SM	*MS*	*SM*	*MS*
SP	*SP*	*SP*	*SP*

In the above system one can see both three figures, of which the first is
divided into two classes, and a distinction between four figures. For
Überweg both interpretations have equal force.[102]

What all the above said boils down to is not that the three figures should
be replaced by four. I have insisted on the latter only because in this way
certain distortions and misinterpretations could be exposed. I do not even
want to exclude the theory of three figures. I think, however, that it should
be relegated to its proper place.

From a modal point of view the following can be said: the advocates of
four figures reckoned with the *possible* instances while those propagating
three figures, with the function of the middle term in mind, tried to find out
what inferences are *actually* used and described figure (IV) as unnatural
upon this consideration.

The followers of these two trends are not antagonistically opposed to
one another just as the possible and the actual do not exclude one another,

102 Überweg, pp. 255–256.

either. At the same time the former had a leaning towards formalism and the latter towards 'practicism'. I think this metaphysical contraposition can be eliminated by advancing from the possible instances to those necessary for our purposes. It must be noted that the advocates of the three figures, led by Aristotle, actually did so in the field of modes: they had stated that each figure allows for 64 modes and then by the method of elimination arrived at the valid inferences.

VI

So far, syllogisms including categorical propositions have been dealt with. Let us consider now other quantified propositions as well. Let us examine, for example, the following mode:

$$M_1, M_2, M_3 \text{ is } P$$
$$\text{Only } M_1, M_2, M_3 \text{ is } P$$
$$\overline{\text{All } S \text{ is } P}$$

Aristotle has already discussed the above inference as a perfect induction. This recognition, however, had no influence whatsoever on the interpretation of this law of syllogism. Works in traditional logic still keep on maintaining that the conclusion in figure (III) must be particular. Then, is the above mode not one of those which belong to figure (III) of the syllogism? This contradiction is due to the fact that *volens-nolens* only categorical propositions are considered in syllogisms.

As we know from the available sources, Occam was the first to include exclusive propositions as well in the analysis of the syllogism. He came to the following conclusions: if both premises are exclusive propositions only, figure (I) will produce an exclusive conclusion while figures (II) and (III) will not (he did not deal with figure (IV)). Conclusions with exceptive propositions are obtained, again, in figure (I) only, namely when the minor premise is an exceptive proposition.[103]

The recognized theory of syllogistic, however, has not been affected by the above analysis in the least. Therefore Tavanets could rightly state even recently:

---

[103] Cf.: Prantl, pp. 408–409.

Two modes of inference prohibited by traditional logic, namely (1) the conclusion drawn from two affirmative premises in the second figure of the categorical syllogism and (2) the conclusion arrived at in the first figure of the categorical syllogism when the minor premise is negative, have proved to be not only possible but even necessary when one of the premises of the categorical syllogism is a universal, singular, or plural exclusive proposition. Examples: (1) "Only the spectrum of sodium contains a yellow line. The spectrum of this material contains a yellow line. This material contains sodium." (2) "Only a party following the most advanced theory may play the role of a vanguard fighter. This party does not follow the most advanced theory. Hence this party cannot play the role of a vanguard fighter.[104]

Menne examined those syllogisms wherein one of the premises is an exclusive particular proposition.[105] Denoting the proposition concerned by u, he got the following valid modes in each figure:

(I)   *aui, euo*
(II)   *auo, euo*
(III) *uai, uao, uau, aui, euo*
(IV)   *uai, euo*

Mode *aui* of figure (I) can be illustrated with the following example:

All horses have four legs.
Only some animals are horses.
_____
Some animals have four legs.

A syllogistical inference can be established also in other cases:

The particular exclusive propositions including the quantifier 'majority' are used in the theory of reasoning in a special way. From two propositions that include the quantifier 'majority' an absolutely sure conclusion can be drawn with regard to some objects which belong to one of the classes mentioned in the predicate of these propositions.

[104] Tavanets, p. 117.
[105] Menne, 'Zur Syllogistik', pp. 91–97.

Let us take these two premises as an example:

(1)   'The majority of animals is ill in this area.'
(2)   'The majority of animals is sheep in this area.'

These two propositions lead to the absolutely sure conclusion that some ill animals are sheep in this area (or: some sheep are ill in this area).[106]

A syllogism can be formed even if an exceptive proposition is involved:

> All M (with the exception of M_1) is P
> All S is M but not M_1
> _____
> All S is P

I think it is unnecessary to demonstrate further modes. What has been said so far is sufficient in order to state that, when syllogisms are discussed, it is reasonable to include all quantified propositions and not only the categorical ones in the analysis.

This wider scope puts the old problems in a new perspective. Let us take, e.g., mode AA of figure (I). The traditional approach accepted only propositions $AEIO$ as conclusion. Let us permit now any quantified proposition in the conclusion. A correct, or valid inference will result from the following among them:

> All S is P
> All S is P or no S is P
> Some but not only this S is P
> This S is P
> Not only this S is P
> Not only some S is P
> Some S is P

In this way (1) we have a complete picture of the correct conclusions from AA, and (2) all such views as the syllogism has (14), (15), (19), or (24) valid modes have been discarded by consistently adopting the above point of view.

---

[106] Tavanets, p. 113.

CHAPTER FOUR

10. THE MODALITY OF TERMS
AND PROPOSITIONS

A

I

Aristotle classified modalities as follows in the 'Hermeneutics'[1]:

It may be.	It cannot be.
It is contingent.	It is not contingent.
It is impossible.	It is not impossible.
It is necessary.	It is not necessary,
It is true.	It is not true.

It is remarkable that true and its negation are enlisted here as modalities. According to Waitz, the Stagirite meant by true in this context what exists. In my opinion this interpretation is unfounded in so far as it merges the modality concerned with other ones, for the possible, the contingent, and the necessary exist just as much. I think Aristotle had in mind *factual* truth, i.e., which expresses facts.—In the *Prior Analytic* Aristotle distinguished necessary, contingent, and simple relations.[2] Commentaries agree that the last of these functions as an expression of facts.

Aristotle, however, failed to consider the factual when analysing the interrelations of the modalities:

> ... we must consider the mutual relation of those affirmations and denials which assert or deny possibility or contingency, impossibility or necessity: for the subject is not without difficulty.[3]

While Aristotle did not draw a distinct line between the objective and the subjective meaning of modalities and considered both of them, the way

---

[1] Aristotle. 1928. 22a12–14.
[2] *Op. cit.*, 29b29–35, 41b30–32.
[3] *Op. cit.*, 21a34–37.

Theophrastus interpreted them was essentially subjective. In his view, for example, possible is what is free of (logical) contradictions.—The Stoics distinguished the following modalities: necessary, possible, impossible, and not necessary.

Shyreswood, a medieval logician (and many others after him), selected six modalities as a basis: true, false, possible, impossible, contingent, and necessary. It was also he who divided propositions into two big groups, namely those which express facts (*de inesse*) and modal propositions.

Opinions on whether or not factual propositions should be regarded as modal ones greatly varied in early modern logic. Following Shyreswood's tradition, Wolff and his school distinguished modal and not-modal propositions. Lambert, on the contrary, made a distinction between three kinds of modal propositions: possible, real, necessary.[4]

Kant shared this latter opinion:

> The *modality* of judgements is a quite peculiar function. Its distinguishing characteristic is that it contributes nothing to the content of the judgement (for, besides quantity, quality, and relation, there is nothing that constitutes the content of a judgment), but concerns only the value of the copula in relation to thought in general. Problematic judgments are those in which affirmation or negation is taken as merely possible (optional). In assertoric judgments affirmation or negation is viewed as real (true) and apodeictic judgments as necessary.[5]

The way Kant interpreted these propositions deserves special attention. A footnote to the text cited above runs as follows:

> Just as if thought were in the problematic a function of the *understanding*; in the assertoric, of the faculty of *judgment*; in the apodeictic, of *reason*.[6]

Elsewhere he declared opinion, belief, and knowledge to be the basis of these very propositions.[7]

In order to clear all possible doubts about his subjectivism, he stated the following:

---

[4] Lambert, p. 89.
[5] Kant, *Kritik* p. 145; Kemp Smith, pp. 109–110.
[6] *Loc. cit.*; Kemp Smith, p. 110.
[7] Cf.: Kant, *Schriften* p. 494.

> This determination of the merely possible or the real or the necessary truth affects only *the judgement itself* but not at all *the thing* about which a judgement is made.[8]

Kant also wanted to emphasize his subjective approach by changing the traditional names of modal propositions. This is why contingent was replaced by problematic and the necessary proposition became apodeictic. According to some historians of logic Kant borrowed the expression 'assertoric' from legal terminology.

Sigwart tried to restrict the scope of modalities even more:

> The so-called assertoric proposition (the simple assertion *A* is *B*) is *not essentially different* from the *apodeictic* proposition (it is necessary that *A* is *B*) in so far as *in every proposition stated with full consciousness, the necessity of stating the proposition is also asserted.*[9]

C. I. Lewis was the first to assess modalities by means of mathematical logic.[10] This aletho-logic which is concerned with assertions studies the following modalities: necessarily true, possibly true, contingently true, and impossible to be true. A. Becker, Łukasiewicz, and Wright should be mentioned among those dealing with modal logic.

All I want to cite here from the above mentioned authors is a statement by Łukasiewicz:

> Under the influence of Plato's theory of ideas Aristotle developed a logic of universal terms and set forth views on necessity which were, in my opinion, disastrous for philosophy ... Modal logic can be described as an extension of the customary logic by the introduction of a 'stronger' and a 'weaker' affirmation; the apodeictic affirmation *Lp* is stronger, and the problematic *Mp* weaker than the assertoric affirmation *p*. If we use the non-committal expressions 'stronger' and 'weaker' instead of 'necessary' and 'contingent', we get rid of some dangerous associations connected with modal terms.[11]

---

[8] *Op. cit.*, p. 540.
[9] Sigwart, p. 190.
[10] Lewis, *Survey*.
[11] Łukasiewicz, 62.

As his work as a whole reveals, these 'dangerous associations' should be understood as the possibility that objective content may be attached to modalities.

For my part I find it equally necessary to assess modalities both objectively and subjectively, with an emphasis on the priority of the objective approach.

As we have seen, the following remarkable classifications have developed during the history of logic:

Aristotle	Stoics	Kant	Logistic
necessary	necessary	apodictical	necessary
impossible	impossible	assertoric	impossible
possible	possible	problematic	possible
contingent	not necessary		accidental
true (fact)			

For orientation's sake let us have a look at Figure 25:

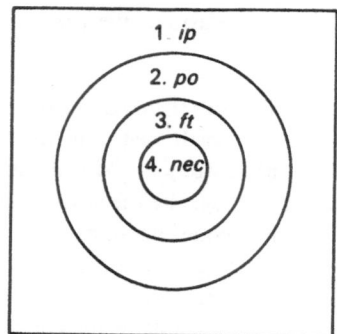

Fig. 25.

The designations are to be understood as follows: *ip* = impossible, *po* = possible, *ft* = fact, *nec* = necessary. Contingent (*c*) is made up of both rings *po* and *ft*. Ring 2 represents 'only possible' and ring 3 'only fact'. Rings 2 and 4 together are what is 'possible but not only fact'. Accordingly, we arrive at the following summary:

0	— void	2 *o* 3	—	*c*
1	— *ip*	2 *o* 4	—	*nec* or only *po*
2	— only *po*	3 *o* 4	—	*ft*
3	— only *ft*	1 *o* 2 *o* 3	—	not *nec*
4	— *nec*	1 *o* 2 *o* 4	—	not only-*ft*
1 *o* 2	— not *ft*	1 *o* 3 *o* 4	—	not only-*po*
1 *o* 3	— only *ft* or *ip*	2 *o* 3 *o* 4	—	*po*
1 *o* 4	— not *c*	1 *o* 2 *o* 3 *o* 4	—	anything

II

Now, let us study Aristotle's following thesis:

> Since there is a difference according as something belongs, necessarily belongs, or may belong to something else (for many things belong indeed, but not necessarily, others neither necessarily nor indeed at all, but it is possible for them to belong), it is clear that there will be different syllogisms to prove each of these relations, and syllogisms with differently related terms, one syllogism concluding from what is necessary, another from what is, a third from what is possible.[12]

This text reveals a duality for it concerns *things* which exist necessarily, simply, and contingently, on the one hand, and necessary, simple, and contingent *propositions* which can be asserted of these things, on the other. With Aristotle, these two aspects have merged into one.

Theophrastus and Eudemus, however, dealt with the second aspect only. Towards the end of the ancient times and at the beginning of the Middle Ages the view (of Boethius, Alfarabi, Petrus Hispanus) that modality is the predicate (*modus est praedicatum*) prevailed. It corresponded to Aristotle's first interpretation but was gradually replaced, from Shyreswood on, by the opinion that two kinds of modal propositions should be distinguished. In one of them modality refers to the terms (*de re*) while in the other to the propositions as a whole (*de dicto*). An example of each of the two kinds of formula is:

(1)   *De re*: S is necessarily P
(2)   *De dicto*: it is necessary that S is P

[12] Aristotle. 1928. 29b29–35.

The former was thought to be of objective nature and the latter of subjective. This distinction, however, is not acceptable. On the one hand, terms may refer to the nonexistent just as much as to the existent, e.g., gods are necessarily immortal. On the other, *de dicto* may refer not to propositions only, but also to a connection in the objective reality, e.g., it is necessary that apple is a fruit. Consequently, to correlate formula (1) with the objective only and formula (2) with the subjective only is unjustified. In reality, the following instances should be reckoned with:

TABLE 23

	Objective	Subjective
De re	(I)	(III)
De dicto	(II)	(IV)

Traditional logic interpreted modal propositions mainly in the spirit of the latter.

> The ground for the classification according to modality is the nature of the interrelation of the object and the property. Our propositions may express whether the connection between the property and the object is actual, possible, or necessary.[13]

This interpretation also has left its mark on works commenting on the Aristotelean modal logic in modern times. That is, it was on the basis of, and in relation to, the *de dicto* interpretation that they have tried to judge whether the Stagirite was right or wrong.

Albrecht Becker deserves credit for having cleared up the relation of *de dicto* and *de re* in his work *Die aristotelische Logik der Möglichkeits-schlüsse* (Berlin, 1933). Most mathematical logicians find it necessary to consider both interpretations. There are, however, some who follow *de dicto* only (e.g., Łukasiewicz).

Sándor Szalai subscribed to this opinion. Aristotle

> ... seems to have taken a position hardly compatible with his views confessed elsewhere, namely that, e.g., the predicate of the proposition '*B* is

---

[13] *Logika*. (Budapest 1956) p. 158.

necessarily A' is '*necessarily A'*. That is, the predicate of the proposition, say, '*All bodies necessarily have volume*' is not 'have volume' but '*necessarily have volume*'. Such a division of the necessary proposition is erroneous because if it were really so the proposition '*All bodies necessarily have volume*' would have nothing to do with the proposition '*All bodies have volume*' since both would have a different predicate; and in this case it would not be conceivable why the latter follows from the former. If understood correctly, the predicate of '*B is necessarily A*' is simply '*A*' and the word 'necessarily' expresses but the necessity of the *connection* between subject *B* and predicate *A*—just as the predicate of '*B may be A*' is simply '*A*' and the word 'may be' expresses the contingency of the *connection* between subject *B* and predicate *A,* or the word '*est*' in the traditional formula '*S est P*' expresses the affirmation of a *connection* between subject *S* and predicate *P*. Nevertheless, it must be understood that Aristotle was *not* of this opinion but, contrary to his posterity as a whole, referred necessity to the predicate of necessary propositions. And having taken *this* position, we shall *not* find Aristotle's following arguments mistaken since they follow from *this* very interpretation of necessary propositions with complete consequentiality.[14]

I have the following objections to the above considerations:

(1) The assertion that Aristotle's position just cited is hardly compatible with his views expressed elsewhere does not hold good since he wrote, among others, the following: "... there are moreover two other pairs, if a term be conjoined with 'not-man', the latter forming a kind of subject. Thus: A″ Not-man is just. B″ Not-man is not-just. C″ Not-man is not-just. D″ Not-man is not not-just."[15] Aristotle made a distinction between negations which refer to the proposition as a whole and those which belong to the predicate only.

(2) Why would it be impossible for two propositions with different predicates to be related to one another? Let us take the following propositions as an example:

All gold is metal.
All gold is precious metal.

The predicates of these two propositions are in a relation of inferiority and superiority respectively which becomes manifest in the following

[14] Aristotelés. *Organon*. Vol. I. (Budapest 1961) p. 245.
[15] Aristotle. 1928. 19b38–41.

proposition: every precious metal is metal. By the knowledge of this middle thesis, we can conclude from the truth of the second proposition to that of the first one. Soon we shall see what an important role this connection plays.

(3) The assertion that Aristotle held the view that necessity belongs to the predicate of the necessary proposition, contrary to his posterity as a whole, is in flat contradiction to the facts of the history of logic. Some of those who agreed with Aristotle in this question have been named at the beginning of this chapter.[16]

(4) The above argument seems to suggest that Aristotle thought necessity belongs to the predicate only. Later on we shall see that he referred necessity also to the proposition as a whole.

Let me quote another Aristotelean text:

> The expression 'it is possible for this to belong to that' may be understood in two senses: 'that' may mean either that to which 'that' belongs or that to which it may belong; for the expression 'A is possible of the subject of B' means that it is possible either of that of which B is stated or of that of which B may possibly be stated. It makes no difference whether we say, A is possible of the subject of B, or all B admits of A. It is clear then that the expression 'A may possibly belong to all B' might be used in two senses.[17]

With regard to the above text Sándor Szalai has observed that Aristotle

> ... finds it necessary to call special attention to the fact that if A is possible of the subject of B, then A may possibly belong to *all* B. But the sentence 'A is possible of the subject of B' may mean not only this but also—at least according to Aristotle—that 'A may belong to that of which B may possibly be stated', and this being a statement of different nature, Aristotle wants to tell it apart from the former one. Even the best commentators seem to find it difficult to understand how this second meaning can be revealed in the sentence 'A is possible of the subject of B' and why Aristotle finds it necessary here to refer to it. Perhaps because in the next paragraph he wants to make a distinction between perfect syllogism that has two contingent premises and imperfect syllogism which is composed of a contingent and an assertoric premise. The formula of the former in the first figure, with

---

[16] Cf.: Bochenski, pp. 260–267.
[17] Aristotle. 1928. 32b26–31.

universal affirmative premises involved, is: *A* may *possibly* belong to *B* which may *possibly* belong to *C*; the formula of the latter is: *A* may *possibly* belong to *B* which *belongs* to *C* (without contingency!). Here, indeed, in the first case *A* is in a relation of contingency with *B* which is also only in a relation of contingency with *C* while in the second case *A* may possibly belong to *B*, but *B* just simply belongs to *C* without any contingency. This interpretation is rather forced, yet the best we can offer for all the others seem to be even more forced.[18]

I think there are three modes mentioned in the Aristotelean quotation. These connections can be described by the signs used in traditional logic as follows:

(1)   *De dicto*: it may be that *S* is *P*.

(2)   *De re*: *S* may be *P*.

(3)   *De dicto* and *de re*: it may be that *S* may be *P*.

Now let us analyse the Aristotelean text under the above aspects:

(1)   "it is possible ... of that of which *B* is stated"

(2)   it belongs to that of which *B* may possibly be stated*

(3)   "it is possible ... of that of which *B* may possibly be stated".

Aristotle must have identified cases (2) and (3) That is why he could declare that the expression "it is possible for this to belong to that" may be understood in two senses.

It is still to be seen whether terms in *de re* interpretation—let us call them *modal terms*—can be assessed as components of propositions only or independently as well. Naturally, they can be treated also in the latter sense, e.g., elements that are necessarily radio-active (more simply: the necessarily radio-active elements). That is, *de re* refers to modal terms while *de dicto* to modal propositions. The *de re* interpretation, as we have seen, is rather different from the *de dicto* interpretation and raises several problems. Theophrastus and his followers found it, of course, more simple to deal with the latter only. Simplification, however, rarely leads to satisfactory solutions.

[18] Aristotelés, *Organon*. Vol. I. (Budapest 1961) p. 264.
* In the Hungarian translation which I find to be closer to the original text of Aristotle this version is obviously present—Gy. T.

The modal term of the above example can be formulated as 'necessarily-*P*'. Since not only the predicate, but also the subject may be a modal term, at least the following cases should be taken into account (to make our survey easier I take only universal affirmative propositions):

(1)   *S* is *P*.
(2)   Necessarily-*S* is *P*.
(3)   *S* is necessarily-*P*.
(4)   Necessarily-*S* is necessarily-*P*.

A *de dicto* interpretation reveals the following modes:

(1)   *S* is *P*.
(2)   It is necessary that *S* is *P*.
(3)   *S* is necessarily *P*.
(4)   It is necessary that *S* is necessarily *P*.

The formula '*S* is necessarily *P*' was interpreted *de re* in traditional logic therefore also identified with the proposition 'all *S* is necessarily-*P*'. As I see it: (1) in the former formula, modality refers to the proposition as a whole and not to the predicate; (2) therefore, the two formulas have different meanings. I shall give my reasons for these statements in the detailed analysis.

The combinations of the above *de re* and *de dicto* modes produce further instance. To enlist all of them is unnecessary since primarily three versions will be discussed. I shall denote by (D1) those instances where modality is at the beginning of the proposition (e.g., it is necessary that *S* is *P*), by (D2) those where it is between the terms (e.g., *S* is necessarily *P*), and by (D3) those where modality refers to the terms themselves (e.g., *S* is necessarily-*P*).

III

Necessity was one of the central categories of ancient Greek philosophy. The atomists, Leucippus and Democritus who took a materialistic position in this respect, paid especially great attention to it. Their approach was, however, one-sided since they treated necessity as an absolute and interpreted it in a fatalistic way.—Protagoras, however, propagated indeterminism ("man is the measure of all things") and denied objective necessity.

As opposed to these views, Aristotle described necessity as follows:

> We say that that which cannot be otherwise is necessarily as it is ...
> Now some things owe their necessity to something other than themselves; others do not, but are themselves the source of necessity in other things. Therefore the necessity in the primary and strict sense is the simple; for this does not admit of more states than one, so that it cannot even be in one state and also in another; for if it did it would already be in more than one.[19]

It is the deterministic standpoint which is manifest in the above quotation: the necessary, though not an absolute, plays a decisive role.

These trends, with some modifications, survived into the Hellenistic period as well. The Stoics took the course of idealistic fatalism, the skeptics that of indeterminism, and the Epicureans that of determinism.

According to the Stoics, necessary is what is true, being either incapable of, or prevented by external circumstances from, being false. What is true and is capable of being false unless external circumstances prevent it from being so is not necessary.[20]

The development of natural sciences drew the attention of modern philosophy to natural necessity. A mechanical determinist school of thought evolved to explain this necessity. In fact, it revived and further developed ancient fatalism. Mechanical determinism was positive in so far as it regarded necessity as primarily objective, but at the same time transformed it into something absolute. Holbach, for example, wrote:

> ... tout en elle [in nature—Gy. T.] est lié par des nœuds invisibles, et tous les effets que nous voyons découlent nécessairement de leurs causes, soit que nous les connaissions, soit que nous ne les connaissions pas. Il peut bien y avoir ignorance de notre part, mais les mots *Dieu, Esprit, Intelligence,* etc., ne remédieront point à cette ignorance; ils ne feront que la redoubler, en nous empêchant de chercher les causes naturelles des effets que nous voyons.[21]

This view gains adherents even nowadays, and also among Marxist philosophers.

---

[19] Aristotle, *Metaphysica.* 1015a34–35, 1015b9–15.
[20] Prantl, Vol. I. p. 463.
[21] Holbach, *Système de la nature.* Paris 1821. Vol. II. p. 124.

Hume's views, on the contrary, reflected a revival of skepticism:

> Our idea, therefore, of necessity and causation arises entirely from the uniformity observable in the operations of nature, where similar objects are constantly conjoined together, and the mind is determined by custom to infer the one from the appearance of the other. These two circumstances form the whole of that necessity, which we ascribe to matter. Beyond the constant *conjunction* of similar objects, and the consequent *inference* from one to the other. we have no notion of any necessity or connexion.[22]

Kant flatly refused the objective interpretation of necessity:

> The apodeictic proposition thinks the assertoric as determined by these laws of the understanding, and therefore as affirming *a priori*; and in this manner it expresses logical necessity.[23]

Marxism has taken the stand of materialist determinism. It means, on the one hand, that this school of thought is equally opposed to fatalism and indeterminism and regards necessity as one of the modalities which is at the same time the basis of possibility and reality.—On the other, it implies its opposition both to vulgar materialism and to idealism.

> Propositions which express necessity emphasize the necessity *expressly*; they *make conscious* the necessity of a phenomenon. In general the differentiation of the modalities of propositions common in school logic does not express the modalities of *objective* relationships but rather the different stages of knowledge.[24]

Now let us turn to the analysis of existential propositions expressing necessity. Aristotle stated, first of all, the following:

> The contradictory of 'it is necessary that it should be' is not 'it is necessary that it should not be', but 'it is not necessary that it should be', and the contradictory of 'it is necessary that it should not be' is 'it is not necessary that it should not be'.[25]

---

[22] Hume, *Enquiries.* p. 82.
[23] See Kant, *Kritik* p. 146. Kemp Smith, p. 110.
[24] Fogarasi, p. 208.
[25] Aristotle. 1928. 22a3–5.

He had three versions in mind here:*

(1)   it is necessary that it should be
(2)   it is necessary that it should not be
(3)   it is not necessary that it should be

According to the text, version (3) denies both of the former two.
In the next chapter the following classification is offered:[26]

(3) it is not necessary that it should be	(2) it is necessary that it should not be
(4) it is not necessary that it should not be	(1) it is necessary that it should be

As compared to the previous list, there is even a fourth version here: it is not necessary that it should not be. Having this in mind, Aristotle gave the following explanation for their relations:

> We must investigate the relation subsisting between these propositions and those which predicate necessity ... In this case, contrary propositions follow respectively from contradictory propositions, and the contradictory propositions belong to separate sequences. For the proposition 'it is not necessary that it should be' is not the negative of 'it is necessary that it should not be', for both these propositions may be true of the same subject; for when it is necessary that a thing should not be, it is not necessary that it should be.
>
> The reason why the propositions predicating necessity do not follow in the same kind of sequence as the rest, lies in the fact that the proposition 'it is impossible' is equivalent, when used with a contrary subject, to the proposition 'it is necessary'. For when it is impossible that a thing should be, it is necessary, not that it should be, but that it should not be, and when it is impossible that a thing should not be, it is necessary that it should be. Thus, if the propositions predicating impossibility or non-impossibility follow without change of subject from those predicating possibility or non-possibility, those predicating necessity must follow with the contrary subject; for the propositions 'it is impossible' and 'it is necessary' are not equivalent, but, as has been said, inversely connected.[27]

[26] Op. cit., 22a23–33.
* My analysis is based on the Hungarian translation — Gy. T.
[27] Op. cit., 22a38–22b9.

In order to understand this text, one has to bear in mind the fact that Aristotle also included the relations of the possible, the impossible, and the contingent in the table from which I have taken the relations of the affirmative and negative forms of the necessary. This is what is implied in the sentence "...the propositions predicating necessity do not follow in the same kind of sequence as the rest". That is, in other cases the propositions that stand in the same line contradicted one another, but in the above table of necessary propositions those in the same line are contrary to one another.

Further on Aristotle corrected himself for the difference at issue had proved to be the result of inconsistent arrangement only.

> Yet perhaps it is impossible that the contradictory propositions predicating necessity should be thus arranged ... the proposition 'it is not necessary that it should not be' follows from the proposition 'it may be'. For this is true also of that which must necessarily be. Moreover the proposition 'it is not necessary that it should not be' is the contradictory of that which follows from the proposition 'it cannot be'; for 'it cannot be' is followed by 'it is impossible that it should be' and by 'it is necessary that it should not be', and the contradictory of this is the proposition 'it is not necessary that it should not be'. Thus in this case also contradictory propositions follow contradictory in the way indicated, and no logical impossibilities occur when they are thus arranged.[28]

That is, if versions (3) and (4) are transposed in the above table this apparent difference will disappear.

Boethius illustrated the relations concerned in the following way:[29]

it is not necessary that it should be   it is necessary that it should not be

it is necessary that it should be

Fig. 26.

[28] *Op. cit.*, 22b10, 23–29.
[29] Prantl, Vol. I. (Berlin 1955) p. 696.

Let us form propositions according to the above versions:

(1)   It is necessary that S should be.
(2)   It is necessary that S should not be.
(3)   It is not necessary that S should be.
(4)   It is not necessary that S should not be.

I shall evaluate them on the basis of Figure 2. The shade-lines mean: it may not be, and the dot in the middle implies: it may be. Under this condition the following values are obtained:

TABLE 24

	(1)	(2)
(1)	f	t/f
(2)	t/f	f
(3)	t	t/f
(4)	t/f	t

Let us modify the conditions. The shade-lines now mean: it is necessary that it should not be, and the dot means: it is necessary that it should be. Then the values of the propositions under discussion are:

TABLE 25

	(1)	(2)
(1)	f	t
(2)	t	f
(3)	t	f
(4)	f	t

When Aristotle arrived at the conclusion that the negative of 'it is necessary that it should be' is not 'it is necessary that it should not be', he had in mind the criterion applied in Table 24. At the same time, Table 25 reveals that this thesis is not universally valid since there is such a criterion under which the propositions concerned negate one another. The same holds true of Boethius' figure.

Let us see now the following propositions: (1) it is necessary that S is P, (2) it is necessary that S is not P, (3) it is not necessary that S is not P, (4) it is not necessary that S is P. The diagrams in Figure 12, as originally interpreted, represent modally indefinite relations (e.g., S is identical with P, S is inferior to P). Just as propositions differ from one another according to their modality so do the relations of terms. These relations may equally be necessary, factual, possible, contingent, etc. If the above propositions were evaluated on the basis of indefinite term relations, the following situation might develop.

Indefiniteness makes it possible to interpret, e.g., the relation 'S is inferior to P' as 'S may possibly be inferior to P'. Under such circumstances the proposition 'it is necessary that S is P', for example, will not produce unambiguous values for it may be both t and f. Therefore, in the interpretation of Figure 15, the indefinite term relations must be replaced by necessary ones everywhere, e.g., 'S is necessarily alien to P'. With this reservation in mind, the values of the propositions concerned coincide with those of Table 6.

In order to give an overall evaluation of necessary relations, we should consider their negations as well, e.g., 'S is not necessarily alien to P'. In this way the number of the instances to be described by diagrams would double. Since all I strive for in the present work is to demonstrate that the propositions under discussion can be evaluated at all, it is sufficient to know what is necessary in order to have an overall picture, without actually considering every instance.

IV

Let us study now the possible and the impossible. In this respect Diodorus should be named first among the still extant sources. He assumed that: (1) what is possible is at the same time real; (2) the impossible cannot follow from the possible; (3) if one of two mutually exclusive instances came into being the other is impossible for if it were possible impossible would follow from possible.[30]

---

[30] Cf.: Überweg–Heinze, Vol. I. p. 138.

Aristotle countered all these three assertions:

(1) There are some who say, as the Megaric school does, that a thing 'can' act only when it is acting, and when it is not acting it 'cannot' act, e.g. that he who is not building cannot build, but only he who is building, when he is building; and so in all other cases. It is not hard to see the absurdities that attend this view. For it is clear that on this view a man will not be a builder unless he is building (for to be a builder is to be able to build), and so with other arts... But we cannot say this, so that evidently potency and actuality are different (but these views make potency and actuality the same, and so it is no small thing they are seeking to annihilate)....[31]

(2) The negative of the proposition 'it is impossible' is consequent upon the proposition 'it may be' and the corresponding positive in the first case upon the negative in the second. For 'it is impossible' is a positive proposition and 'it is not impossible' is negative.[32]

(3) ... if that which is deprived of potency is incapable, that which is not happening will be incapable of happening; but he who says of that which is incapable of happening either that it is or that it will be will say what is untrue; for this is what incapacity meant. Therefore these views do away with both movement and becoming. For that which stands will always stand, and that which sits will always sit, since if it is sitting it will not get up; for that which, as we are told, cannot get up will be incapable of getting up.[33]

He formulated it also in a positive way:

Potent or capable ... in one sense will mean that which can begin a movement (or a change in general, for even that which can bring things to rest is a 'potent' thing) in another thing or in itself *qua* other; and in one sense that over which something else has such a potency; and in one sense that which has a potency of changing into something, whether for the worse or for the better for even that which perishes is thought to be 'capable' of perishing, for it would not have perished if it had not been capable of it; but, as a matter of fact, it has a certain disposition and cause and principle which fits it to suffer this.[34]

---

[31] Aristotle, *Metaphysica*. 1046b28–35, 1047a18–20.
[32] Aristotle. 1928. 22a35–38.
[33] Aristotle, *Metaphysica*. 1047a10–18.
[34] *Op. cit.*, 1019a33–1019b5.

He stated the following about the modality contradictory to capacity, or possibility:

> Incapacity is privation of capacity . . . either in general or in the case of something that would naturally have the capacity, or even at the time when it would naturally already have it . . . The impossible is that of which the contrary is of necessity true, e.g. that the diagonal of a square is commensurate with the side is impossible, because such a statement is a falsity of which the contrary is not only true but also necessary; that it is commensurate, then, is not only false but also of necessity false.[35]

While Aristotle treated possibility as a primarily objective category, Theophrastus strived for making it subjective. In his view possible is what is free of contradiction. He had, of course, the logical, i.e. subjective, contradiction in mind. It is easy to see that contradictory is one of the subjective expressions of impossible.

The Stoics, too, tried to investigate this problem with regard to the mind: possible is what is capable of being true unless the external circumstances prevent it from being so.—Impossible is what is incapable of this.[36]

Kant, too, was against the objective interpretation of possibility:

> The problematic proposition is therefore that which expresses only logical (which is not objective) possibility — a free choice of admitting such a proposition, and a purely optional admission of it into the understanding.[37]

Klaus, on the contrary, propagates the materialistic view:

> First of all of course it must be established that the category, possibility, is grounded in the real world. It has nothing to do with believing something to be possible. If I say, "It is possible that I will win this chess match," I often mean that I am of the opinion that my chances are good. Here possibility is meant in a subjective sense.
>
> It is different when I say, "It is possible to checkmate the opponent's king with two knights." This means, namely, that such checkmate constellations exist. . .[38]

---

[35] Op. cit., 1019b15–18, 24–28.
[36] Cf.: Prantl, Vol. I. p. 463.
[37] Kant, Kritik p. 146; Kemp Smith, p. 110.
[38] Klaus, p. 77.

The examples illustrating the distinction reflect *de dicto* and *de re* respectively the way they were interpreted in the Middle Ages. In my view both propositions may imply both objective and subjective possibility respectively. Whether the linguistic expression of modality stands at the beginning of the sentence or between the terms is a factor unsuitable for distinguishing these two meanings. It would be reasonable to use the words 'possible' or 'may be' to express an objective relation, and 'I find it possible' or 'it may be assumed' for subjective relations.

By possibility I mean what allows for the necessary but excludes the impossible. It is, of course, implied that the same relation is kept in mind. E.g., the proposition 'it is possible that *S* is *P*' does not allow for 'it is necessary that *S* is not *P*' and does not exclude 'it is impossible that *S* is not *P*'.

What is the relation between the possible and the impossible, on the one hand, and the existent and the nonexistent, on the other? In everyday thought, the nonexistent is often identified with the impossible. When the train did not exist, many people thought it also to be impossible. This identification was based on the assumption that what does not exist is impossible.

The expression 'it is not' is often used instead of 'it cannot be'. It is, indeed, the impossible that follows from 'it cannot be' (it is not possible that it should be). 'It is not', however, is not identical with the impossible and, in fact, does not even lead to it as a consequence. Yet the impossible implies the nonexistent.

Aristotle wrote the following on the relation of the existent and the possible:

> ... if, when *A* is real, *B* must be real, then when *A* is possible, *B* also must be possible. For if *B* need not be possible, there is nothing to prevent its not being possible. Now let *A* be supposed possible. Then, when *A* was possible, we agreed that nothing impossible followed if *A* were supposed to be real; and then *B* must of course be real. But we supposed *B* to be impossible. Let it be impossible, then. If, then, *B* is impossible, *A* also must be so. But the first *was* supposed impossible; therefore the second also is impossible. If, then, *A* is possible, *B* also will be possible, if they were so related that if *A* is

real, *B* must be real. If, then, *A* and *B* being thus related, *B* is not possible on this condition, *A* and *B* will not be related as was supposed. And if when *A* is possible, *B* must be possible, then if *A* is real, *B* also must be real. For to say that *B* must be possible, if *A* is possible, means this, that if *A* is real both at the time when and in the way in which it was supposed capable of being real, *B* also must then and in that way be real.[39]

In accordance with the above the following connection can be established: what exists is possible.

Now we should find out whether or not this relation holds true also in an inverse form, i.e., what is impossible exists. If this thesis held true, then the connection that what does not exist is impossible ought to be true as well. Since the latter is false the former cannot be accepted, either.

V

Now let us analyse the propositions expressing possibility. Aristotle discussed this question proceeding from the fact that the negation of '*is*' is '*is not*'. Having illustrated it with several examples, he continued as follows:

If then this rule is universal, the contradictory of 'it may be' is 'it may *not* be', not 'it cannot be'...

But since it is impossible that contradictory propositions should both be true of the same subject, it follows that 'it may *not* be' is not the contradictory of 'it may be'. For it is a logical consequence of what we have said, either that the same predicate can be both applicable and inapplicable to one and the same subject at the same time, or that it is not by the addition of the verbs 'be' and 'not be', respectively, that positive and negative propositions are formed. If the former of these alternatives must be rejected, we must choose the latter. The contradictory, then, of 'it may be' is 'it cannot be'.[40]

There are three versions here:

(1)   it may be
(2)   it may not be
(3)   it cannot be

[39] Aristotle, *Metaphysica*. 1047b14–30.
[40] Aristotle. 1928. 21b10–11, 18–23.

Later on, in the chapter quoted above, the following can be read:

> The contradictory, then, of 'it may *not* be' is not 'it cannot be', but 'it cannot not be', and the contradictory of 'it may be' is not 'it may *not* be', but 'it cannot be'. Thus the propositions 'it may be' and 'it may *not* be' appear each to imply the other: for, since these two propositions are not contradictory, the same thing both may and may *not* be. But the propositions 'it may be' and 'it cannot be' can never be true of the same subject at the same time, for they are contradictory. Nor can be propositions 'it may not be' and 'it cannot not be' be at once true of the same subject.[41]

As compared to the previous quotation, there is even a fourth version here: it cannot not be.

In the next chapter the following classification is introduced:[42]

(1) it may be	(3) it cannot be
(2) it may not be	(4) it cannot not be

In order to analyse these versions propositions have to be formed:

(1) S may be
(2) S may not be
(3) S cannot be
(4) S cannot not be

I shall evaluate them on the basis of Figure 2. The shade-lines here mean: it may not be, the dot implies: it may be. Under this condition the following values are arrived at (see Table 26):

TABLE 26

	(1)	(2)
(1)	t/f	t
(2)	t	t/f
(3)	t/f	f
(4)	f	t/f

[41] *Op. cit.*, 21b34–22a3.
[42] *Op. cit.*, 22a24–34.

Now let us modify the conditions. The shade-lines mean: it is necessary that it should not be, and the dot implies: it is necessary that it should be. Then the values of the propositions concerned are·

TABLE 27

	(1)	(2)
(1)	*f*	*t*
(2)	*t*	*f*
(3)	*t*	*f*
(4)	*f*	*t*

Now let us see Aristotle's dilemma. He proceeded from the opposition of one affirmation and one negation. However, he ended up with two negations for the proposition 'it may be'. Which of them is the real negation? The above quotations show that he could not make a definite decision. His indecision was due to the fact that both solutions seemed to be plausible. A definite decision, in lack of adequate criteria, would have led to one-sidedness.

Under the conditions given during the evaluation, I see the solution of this problem as follows: if the possible is taken as a criterion, then the negation of 'it may be' can only be 'it cannot be', according to Table 26. If, however, our criterion is the necessary, then 'it may not be' and 'it cannot be' become equivalent. In this way both propositions prove to be equally a negation of 'it may be' without breaking the rule that one proposition has but one negation, since this one negation may appear in various forms.

Let us analyse the following propositions: (1) S may be P, (2) S may not be P, (3) S cannot be P, (4) S cannot not be P. On the ground of necessary term relations we get the following values:

TABLE 28

	(1)	(2)	(3)	(4)	(5)	(6)	(7)	(8)	(9)	(10)	(11)	(12)
(1)	*t*	*t*	*t*	*t*	*f*	*f*	*f*	*f*	*t*	*t*	*t*	*t*
(2)	*t*	*t*	*f*	*f*	*t*	*t*	*t*	*t*	*f*	*f*	*t*	*t*
(3)	*f*	*f*	*f*	*f*	*t*	*t*	*t*	*t*	*f*	*f*	*f*	*f*
(4)	*f*	*f*	*t*	*t*	*f*	*f*	*f*	*f*	*t*	*t*	*f*	*f*

Aristotle had the following opinion of facts (reality):

> Actuality, then, is the existence of a thing not in the way which we express by 'potentially'; we say that potentially, for instance, a statue of Hermes is in the block of wood and the half-line is in the whole, because it might be separated out, and we call even the man who is not studying a man of science, if he is capable of studying; the thing that stands in contrast to each of these exists actually.[43]

Let us take existential propositions in the actual mode: (1) it is a fact that S is, (2) it is a fact that S is not, (3) it is not a fact that S is, (4) it is not a fact that S is not. Their evaluation is based on Figure 2. The shade-lines mean: 'it is a fact that it is not', and the dot means: 'it is a fact that it is'.

TABLE 29

	(1)	(2)
(1)	f	t
(2)	t	f
(3)	t	f
(4)	f	t

Let us continue with the following propositions: (1) it is a fact that S is P, (2) it is a fact that S is not P, (3) it is not a fact that S is P, (4) it is not a fact that S is not P. On the ground of necessary term relations the following values are arrived at:

TABLE 30

	(1)	(2)	(3)	(4)	(5)	(6)	(7)	(8)	(9)	(10)	(11)	(12)
(1)	t/f	t/f	t	t	f	f	f	f	t	t	t/f	t/f
(2)	t/f	t/f	f	f	t	t	t	t	f	f	t/f	t/f
(3)	t/f	t/f	f	f	t	t	t	t	f	f	t/f	t/f
(4)	t/f	t/f	t	t	f	f	f	f	t	t	t/f	t/f

---

[43] Aristotle, *Metaphysica.* 1048a31–35.

16*

B

I

The 'only possible' corresponds to ring 2 in the figure of modalities. It should be understood as 'it is possible that *p* and it is not a fact that *p*'. Aristotle pronounced the following view of this modality:

> But also the infinite and the void and all similar things are said to exist potentially and actually in a different sense from that which applies to many other things, e.g. to that which sees or walks or is seen. For of the latter class these predicates can at some time be also truly asserted without qualification; for the seen is so called sometimes because it is being seen, sometimes because it is capable of being seen. But the infinite does not exist potentially in the sense that it will ever actually have separate existence, it exists potentially only for knowledge. For the fact that the process of dividing never comes to an end ensures that this activity exists potentially, but not that the infinite exists separately.[44]

The 'only possible' is at issue, for example, when the cognizability of the world is discussed. Agnosticism, as everyone knows, denies this possibility. However, the history and experience of mankind as a whole proves that science has never reached a point in its development where unsurmountable barriers blocked further research. Each discovery raised a whole series of new problems but these problems have always been solved. Scientific cognition may be hindered by technical and social difficulties but there are no unsurmountable barriers which could block this progress.

Agnosticism was opposed by another mistaken extreme. According to this latter view the development of cognition will reach a point where nothing unknown is left for it. Laplace formulated this view as follows: A mind which at some given moment would know all the forces at work in nature and the position of all the beings that make up nature in relation to one another, on the one hand, and would be comprehensive enough to analyse these data, on the other, would sum up the movements of the biggest bodies of the universe and the lightest atom in the same formula:

[44] *Op. cit.*, 1048b9–18.

nothing would be uncertain for it, and it could see the future just as clearly as the past.[45]

This metaphysical approach, on the one hand, disregards the fact that cognition is an endless process of development. Whoever presupposes that some time everything will be known regards the evolution of cognition as something finite which will come to an end at a certain point. On the other hand, reality itself, as we have already learned to know, keeps on evolving without end. Therefore our knowledge can never completely reflect reality.

Marxism rejects both agnosticism and metaphysical materialism. Its objection to agnosticism is that it puts barriers of nonexistent principles in the way of cognition while metaphysical materialism erroneously asserts that it is only a question of time and ability in order to learn to know everything.

Let us examine the following versions:

(1)   It is only possible that S is P.
(2)   It is only possible that S is not P.

In case of necessary term relations, their values are as follows:

TABLE 31

	(1)	(2)	(3)	(4)	(5)	(6)	(7)	(8)	(9)	(10)	(11)	(12)
(1)	t/f	t/f	f	f	f	f	f	f	f	f	t/f	t/f
(2)	f/t	f/t	f	f	f	f	f	f	f	f	f/t	f/t

Ring 3 illustrates "only a fact". This modality holds true, e.g., of not-recurring phenomena, for instance: Marx was born on the 5th of May, in 1818, in Trier. 'Only a fact' implies that it is a fact that p and it is possible that not p.

Let us compare what is said above with what Aristotle thought, among others, of this modality:

> ... that which is of necessity is actual. Thus, if that which is eternal is prior, actuality is also prior to potentiality. Some things are actualities without

---

[45] Cf.: Laplace, *Essai* p. 3.

potentiality, namely, the primary substances; a second class consists of those things which are actual but also potential, whose actuality is in nature prior to their potentiality, though posterior in time; a third class comprises those things which are never actualized, but are pure potentialities.[46]

Sándor Szalai commented on this paragraph in the following way:

> This whole argumentation at the end of the chapter is essentially an 'ontological appendix' to the above modal logical considerations where Aristotle arranges the modalities of propositions in a certain order according to the various levels of existence, namely: the necessary actually exists. Since, in accordance with the foregoing, possibility follows from the necessary (if something is necessary it is also possible, but it does not hold true the other way round) and therefore actuality (ἐνέργεια, actus) is prior to potentiality (δύναμις, potentia). There is such an actually existing thing (ἐνέργεια) where no potentiality (δύναμις) is involved but it is already all pure actuality — this is 'actus purus', the pure form, the pure essence, which is a primary substance but—as explained in the Metaphysics—is the basis of all being; there are also existing things composed of actuality and potentiality: the existing things of the physical world; and finally, there are potentialities which never become actualities (e.g., the least possible quantity, the biggest possible number, etc.).[47]

The modality at issue was treated by Aristotle indeed only from an ontological point of view. In this way, however, he has created a basis for logical analysis. Let us examine the following propositions:

(1)   It is only a fact that S is P.
(2)   It is only a fact that S is not P.

Considering necessary term relations, one obtains the following values:

TABLE 32

	(1)	(2)	(3)	(4)	(5)	(6)	(7)	(8)	(9)	(10)	(11)	(12)
(1)	f/t	f/t	f	f	f	f	f	f	f	f	f/t	f/t
(2)	t/f	t/f	f	f	f	f	f	f	f	f	t/f	t/f

[46] Aristotle. 1928. 23a22–27.
[47] Aristotelés, Organon. Vol. I. (Budapest 1961). p. 171.

II

Aristotle regarded the contingent as equivalent to the possible in the 'Hermeneutics'.[48] Elsewhere, however, he distinguished them: "... a thing is capable of doing something if there will be nothing impossible in its having the actuality of that of which it is said to have the capacity."[49]— "I use the terms 'to be possible' and 'the possible' of that which is not necessary but, being assumed, results in nothing impossible."[50] Thus, by the possible he meant whatever allows the necessity of the given connection and excludes its impossibility only, e.g., it is possible that man is a living being. He regarded as contingent whatever excludes both the necessity and the impossibility of the given connection, e.g., it is contingent that a swan is white.

Theophrastus was of the opinion that this distinction is senseless and unnecessary. He regarded the modalities concerned as synonyms in the sense of a proposition expressing possibility.

The dominant school of thought in the Middle Ages distinguished between the possible (*possibilis*) and the contingent (*contingens*). This view was asserted, among others, by Shyreswood, Petrus Hispanus, Albertus Magnus, and Occam.[51]

Sigwart discredited the statement that the modality at issue is a proposition:

> Taken as a proposition about *A*, the so-called problematical proposition is not a proposition but merely the thought of a proposition, the uncompleted attempt at a proposition. The veritable proposition that lies in the formula: *A* is perhaps *B*, is only the statement about myself, the doubter, that I do not know whether *A* is *B* or not...[52]

Albrecht Becker who accurately distinguished between the contingent and the possible in his work played an outstanding role in clearing up this confusion.

---

[48] Aristotle, 1928. 22a15–17.
[49] Aristotle, *Metaphysica*. 1047a24–26.
[50] Aristotle. 1928. 32a18–20.
[51] Cf.: Prantl, Vol. III. pp. 4, 44, 105, 380.
[52] Sigwart, Vol. I. p. 192.

Aristotle wrote the following with regard to affirmative and negative contingent propositions:

> Now it appears that the same thing both may and may not be; for instance, everything that may be cut or may walk may also escape cutting and refrain from walking; and the reason is that those things that have potentiality in this sense are not always actual. In such cases, both the positive and the negative propositions will be true; for that which is capable of walking or of being seen has also a potentiality in the opposite direction.[53]

I shall analyse the following versions of the above modality:

(1)   It is contingent that S should be.
(2)   It is contingent that S should not be.
(3)   It is not contingent that S should be.
(4)   It is not contingent that S should not be.

I shall evaluate them on the basis of Figure 2. The shade-lines mean: it is contingent that it should not be, and the dot implies: it is contingent that it should be. Under this condition we get the following values:

TABLE 33

	(1)	(2)
(1)	t	t
(2)	t	t
(3)	f	f
(4)	f	f

Further versions are: (1) it is contingent that S is P, (2) it is contingent that S is not P, (3) it is not contingent that S is P, (4) it is not contingent that S is not P. Their values on the basis of necessary term relations are shown on Table 34.

It is a well-known fact that according to the fatalists (e.g., Democritus) all that exists does so necessarily. This view was naturally accompanied by denial of the accidental, at least the objective accidental. They regarded as accidental whatever there is whose cause we do not know.—The

[53] Aristotle. 1928. 21b13–17.

TABLE 34

	(1)	(2)	(3)	(4)	(5)	(6)	(7)	(8)	(9)	(10)	(11)	(12)
(1)	t	t	f	f	f	f	f	f	f	f	t	t
(2)	t	t	f	f	f	f	f	f	f	f	t	t
(3)	f	f	t	t	t	t	t	t	t	t	f	f
(4)	f	f	t	t	t	t	t	t	t	t	f	f

indeterminists (e.g., Protagoras), on the contrary, denied the necessity and overestimated the role of the accidental.

Aristotle, rejecting both extreme views, took the determinist position and, accordingly, considered necessity as fundamental but admitted also the objective accidental. In the *Metaphysics*, he still used 'accidental' as the synonym of 'contingent':

> 'Accident' means that which attaches to something and can be truly asserted, but neither of necessity nor usually, e.g. if some one in digging a hole for a plant has found treasure. This—the finding of treasure—is for the man who dug the hole an accident; for neither does the one come of necessity from the other, nor, if a man plants, does he usually find treasure. And a musical man *might* be pale; but since this does not happen of necessity nor usually we call it an accident.[54]

In the *Organon*, however, he expressed a more differentiated view:

> ... the expression 'to be possible' is used in two ways. In one it means to happen generally and fall short of necessity, e.g. man's turning grey or growing or decaying, or generally what naturally belongs to a thing (for this has not its necessity unbroken, since man's existence is not continuous forever, although if a man does exist, it comes about either necessarily or generally). In another sense the expression means the indefinite, which can be both thus and not thus, e.g. an animal's walking or an earthquake's taking place while it is walking, or generally what happens by chance: for none of these inclines by nature in the one way more than in the opposite.[55]

---

[54] Aristotle, *Metaphysica*. 1025a13–21.
[55] Aristotle. 1928. 32b5–14.

In my opinion the latter view is the correct one since it makes a distinction between the two versions of contingent propositions. I shall designate the former version by the term 'as a rule' because as far as I know no special logical term has been developed and accepted for it. The expression 'accidental' is already used for denoting the second one.

Medieval logic used accidental mainly in the sense of contingent and discussed it in the study of predicables (what can be predicated) primarily with regard to properties.

The category of accident became a peripheral question of logic, but at the same time one of the important subjects of philosophers in the modern age. Here only two trends, which are contrary to one another, can be mentioned. The adherents of mechanical determinism denied the accidental:

> ... tout en elle [nature—Gy. T.] est lié par des nœuds invisibles, et tous les effets que nous voyons découlent nécessairement de leurs causes, soit que nous les connaissions, soit que nous ne les connaissions pas.[56]

Hume, on the contrary, maintained:

> It seems evident, that, when the mind looks forward to discover the event, which may result from the throw of such a dye, it considers the turning up of each particular side as alike probable; and this is the very nature of chance, to render all the particular events, comprehended in it, entirely equal.[57]

Marxism countered both extremes:

> Chance overthrows necessity, as conceived hitherto. The previous idea of necessity breaks down. To retain it means dictatorially to impose on nature as a law a human arbitrary determination that is in contradiction to itself and to reality, it means to deny thereby all inner necessity in living nature, it means generally to proclaim the chaotic kingdom of chance to be the sole law of living nature.[58]

The evaluation of accidental existential propositions is based on Figure 2 where the shade-lines mean: it is accidental that it is not, and the dot means: it is accidental that it is.

[56] Holbach, Vol. II. p. 124.
[57] Hume, *Enquiries* p. 57.
[58] Engels, *Dialektik der Natur.* p. 235; Clemens Dutt tr. p. 293.

TABLE 35

	(1)	(2)
(1) It is accidental that S is.	t	t
(2) It is accidental that S is not.	t	t
(3) It is not accidental that S is.	f	f
(4) It is not accidental that S is not.	f	f

III

That degree of possibility which is already close to necessity was called probability by Aristotle.

> ... those opinions are 'generally accepted' which are accepted by every one or by the majority or by the philosophers—i.e. by all, or by the majority, or by the most notable and illustrious of them.[59]

Later on another definition, namely that probability is the measure of possibility, became widely accepted.

Probability is often considered as being of the same kind as chance.

> Though there be no such thing as *Chance* in the world; our ignorance of the real cause of any event has the same influence on the understanding, and begets a like species of belief or opinion.
>
> There is certainly a probability, which arises from a superiority of chances on any side; and according as this superiority increases, and surpasses the opposite chances, the probability receives a proportionable increase, and begets still a higher degree of belief or assent to that side, in which we discover the superiority. If a dye were marked with one figure or number of spots on four sides, and with another figure or number of spots on the two remaining sides, it would be more probable, that the former would turn up than the latter; though, if it had a thousand sides marked in the same manner, and only one side different, the probability would be much higher, and our belief or expectation of the event more steady and secure. This process of thought or reasoning may seem trivial and obvious; but to those who consider it more narrowly, it may, perhaps, afford matter for curious speculation.[60]

---

[59] Aristotle. 1928. 100a22–24.
[60] *Enquiries* pp. 56–57.

The modalities discussed here were distinguished as early as by Aristotle in the following way: the accidental excludes the necessary while the probable does not.

Kant made the following statement:

> To the study of the certainty of our knowledge also belongs the study of the knowledge of the probable, which is to be seen as an approximation to certainty.
>
> Under probability is to be understood a belief based on insufficient reasons, which however have a greater relation to sufficient reason than do the reasons for the opposite position.[61]

Fogarasi, in opposition to the subjectivist views, wrote the following while investigating induction:

> But probability, too, has two senses, an objective and a subjective. By way of induction one ascertains not whether a *consciousness considers* the occurrence of certain phenomena to be *probable*, but rather whether the occurrence of these phenomena *is* probable in an objective sense. If, for instance, we draw inductive conclusions about the probable weather situation on the basis of observations (century long experience with successive changes in weather conditions), we obviously want to determine the objective probability of weather conditions and not the probability states of our consciousness, as present day subjectivism maintains. The struggle between materialism and idealism on the question of probability is being pursued in our times with increasing acrimony. The idealism of today in physics and biology attributes a subjective sense to probability and thus brings about a total confusion which has become an impediment to scientific progress itself in entire areas of science....[62]

In accordance with the previous procedure, I shall evaluate probable existential propositions on the basis of Figure 2. The shade-lines mean: it is necessary that it should not be, and the dot means: it is necessary that is should be.

---

[61] Kant, *Schriften* p. 512.
[62] Fogarasi, pp. 269–270.

TABLE 36

	(1)	(2)
(1) It is probable that S is.	f	t
(2) It is probable that S is not.	t	f
(3) It is not probable that S is.	t	f
(4) It is not probable that S is not.	f	t

Now let us see the summary of modal propositions 'S is P':

0	— S is P and not S is P
1	— It is impossible that S is P (it is possible but not only-possible that S is not P)
2	— It is only possible that S is P (it is possible but not a fact that S is P)
3	— It is only a fact that S is P (it is a fact but not necessary that S is P)
4	— It is necessary that S is P (it is a fact but not only-fact that S is P)
1 o 2	— It is not a fact that S is P (it is possible that S is P or it is impossible that S is P)
1 o 3	— It is only a fact that S is P or it is impossible that S is P
1 o 4	— It is not contingent that S is P (it is necessary that S is P or it is impossible that S is P)
2 o 3	— It is contingent that S is P (it is possible but not necessary that S is P)
2 o 4	— It is possible but not only-fact that S is P
3 o 4	— It is a fact that S is P (it is only a fact or necessary that S is P)
1 o 2 o 3	— It is not necessary that S is P (it is possible that S is not P)
1 o 2 o 4	— It is not a fact only that S is P (it is not only-possible that S is P)
1 o 3 o 4	— It is not only-possible that S is P (it is a fact that S is P or it is impossible that S is P)

2 *o* 3 *o* 4   — It is possible that *S* is *P* (it is not impossible that *S* is *P*)

1 *o* 2 *o* 3 *o* 4   — *S* is *P* or not *S* is *P*

The following propositions successively will have the values listed in Table 37:

(1)   *S* is not *P* and not *S* is not *P*
(2)   It is necessary that *S* is not *P*
(3)   It is only a fact that *S* is not *P*
(4)   It is only possible that *S* is not *P*
(5)   It is impossible that *S* is not *P*
(6)   It is a fact that *S* is not *P*
(7)   It is possible but not only-fact that *S* is not *P*
(8)   It is not contingent that *S* is not *P*
(9)   It is contingent that *S* is not *P*
(10)   It is only a fact or impossible that *S* is not *P*

[continued on page 241]

TABLE 37

	(1)	(2)	(3)	(4)	(5)	(6)	(7)	(8)	(9)	(10)	(11)	(12)
(1)	*f*	*f*	*f*	*f*	*f*	*f*	*f*	*f*	*f*	*f*	*f*	*f*
(2)	*f*	*f*	*f*	*f*	*t*	*t*	*t*	*t*	*f*	*f*	*f*	*f*
(3)	*t/f*	*t/f*	*f*	*f*	*f*	*f*	*f*	*f*	*f*	*f*	*t/f*	*t/f*
(4)	*f/t*	*f/t*	*f*	*f*	*f*	*f*	*f*	*f*	*f*	*f*	*f/t*	*f/t*
(5)	*f*	*f*	*t*	*t*	*f*	*f*	*f*	*f*	*t*	*t*	*f*	*f*
(6)	*t/f*	*t/f*	*f*	*f*	*t*	*t*	*t*	*t*	*f*	*f*	*t/f*	*t/f*
(7)	*f/t*	*f/t*	*f*	*f*	*f*	*f*	*f*	*f*	*f*	*f*	*f/t*	*f/t*
(8)	*f*	*f*	*t*	*t*	*t*	*t*	*t*	*t*	*t*	*t*	*f*	*f*
(9)	*t*	*t*	*f*	*f*	*f*	*f*	*f*	*f*	*f*	*f*	*t*	*t*
(10)	*t/f*	*t/f*	*t*	*t*	*f*	*f*	*f*	*f*	*t*	*t*	*t/f*	*t/f*
(11)	*f/t*	*f/t*	*t*	*t*	*f*	*f*	*f*	*f*	*t*	*t*	*f/t*	*f/t*
(12)	*t*	*t*	*f*	*f*	*t*	*t*	*t*	*t*	*f*	*f*	*t*	*t*
(13)	*t/f*	*t/f*	*t*	*t*	*t*	*t*	*t*	*t*	*t*	*t*	*t/f*	*t/f*
(14)	*f/t*	*f/t*	*t*	*t*	*t*	*t*	*t*	*t*	*t*	*t*	*f/t*	*f/t*
(15)	*t*	*t*	*t*	*t*	*f*	*f*	*f*	*f*	*t*	*t*	*t*	*t*
(16)	*t*	*t*	*t*	*t*	*t*	*t*	*t*	*t*	*t*	*t*	*t*	*t*

(11)   It is not a fact that S is not P
(12)   It is possible that S is not P
(13)   It is not only-possible that S is not P
(14)   It is not only a fact that S is not P
(15)   It is not necessary that S is not P
(16) S is not P or not S is not P

IV

What is the function of the essential properties of objects in logic? Aristotle devoted a whole chapter to this subject both in the *Metaphysics*[63] and in the *Posterior Analytics*.[64] Let me quote some of his significant assertions from the latter work:

> Essential attributes are (1) such as belong to their subject as elements in its essential nature (e.g. line thus belongs to triangle, point to line; for the very being or 'substance' of triangle and line is composed of these elements, which are contained in the formulae defining triangle and line): (2) such that, while they belong to certain subjects, the subjects to which they belong are contained in the attribute's own defining formula...Extending this classification to all other attributes, I distinguish those that answer the above description as belonging essentially to their respective subjects; whereas attributes related in neither of these two ways to their subjects I call accidents or 'coincidents'; e.g. musical or white is a 'coincident' of animal ... So, since any given predicate must be either affirmed or denied of any subject, essential attributes must inhere in their subjects of necessity.

This view has essentially been accepted for two thousand years. Let us see, for example, Spinoza's definition:

> I say that appertains to the essence of a thing which, when granted, necessarily involves the granting of the thing, and which, when removed, necessarily involves the removal of the thing; or that without which the thing, or on the other hand, which without the thing can neither exist nor be conceived.[65]

[63] Aristotle, *Metaphysics*. 1017b15–43.
[64] Aristotle. 1928. 73a25–74a4.
[65] Spinoza, p. 37.

Mill wrote the following about the way a contrary opinion had developed:

> Almost all metaphysicians prior to Locke, as well as many since his time, have made a great mystery of Essential Predication, and of predicates which are said to be of the *essence* of the subject. The essence of a thing, they said, was that without which the thing could neither be, nor be conceived to be. Thus, rationality was of the essence of man, because without rationality, man could not be conceived to exist ...
>
> it was reserved for Locke at the end of the seventeenth century, to convince philosophers that the supposed essences of classes were merely the signification of their names; nor, among the signal services which his writings rendered to philosophy, was there one more needful or more valuable.[66]

Marxist philosophers oppose the positivistic attempts at neglecting the essence and defend this category as justified:

> Under essential properties we understand the following:
>
> (a) A property, which is inseparably joined to the thing in question and so determines it that the thing stops being itself if the property in question disappears. Among the essential properties of a fish, for example, is that of breathing through gills. If this property does not exist, then the creature in question is not a fish, even if its external appearance (e.g. in whales) seems to indicate that it is.
>
> (b) Not every property is essential. Every thing has infinitely many properties. The non-essential properties are distinguished by the fact that they are variable within certain limits, that they may even disappear completely without making the thing to which they are attributed cease to be what it is. The essential properties of humans consist in the ability to work, to possess consciousness and language, etc. Besides these there are, however, countless other properties which humans have, e.g., teeth, a particular hair colour, etc. But a change in these properties, even their disappearance, does not change the fact that the individual human in question remains in essence a human.
>
> (c) Finally, under the essential properties of a thing we understand two separate groups of properties. The one group, which we also denote as *qualities,* belong necessarily to the thing in question, and they belong *only* to

[66] Mill, *A System of Logic Ratiocinative and Inductive.* pp. 110, 112.

this thing and this phenomenon. They thus constitute at the same time the specific difference which distinguishes this thing and this phenomenon from other things and phenomena. The essential properties of the second group also necessarily belong to the thing in question, but they are not limited to this thing and are also to be found in other things.[67]

Now let us compare the designations of properties with the types of modalities. I shall build this analysis on Figure 25 and the explanation attached to it. Version 2 *o* 3 will be the easiest to solve since during the history of logic there has been an agreement on the fact that both properties and propositions can be contingent. Contingent is what equally excludes both necessary and impossible.

The texts cited so far have revealed that Aristotle used the essential in the same sense as the necessary. Later, however, it became general practice to speak of essential properties and necessary propositions. This is why the expression 'essential proposition' sounds so strange (it is less unusual to speak of necessary properties). This strangeness, however, can be overcome by an adequate formulation, e.g., 'man necessarily lives in society' = 'it belongs to the essence of man to live in society'. That is, it can be stated with regard to propositions that a necessary relation is at the same time also an essential relation, and the other way round. It is no accident that the positivists reject both the necessary and the essential.

It follows from the above said that the inessential (not-essential) corresponds to the not-necessary. This correlation contributes to the elimination of an inaccuracy. The inessential is often used in the same sense as the contingent (additional, accidental). Figure 25, however, reveals that the contingent (2 *o* 3) is not identical with the not-necessary (1 *o* 2 *o* 3). The difference is that the contingent excludes the impossible while the not-necessary does not. Consequently, inessential is not identical with contingent.

To the best of my knowledge there are no more names used for properties which could be compared with the modalities discussed so far. Thus, it is reasonable to use the existing modal names for the remaining properties. Let us take the following versions as an example:

[67] Klaus, pp. 141–142.

17

2 *o* 3 *o* 4   e.g., 'baldness is a possible property of man'
1   e.g., 'it is impossible that a stone should have the property of nutrition'
1 *o* 2   e.g., 'it is a not-actual property of France that it is socialist'

The analogy revealed above makes it possible (1) to treat properties, too, in modal logic, and (2) to use the results of modal logic in the classification of properties.

V

Let us examine the relation of modal propositions, proceeding from the following table of Aristotle:

TABLE 38

(1) It may be.	(I)	(9) It cannot be.	(III)
(2) It is contingent.		(10) It is not contingent.	
(3) It is not impossible that it should be.		(11) It is impossible that it should be.	
(4) It is not necessary that it should not be.		(12) It is necessary that it should not be.	
(5) It may not be.	(II)	(13) It cannot not be.	(IV)
(6) It is contingent that it should not be.		(14) It is not contingent that it should not be.	
(7) It is not possible that it should not be.		(15) It is impossible that it should not be.	
(8) It is not necessary that it should be.		(16) It is necessary that it should be.	

(In Aristotle's table the expression 'it is not necessary that it should not be' is in the place of 'it is not necessary that it should be', and the other way round. In his further argumentation, however, he modified this sequence, and I publish his table here accordingly.)

He established the following relations between the propositions classified above:

(1) The statements in each group marked with a Roman number are equivalent, e. g., 'it is contingent', 'it is not impossible that it should be', and 'it is not necessary that it should not be' all follow from the statement

'it may be'. Later, in Chapter 13 of the *Prior Analytic*, Aristotle made a distinction between the possible and the contingent and in this way arrived at the correct conclusion that 'it is contingent' is not equivalent to the other three versions. Therefore I disregard the contingent in the analysis of the above table.

(2) Now the propositions 'it is impossible that it should be' and 'it is not impossible that it should be' are consequent upon the propositions 'it may be', 'it is contingent', and 'it cannot be', 'it is not contingent', the contradictories upon the contradictories. But there is inversion. The negative of the proposition 'it is impossible' is consequent upon the proposition 'it may be' and the corresponding positive in the first case upon the negative in the second. For 'it is impossible' is a positive proposition and 'it is not impossible' is negative.'[68]

That is, there is a contradiction between the modes in groups (I) and (III), and in groups (II) and (IV) respectively.

(3)   We must investigate the relation subsisting between these propositions and those which predicate necessity. That there is a distinction is clear. In this case, contrary propositions follow respectively from contradictory propositions, and the contradictory propositions belong to separate sequences. For the proposition 'it is not necessary that it should be' is not the negative of 'it is necessary that it should not be', for both these propositions may be true of the same subject; for when it is necessary that a thing should not be, it is not necessary that it should be.[69]

Accordingly, there is a subcontrary relation between the modes of groups (I) and (II) and a contrariety between those of groups (III) and (IV).

(4) ... when it is necessary that a thing should be, it is possible that it should be. (For if not, the opposite follows, since one or the other must follow; so, if it is not possible, it is impossible, and it is thus impossible that a thing should be, which must necessarily be; which is absurd.) ...

But again, the proposition 'it is necessary that it should be' does not follow from the proposition 'it may be', nor does the proposition 'it is necessary that it should not be'. For the proposition 'it may be' implies a

[68] Aristotle. 1928. 22a32—37.
[69] *Op. cit.*, 22a38—22b3.

twofold possibility, while, if either of the two former propositions is true, the twofold possibility vanishes. For if a thing may be, it may also not be, but if it is necessary that it should be or that it should not be, one of the two alternatives will be excluded."[70]

There is a relation of subordination between the modes of groups (IV) and (I), and those of groups (III) and (II), respectively.

Shyreswood developed the above Aristotelean table into what is called the logical square.[71] [See p. 247—Ed.]

With regard to the contingent only a relation of contradiction can be correctly stated.

The mode 'it is a fact' has not occurred in the foregoing. It is due to the fact that Aristotle dealt with the actual when he was investigating categorical propositions and syllogisms respectively, and not modalities. Later on, however, many logicians also ranked the actual among the modalities, and having taken it into consideration, stated the following these:

(1)   One can conclude from the necessary to the actual.
(2)   One can conclude from the necessary to the possible.
(3)   One can conclude from the actual to the possible.

In traditional logic the study of modal immediate inferences was basically confined to these very theses.

Let us proceed from the following formulas:

(1)   It is necessary that p
(2)   It is necessary that not p
(3)   It is not necessary that p
(4)   It is not necessary that not p

The relations of these formulas and the immediate inferences based on them can be seen from the comprehensive tables which contain the values of modal propositions. The fourth is subordinated to the first, the third to

---

[70] *Op. cit.*, 22b11—15, 18—22.
[71] Prantl, Vol. III. p. 14.

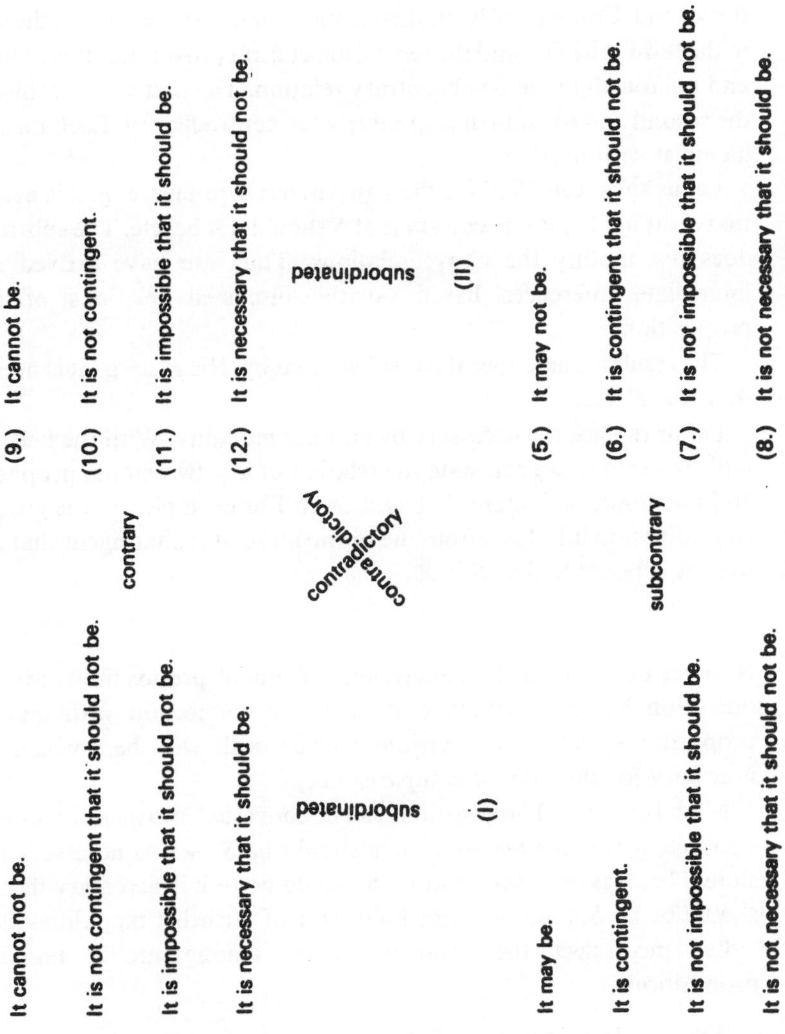

(III)

(9.) It cannot be.

(10.) It is not contingent.

(11.) It is impossible that it should be.

(12.) It is necessary that it should not be.

subordinated

(II)

(5.) It may not be.

(6.) It is contingent that it should not be.

(7.) It is not impossible that it should not be.

(8.) It is not necessary that it should be.

contrary

contradictory

contradictory

subcontrary

(IV)

(13.) It cannot not be.

(14.) It is not contingent that it should not be.

(15.) It is impossible that it should not be.

(16.) It is necessary that it should be.

subordinated

(I)

(1.) It may be.

(2.) It is contingent.

(3.) It is not impossible that it should be.

(4.) It is not necessary that it should not be.

Fig. 27

the second. Consequently, the first is superior to the fourth, and the second to the third. The first and the second are contrary to one another. The third and the fourth are in a subcontrary relation. The first and the third, and the second and the fourth respectively are contradictory. Each formula is identical with itself.

Let us substitute 'S is' for the sign p in our formulas, e. g. it is necessary that S should be; it is necessary that S should not be, etc. The substitution does not modify the above relations. Thus, we have arrived at the immediate inferences based on the discussed relations of modal propositions.

The result is analogous if p is substituted by 'S is P', e. g., it is necessary that S is P, etc.

Let us replace the necessary by another modality. With the help of the table referred to we can state the relation of any two modal propositions and the immediate inference based on it. For example, on the ground of subordination it follows from the proposition 'it is contingent that S is P' that 'it is possible that S is P'.

Now let us examine the conversions of modal propositions. Since this operation has been analysed in detail in connection with quantified propositions, only those versions will be dealt with here which I find necessary for the sake of comparison.

Modal existential propositions can be converted into identical ones. For example, it is necessary that there should be $S \leftrightarrow$ it is necessary that S should be; it is necessary that no S should be $\leftrightarrow$ it is necessary that there should be no S. And the same holds true of the other modalities as well.

Let me select the following from among modal unqualified propositions:

(1)	It is necessary that S is P	must not be converted
(2)	It is necessary that S is not P	$\leftrightarrow$ It is necessary that P is not S

The restriction 'it must not be converted' means with regard to the proposition 'it is necessary that S is P' that it cannot be converted into identical (simply). It can be converted, however, by subordination: it is possible that *P* is *S*.

(1)	It is possible that S is P	↔	It is possible that P is S
(2)	It is possible that S is not P		must not be converted

Other modal propositions should be converted neither in their affirmative nor in their negative forms.

The operation of obversion has already been mentioned in connection with categorical propositions. Obversion is an immediate inference in the course of which the quality of the obvertend is changed and its predicate is replaced by a contradictory term. Let us perform this operation on modal propositions.

(1)	It is necessary that S is P	↔	It is necessary that S is not not-P
(2)	It is necessary that S is not P	↔	It is necessary that S is not-P

These relations can be formulated also in a different way:

(1)	It is necessary that S is P	↔	It is impossible that S is not P
(2)	It is necessary that S is not P	↔	It is impossible that S is P

Necessity can be replaced by any other modality and the operation of obversion can still be performed according to models (1) and (2).

The following modal obversions correspond to the obversion of categorical propositions (in the sequence: *A*, *E*, *I*, *O*):

(1)	It is necessary that S is P	↔	It is impossible that S is not-P
(2)	It is impossible that S is P	↔	It is necessary that S is not-P
(3)	It is possible that S is P	↔	It is possible that S is not-P

(4)   It is possible that   ↔ It is possible that
         S is not P   S is not-P

Contraposition is an immediate inference in the course of which the terms are transported and at least one of them is replaced by a contradictory term. Let us begin with those versions where both terms are replaced by their contradictory:

(1)   It is necessary that   ↔ It is necessary that
         S is P   not-P is not-S
(2)   It is necessary that   must not be
         S is not P   contraposed
(3)   It is possible that   must not be
         S is P   contraposed
(4)   It is possible tnat   ↔ It is possible that
         S is not P   not-P is not not-S

Now only one of the terms is replaced by its contradictory. The following results are obtained by the contraposition of the predicate:

(1)   It is necessary that   ↔ It is impossible that
         S is P   not-P is S
(2)   It is impossible that   ↔ It is possible that
         S is P   not-P is S
(3)   It is possible that   must not be
         S is P   contraposed
(4)   It is possible that   ↔ It is possible that
         S is not P   not-P is S

The contraposition of the subject leads to the following versions:

(1)   It is necessary that   ↔ It is possible that
         S is P   P is not not-S
(2)   It is impossible that   ↔ It is necessary that
         S is P   P is not-S
(3)   It is possible that   ↔ It is possible that
         S is P   P is not not-S
(4)   It is possible that   must not be
         S is not P   contraposed

Of modal propositions syllogisms can be established, e. g.:

> It is necessary that M is P
> It is possible that S is M
> ————————————————
> It is possible that S is P

It needs no further explanation that a modal syllogism can be correlated to any categorical syllogism if 'all' is replaced by 'necessary' and 'some' by 'possible'.

11. MODALITY AND QUANTIFICATION

A

I

The relation of the general and the necessary has attracted more attention than the interrelations of any other categories implied in the title. Aristotle wrote, among others, the following on this issue:

> Some things occur of necessity, others usually, others however it may chance; if therefore a necessary event has been asserted to occur usually, or if a usual event (or, failing such an event itself, its contrary) has been stated to occur of necessity, it always gives an opportunity for attack. For if a necessary event has been asserted to occur usually, clearly the speaker has denied an attribute to be universal which is universal, and so has made a mistake; and so he has if he has declared the usual attribute to be necessary: for then he declares it to belong universally when it does not so belong"[72]

Elsewhere he wrote as follows:

> I term 'commensurately universal' an attribute which belongs to every instance of its subject, and to every instance essentially and as such; from which it clearly follows that all commensurate universals inhere *necessarily* in their subjects.[73]

---

[72] Aristotle. 1928. 112 bl—10.
[73] *Op. cit.*, 73b27—28.

Avicenna ranked the expressions 'all' and 'nothing' among the designations of modality. His view was shared by Thomas Aquinas.[74]

Kant took the following position as to the relation of the categories at issue:

> Necessity and strict universality are ... sure criteria of *a priori* knowledge, and are inseparable from one another. But since in the employment of these criteria the contingency of judgments is sometimes more easily shown than their empirical limitation, or, as sometimes also happens, their unlimited universality can be more convincingly proved than their necessity, it is advisable to use the two criteria separately, each by itself being infallible.[75]

What had been only a practical consideration for Kant became a question of principle for the indeterminists. The tendency to separate these two categories runs right through the modal logic of Łukasiewicz as a whole. Why did he need this distinction? He himself has answered this question:

> By determinism I understand a theory which states that if an event E happens at the moment t, then it is true at any moment earlier than t that E happens at the moment t. The strongest argument in defence of this theory is based on the law of causality which states that every event has a cause in some earlier event. If so, it seems to be evident that all future events have causes which exist today, and existed from eternity, and therefore all are predetermined.
>
> The law of causality, however. understood in its full generality should be regarded as merely a hypothesis. It is true, of course, that astronomers, relying on some laws known to govern the universe, are able to predict for years in advance the positions and motions of heavenly bodies with considerable accuracy. Just at the moment I finished writing the previous sentence a bee flew humming past my ear. Am I to believe that this event too has been predetermined from all eternity and by some unknown laws governing the universe? ... Tomorrow's sea-fight is a contingent event, and if there are such events, determinism is refuted.[76]

---

[74] Cf.: Prantl, Vol. III. p. 117.
[75] Kant, *Kritik* pp. 49—51; Kemp Smith, p. 44.
[76] Łukasiewicz, pp. 207—208.

The above quotation makes it clear that by determinism Łukasiewicz meant mechanical determinism. It does not require further evidence that it is not this kind of determinism which I want to defend.

There are basically two false extremes: one of them separates the necessary and the general from one another, the other identifies them. While these extreme views occur relatively rarely, those which lay an emphasis on one of them at the cost of the other are very frequent. Logic should reckon both with their difference and with their identity. Sándor Szalai was right.in asserting:

> Let us say a few words about the necessary (apodictic) propositions first. What predicates can be related to a subject either in affirmative or in negative is an epistomological and ontological question of great importance which Aristotle fails to discuss in detail in the *Prior Analytics*. There is no doubt, however, that such propositions exist, and they are usually distinguished from the usual assertoric propositions. For example, that all children of Callias are boys can be asserted assertorically only, and we would not say that 'all children of Callias are necessarily boys' since we do not know of such a logical natural law that makes necessary for Callias to have only boys as children. Naturally, if such a law — of genetics, for instance — which would make it necessary for Callias to have sons only became known, then we would be willing to assert the above necessary (apodictic) proposition, just as we say, for example, such things as: 'All numbers divisible by four are necessarily divisible by two as well', or 'The speed of all freely falling bodies necessarily increases in proportion to the time of fall'.[77]

The general and the necessary are in a relation of the subalternant and the subalternate. All that is necessary is at the same time general, or universal, but it does not hold true the other way round. For a detailed survey of their relations should be based on the following propositions: (1) it is necessary that S is P, (2) it is necessary that S is not P, (3) all S is P, (4) no S is P. The third is subordinated to the first and the fourth to the second. The rest of the instances, with the exception of identities, can be characterized by a relation of contrariety.

[77] Aristotelés, *Organon*. Vol. I. (Budapest 1961) pp. LXXXVII—LXXXVIII.

II

Let us examine the relation of the individual and the essence. In the *Metaphysics* we can read the following:

> Each thing itself, then, and its essence are one and the same in no merely accidental way... to *know* each thing, at least, is just to know its essence, so that even by the exhibition of instances it becomes clear that both must be one. (But of an accidental term, e.g. 'the musical' or 'the white', since it has two meanings, it is not true to say that it itself is identical with its essence; for both that to which the accidental quality belongs, and the accidental quality, are white, so that in a sense the accident and its essence are the same, and in a sense they are not ...)[78]

Later on Aristotle wrote as follows:

> ... there is neither definition of nor demonstration about sensible individual substances, because they have matter whose nature is such that they are capable both of being and of not being; for which reason all the individual instances of them are destructible. If then demonstration is of necessary truths and definition is a scientific process, and if, just as knowledge cannot be sometimes knowledge and sometimes ignorance, but the state which varies thus is opinion, so too demonstration and definition cannot vary thus, but it is opinion that deals with that which can be otherwise than it is, clearly there can neither be definition of nor demonstration about sensible individuals ...
> ... e.g. if one were defining you, he would say 'an animal which is lean' or 'pale', or something else which will apply also to some one other than you. If any one were to say that perhaps all the attributes taken apart may belong to many subjects, but together they belong only to this one, we must reply first that they belong also to both the elements; e.g. 'two-footed animal' belongs to animal and to the two-footed.[79]

In the former quotation Aristotle admitted that individual things also have essence while in the latter he denied such a relation. This duality survived into the Middle Ages. In modern logic the first view was rejected in the attempt to overcome this duality.

---

[78] Aristotle, *Metaphysica*, 1031b18–25.
[79] *Op. cit.*, 1039b28–1040a2, 1040a13–17.

Individuals have no essence. When the schoolmen talked of the essence of an individual, they did not mean the properties implied in its name, for the names of individuals imply no properties. They regarded as of the essence of an individual, whatever was of the essence of the species in which they were accustomed to place that individual; i.e. of the class to which it was most familiarly referred, and to which, therefore, they conceived that it by nature belonged... If *man* was a substance inhering in each individual man, the *essence* of man (whatever that might mean) was naturally supposed to accompany it; to inhere in John Thompson, and to form the *common essence* of Thompson and Julius Caesar. It might then be fairly said, that rationality, being of the essence of Man, was of the essence also of Thompson. But if Man altogether be only the individual men and a name bestowed upon them in consequence of certain common properties, what becomes of John Thompson's essence?[80]

Whoever accepts this argument will surely agree with the following reasoning as well:

The definition of a word being the proposition which enunciates its meaning, words which have no meaning are unsusceptible of definition. Proper names, therefore, cannot be defined. A proper name being a mere mark put upon an individual, and of which it is the characteristic property to be destitute of meaning, its meaning cannot of course be declared; though we may indicate by language, as we might indicate still more conveniently by pointing with the finger, upon what individual that particular mark has been, or is intended to be, put.[81]

According to this view the definition of individual, or singular, terms is impossible mainly because of the presupposition that individuals have no essence. Here the essential properties are regarded as being general and nothing but general. It is true that essential properties are characteristic not only of individuals but that does not imply that they are not characteristic of individuals. The general and the individual are not mutually exclusive opposites. Consequently, the essence is not only general but at the same time also of an individual nature.

---

[80] Mill, p. 114.
[81] *Op. cit.*, p. 133.

Let me also recall the assertion that essential is that without which a thing could not exist. So, if individuals had no essence it would follow that individual things did not exist—an obvious absurdity.

I think there are logical interrelations not only between the above modalities and quantifiers but in the sense that any modal proposition can be compared to any quantified proposition. To demonstrate this thesis let me quote the following propositions: (1) some S is P, (2) some S is not P, (3) it is possible that S is P, (4) it is possible that S is not P. The third is subordinated to the first, and the fourth to the second. In the rest of cases, with the exception of identities, a subcontrary relation can be revealed.

Modal propositions can be converted into quantified ones just as well as into modal ones, e.g., it is necessary that S is P; some P is S. Quantified propositions, on the other hand, can be converted into modal ones: all S is P, it is possible that P is S.

Syllogisms can also combine modal and quantified propositions, e.g.:

> It is necessary that M is P
> All S is M
> _____
> It is possible that S is P

B

I

In the previous chapter we have studied the relations of propositions such that one was modal and the other quantified. Let us now see propositions wherein both modality and quantification occur together. Let us begin with the quantified propositions predicating possibility. First of all, the following instances will occur, according to D1:

> *Apo*: It is possible that all S is P
> *Epo*: It is possible that no S is P
> *Ipo*: It is possible that some S is P
> *Opo*: It is possible that some S is not P

TABLE 39

	(1)	(2)	(3)	(4)	(5)	(6)	(7)	(8)	(9)	(10)	(11)	(12)	(13)
Apo	*f*	*f*	*t*	*t*	*f*	*f*	*f*	*f*	*t*	*t*	*f*	*f*	*t*
Epo	*f*	*f*	*f*	*f*	*t*	*t*	*t*	*t*	*f*	*f*	*f*	*f*	*t*
Ipo	*t*	*t*	*t*	*t*	*f*	*f*	*f*	*f*	*t*	*t*	*t*	*t*	*t*
Opo	*t*	*t*	*f*	*f*	*t*	*t*	*t*	*t*	*f*	*f*	*t*	*t*	*t*

Numbers (1)–(12) refer to not-contingent term relations while number (13) implies the criterion 'it is contingent that *S* is *P*'.

According to D2 the following propositions can be formed:

(1)   All *S* may be *P*
(2)   No *S* can be *P*
(3)   Some *S* may be *P*
(4)   Some *S* cannot be *P*

Their values are:

TABLE 40

	(1)	(2)	(3)	(4)	(5)	(6)	(7)	(8)	(9)	(10)	(11)	(12)	(13)
(1)	*f*	*f*	*t*	*t*	*f*	*f*	*f*	*f*	*t*	*t*	*f*	*f*	*t*
(2)	*f*	*f*	*f*	*f*	*t*	*t*	*t*	*t*	*f*	*f*	*f*	*f*	*f*
(3)	*t*	*t*	*t*	*t*	*f*	*f*	*f*	*f*	*t*	*t*	*t*	*t*	*t*
(4)	*t*	*t*	*f*	*f*	*t*	*t*	*t*	*t*	*f*	*f*	*t*	*t*	*t*

The comparison of Tables 39 and 40 reveals that there is a difference only in case of proposition 2. What is the reason? Propositions (1), (3), and (4) in Table 39 are successively equivalent to propositions (1), (3), and (4) in Table 40. For example, the proposition 'it is possible that every man is a living being' means the same as 'every man may be a living being'. However, the proposition 'it is possible that no table is blue' means something different from 'no table can be blue'.

Why is it exactly the universal negative proposition which displays this difference? The proposition 'it is possible that no *S* is *P*' can be transcribed like this as well: 'it is possible that all *S* is not *P*'. Its equivalent is 'all *S* may be not *P*', while the equivalent of 'no *S* can be *P*' is 'all *S* cannot (is impossible to) be *P*'.

Accordingly, the D2 propositions are equivalent to the corresponding D1 ones:

It is possible that all S is P	↔	All S may be P
It is possible that no S is P	↔	All S may not be P
It is possible that some S is P	↔	Some S may be P
It is possible that some S is not P	↔	Some S may not be P

Thus, it has become clear that the *de re* interpretation of D2 versions is an error. Let us compare, e.g., the following propositions:

All S may be P
All S is possible-P

There are indefinite P's in the former version, and possible P's in the latter one.

Now comes the necessary quantified proposition. Let us see first the D1 versions:

Anec: It is necessary that all S is P
Enec: It is necessary that no S is P
Inec: It is necessary that some S is P
Onec: It is necessary that some S is not P

The criteria have not changed.

TABLE 41

	(1)	(2)	(3)	(4)	(5)	(6)	(7)	(8)	(9)	(10)	(11)	(12)	(13)
Anec	*f*	*f*	*t*	*t*	*f*	*f*	*f*	*f*	*t*	*t*	*f*	*f*	*f*
Enec	*f*	*f*	*f*	*f*	*t*	*t*	*t*	*t*	*f*	*f*	*f*	*f*	*f*
Inec	*t*	*t*	*t*	*t*	*f*	*f*	*f*	*f*	*t*	*t*	*t*	*t*	*f*
Onec	*t*	*t*	*f*	*f*	*t*	*t*	*t*	*t*	*f*	*f*	*t*	*t*	*f*

The propositions according to D2 are:

(1)   All S is necessarily P
(2)   No S is necessarily P
(3)   Some S is necessarily P
(4)   Some S is necessarily not P

Their values are:

TABLE 42

	(1)	(2)	(3)	(4)	(5)	(6)	(7)	(8)	(9)	(10)	(11)	(12)	(13)
(1)	f	f	t	t	f	f	f	f	t	t	f	f	f
(2)	f	f	f	f	t	t	t	t	f	f	f	f	t/f
(3)	t	t	t	t	f	f	f	f	t	t	t	t	f
(4)	t	t	f	f	t	t	t	t	f	f	t	t	f

Let us continue with the following propositions: (1) it is contingent that all S is P, (2) it is contingent that no S is P, (3) it is contingent that some S is P, (4) it is contingent that some S is not P. To state their values I start with Aristotle's following remarks:

> ... all premisses in the mode of possibility are convertible into one another. I mean not that the affirmative are convertible into the negative, but that those which are affirmative in form admit of conversion by opposition, e.g. 'it is possible to belong' may be converted into 'it is possible not to belong', and 'it is possible for A to belong to all B' into 'it is possible for A to belong to no B' or 'not to all B', and 'it is possible for A to belong to some B' into 'it is possible for A not to belong to some B'. And similarly the other propositions in this mode can be converted. For since that which is possible is not necessary, and that which is not necessary may possibly not belong, it is clear that if it is possible that A should belong to B, it is possible also that it should not belong to B: and if it is possible that it should belong to all, it is also possible that it should not belong to all. The same holds good in the case of particular affirmations: for the proof is identical.[82]

Let us transcribe the propositions concerned in an expounded form:

Ac: It is possible that all S is P and it is possible that no S is P

Ec: It is possible that no S is P and it is possible that all S is P

Ic: It is possible that some S is P and it is possible that some S is not P

Oc: It is possible that some S is not P and it is possible that some S is P

---

[82] Aristotle. 1928. 32a30–32b1.

18

Accordingly, *Ac* is equivalent to *Ec* and *Ic* to *Oc,* respectively. The question what relations occur in the other combinations has been left unanswered.

Aristotle failed to act consistently in his above reasoning. He, correctly, proceeded from the fact that 'it is possible to belong' is equivalent to 'it is possible not to belong'. Accordingly, however, 'it is possible to belong to all', is not equivalent to 'it is possible to belong to none' but to 'it is possible not to belong to some'. In accordance with the above said the other equivalencies have to be modified as well. In expounded form we get the following versions:

Ac: It is possible that all *S* is *P* and it is possible that some *S* is not *P*

Ec: It is possible that no *S* is *P* and it is possible that some *S* is *P*

Ic: It is possible that some *S* is *P* and it is possible that no *S* is *P*

Oc: It is possible that some *S* is not *P* and it is possible that all *S* is *P*

Accordingly, *Ac* is equivalent to *Oc* and *Ec* to *Ic*, respectively. In order to reveal the remaining relations, let us evaluate the propositions concerned on the ground of unchanged criteria.

TABLE 43

	(1)	(2)	(3)	(4)	(5)	(6)	(7)	(8)	(9)	(10)	(11)	(12)	(13)
Ac	*f*	*f*	*f*	*f*	*f*	*f*	*f*	*f*	*f*	*f*	*f*	*f*	*t*
Ec	*f*	*f*	*f*	*f*	*f*	*f*	*f*	*f*	*f*	*f*	*f*	*f*	*t*
Ic	*f*	*f*	*f*	*f*	*f*	*f*	*f*	*f*	*f*	*f*	*f*	*f*	*t*
Oc	*f*	*f*	*f*	*f*	*f*	*f*	*f*	*f*	*f*	*f*	*f*	*f*	*t*

Each of the propositions discussed is equivalent to the others. Thus, Aristotle correctly recognized certain relations, yet failed to substantiate them in the right way.

II

Aristotle studied only modal categorical propositions. Let us extend the analysis to all quantified modes. Let us take the necessary as modality and join it with the quantifiers, e.g., it is necessary that only this *S* is *P*.

Following the previous sequence of the quantifiers, we arrive at a comprehensive table of values:

TABLE 44

	(1)	(2)	(3)	(4)	(5)	(6)	(7)	(8)	(9)	(10)	(11)	(12)	(13)
(1)	f	f	f	f	f	f	f	f	f	f	f	f	f
(2)	f	f	f	f	t	t	t	t	f	f	f	f	f
(3)	t/f	t/f	f	f	f	f	f	f	f	f	t/f	t/f	f
(4)	f/t	f/t	f	f	f	f	f	f	f	f	f/t	f/t	f
(5)	f	f	t⁻	t	f	f	f	f	t	t	f	f	f
(6)	t/f	t/f	f	f	t	t	t	t	f	f	t/f	t/f	f
(7)	f/t	f/t	f	f	t	t	t	t	f	f	f/t	f/t	f
(8)	f	f	t	t	t	t	t	t	t	t	f	f	f
(9)	t	t	f	f	f	f	f	f	f	f	t	t	f
(10)	t/f	t/f	t	t	f	f	f	f	i	t	t/f	t/f	f
(11)	f/t	f/t	t	t	f	f	f	f	t	t	f/t	f/t	f
(12)	t	t	f	f	t	f	f	f	f	f	t	t	f
(13)	t/f	t/f	t	t	t	t	t	t	t	t	t/f	t/f	f
(14)	f/t	f/t	t	t	t	t	t	t	t	t	f/t	f/t	f
(15)	t	t	t	t	f	f	f	f	t	t	t	t	f
(16)	t	t	t	t	t	t	t	t	t	t	t	t	f

In the foregoing I have investigated necessary propositions only. An overall survey, however, would be possible only if the analysis covered all modalities. Since the aspects of the operations are given, I think further details can be dispensed with.

So far such propositions have been under discussion where the terms were modally indefinite. Let us see now the following propositions:

A: All S is necessarily-P
E: No S is necessarily-P
I: Some S is necessarily-P
O: Some S is not necessarily-P

If we insisted on using modally indefinite S's and P's in the diagrams we could not decide unambiguously in all cases the value of the propositions concerned. Proposition A, for example, is t/f in case of subordination, e.g.:

18*

> All men are necessarily-living beings *(t)*.
> All accidents are necessarily-events *(f)*.

Let us modify our starting point by taking indefinite *S*'s and necessarily-*P*'s. Upon this condition the values of the propositions concerned coincide with the values in Table 6.

Now let us turn our attention to the theme of immediate inferences based on modal quantified propositions. Occam constructed the following scheme: [83]

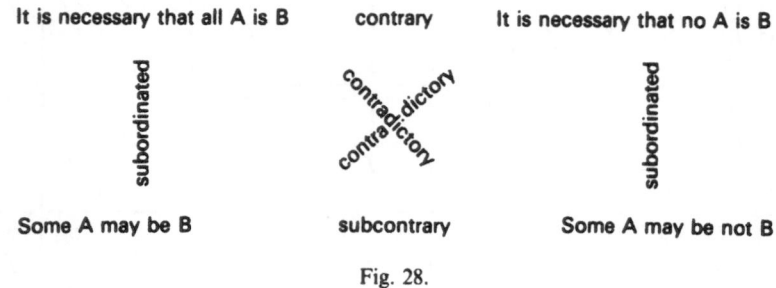

Fig. 28.

Buridan tried to illustrate the relations at issue with a rather complicated figure. [84]

All these analyses, however, covered a narrow field only: of the quantifiers all, none, and some, while among the modalities necessary, possible, and impossible were considered. Why did they not go beyond this scope? On the one hand, because they failed to recognize several combination possibilities. On the other, because having discovered other versions, they were confronted with difficulties.

> ... in the field of modal propositions very strange relations prevail.
> Namely:
> 'All *A* may be *B*' and 'All *A* may be not *B*' are not at all contrary
> propositions since, if they are possible propositions, both of them can be

[83] Prantl, Vol. III. p. 417.
[84] *Op. cit.*, Vol. IV. p. 23.

true at the same time. In fact, in this case these two propositions together, i.e. 'All *A* may be *B* and all *A* may be not *B*', simply constitute a contingent proposition, namely 'All *A* may contingently be *B*'. On the ground of the above said, it can be also stated that from this contingent proposition 'All *A* may contingently be *B*' (it may and may not be *B*) each of the former possible propositions follows, i.e. also that 'All *A* may be *B*' and also that 'All *A* may be not *B*'. Furthermore, due to its strange duality, the affirmative proposition 'All *A* may contingently be *B*' can be also easily replaced by the negative proposition 'All *A* may contingently not be *B*' for this contingent proposition both in the affirmative and in the negative conveys the same, namely that all *A* may be *B* but it is just as much possible that no *A* is *B*.[85]

Similarly strange things occur also in the case of 'only possible', 'only a fact', etc. All these problems can be solved, however, if one succeeds in defining the values of modal quantified propositions, because the comparison of the values of two such propositions reveals what immediate conclusion can be drawn from them. Let us take these propositions as an example: it is necessary that no *S* is *P*—it is necessary that only some *S* is *P*. Table 46 reveals that they are in a contrary relation.

III

Now I turn to the conversion of modal quantified propositions. Referring to the conversion of categorical propositions, Aristotle wrote the following:

> The same manner of conversion will hold good also in respect of necessary premisses. The universal negative converts universally; each of the affirmatives converts into a particular.
>
> If it is necessary that no *B* is *A*, it is necessary also that no *A* is *B*. For if it is possible that some *A* is *B*, it would be possible also that some *B* is *A*. If all or some *B* is *A* of necessity, it is necessary also that some *A* is *B*: for if there were no necessity, neither would some of the *B*s be *A* necessarily.
>
> But the particular negative does not convert, for the same reason which we have already stated [i.e. with regard to categorical propositions—Gy. T.].[86]

---

[85] Aristotelés, *Organon*. Vol. I. (Budapest 1961). pp. LXXXIX–CX.
[86] Aristotle. 1928. 25a26–36.

Accordingly, there are the following instances formulated in the traditional way:

(1)   It is necessary that all S is P
            It is necessary that some P is S
(2)   It is necessary that no S is P
            It is necessary that no P is S
(3)   It is necessary that some S is P
            It is necessary that some P is S
(4)   It is necessary that some S is not P
            It must not be converted

The analysis is correct both according to *Nec*1 and *Nec*2.

As to the conversion of propositions predicating possibility, the following can be read with Aristotle:

> ... if it is possible that all or some B is A, it will be possible that some A is B. For if that were not possible, then no B could possibly be A ... if it is possible for no man to be a horse, it is also admissible for no horse to be a man; and if it is admissible for no garment to be white, it is also admissible for nothing white to be a garment. For if any white thing must be a garment, then some garment will necessarily be white... The particular negative also must be treated like those dealt with above [i.e., as Aristotle has stated in connection with categorical propositions—Gy. T.].[87]

These statements hold good both with regard to P1 and P2.

Aristotle had the following opinion of the conversion of contingent propositions:

> In respect of possible premisses, since possibility is used in several senses (for we say that what is necessary and what is not necessary and what is potential is possible), affirmative statements will all convert in a manner similar to those described [i.e., to necessary propositions—Gy. T.]...
>
> But in negative statements the case is different. Whatever is said to be possible, either because B necessarily is A, or because B is not necessarily A, admits of conversion like other negative statements...
>
> But if anything is said to be possible because it is the general rule and natural (and it is in this way we define the possible), the negative premisses

[87] *Op. cit.*, 25a40–25b2, 9–13.

can no longer be converted like the simple negatives; the universal negative premiss does not convert, and the particular does.[88]

Accordingly, there are the following cases in traditional formulation:

(1)   It is contingent that all S is P
        It is contingent that some P is S

(2)   It is contingent that no S is P
        It must not be converted

(3)   It is contingent that some S is P
        It is contingent that some P is S

(4)   It is contingent that some S is not P
        It is contingent that some P is not S

Let us examine these versions one after the other. Aristotle gave the following justification for the conversion of cases (1) and (3): "... if it is possible that all or some B is A, it will be possible that some A is B. For if that were not possible, then no B could possibly be A. This has been already proved."[89] With regard to affirmative propositions, Aristotle found it unnecessary to make a distinction between the possible and the contingent.

Aristotle declared the universal negative proposition (2) impossible to convert. He tried to prove this assertion in three different ways[90]. I think it is sufficient to analyse the first evidence, or proof, of the three.

> First we must point out that the negative problematic proposition is not convertible, e.g. if A may belong to no B, it does not follow that B may belong to no A. For suppose it to follow and assume that B may belong to no A. Since then problematic affirmations are convertible with negations, whether they are contraries or contradictories, and since B may belong to no A, it is clear that B may belong to all A. But this is false: for if all this can be that, it does not follow that all that can be this: consequently the negative proposition is not convertible.[91]

[88] *Op. cit.*, 25a36–39, 25b3–6, 14–18.
[89] *Op. cit.*, 25a40–25b2.
[90] *Op. cit.*, 36b35–37a31.
[91] *Op. cit.*, 36b35–37a4.

Here Aristotle proceeded from the following equivalence:

(1) A may belong to no $B \leftrightarrow A$ may belong to all B
(2) B may belong to no $A \leftrightarrow B$ may belong to all A

Then he argued as follows: if a universal negative proposition could be simply converted, then the universal affirmative proposition equivalent to it could also be simply converted. A universal affirmative proposition, however, can be converted by limitation only. Consequently, the universal negative proposition equivalent to it must not be simply converted.

Theophrastus, in opposition to this view, maintained that there is no need to distinguish between possible and contingent negative propositions. That is, in his opinion a universal negative proposition implies that the predicate does not belong to the subject, and, be this connection possible or contingent, the subject will not belong to the predicate in any of these cases. Therefore, universal negative propositions can be simply converted.

Theophrastus made a double mistake. On the one hand, he ignored those universal negative propositions where the predicate belongs to the subject, e.g., it is contingent that none of the people in this room is a man. The connection becomes obvious as soon as we transcribe it in the usual way: it is possible that none of the people in this room is a man and *it is possible that some of the people in this room are men*. And on the other hand, he erroneously identified the possible and the contingent.

Drobisch wrote the following:

> We cannot convince ourselves of the inadmissibility of the pure conversion of the universal negative problematic proposition which Aristotle asserted, even though we are essentially in agreement with his concept of possibility. The proposition: it is possible that all S are *not P*, expresses the fact that neither the proposition: all S are *not P*, nor one of its opposites: some (all) S *are P*, is self-contradictory. Therefore neither the converse of the first proposition: all P are *not S*, nor that of the opposed proposition: some P *are S*, is self-contradictory. This means, however that the proposition: all P are *not S*, is possible and thus that the proposition: it is possible that all S are *not P*, is purely convertible. Or to make it short: if it is neither impossible nor necessary that *all S are not P*, that is, that all S are excluded from all P, then it is also neither impossible nor necessary, that *all*

P are not S. Thus there is no contradiction if in fact *some P* are necessarily *not S*; for the conversion asserts only that it is not for *all P* necessary *not* to be *S*.[92]

This argument is a rather confused defence of Theophrastus's view.

H. Maier raised the following question:[93] Why did Aristotle declare that proposition *E* must not be converted? He could have done the following: proposition *E* is equivalent to proposition *A*, and the latter can be converted by limitation: therefore the former can be, too: it is contingent that some *P* is not *S*. The most plausible of the explanations Maier has given seems to be that Aristotle could not prove the above operation by the means known to him.

Łukasiewicz was of the opinion that the contingent proposition *E* can be converted. Since it would require too much space to describe and explain his evidence let me refer to the source only.[94]

To disprove Aristotle's thesis about the nonconvertibility of proposition *E* let us take the following example to the contrary: it is contingent that no high-jumper is a broad-jumper (*t*)—it is contingent that no broad-jumper is a high-jumper (*t*). He proceeded from a wrong supposition in order to prove the thesis at issue. He thought the contingent proposition *A* could be converted by limitation only, in accordance with other modal propositions *A*. If, however, the considerations concerning evaluation hold true, the contingent proposition *A*, too, can be simply converted. In this way the convertibility of the contingent proposition *E* does not contradict that of the contingent proposition *A*.

As a summary let us compare the two main positions which have developed in the history of logic:

Aristotle	Theophrastus
(1) convertible by limitation	convertible by limitation
(2) not convertible	simply convertible
(3) simply convertible	simply convertible
(4) simply convertible	not convertible

[92] Drobisch pp. 88–89.
[93] Maier, pp. 38–41.
[94] Łukasiewicz, p. 59.

In my view the propositions concerned can be simply converted.

Traditional logic restricted itself to the study of the conversion of the propositions discussed above. According to what has been said in the previous paragraph, one can go beyond this narrow scope. A table of values should be made for modal quantified propositions (Table 44 serves as a part of it) as a basis for deciding whether or not they can be converted.

C

I

Aristotle dealt with the following modal quantified syllogisms:

TABLE 45[95]

	Modality of premises		Figure	The part of the *Prior Analytic* in which it is discussed
	major premise	minor premise		
(1)	apodictic	apodictic	I—III	chapter 8
(2)	apodictic	assertoric ⎫	I—III	chapters 9–11
(3)	assertoric	apodictic ⎭		
(4)	contingent	contingent ⎫		
(5)	contingent	assertoric	I	chapters 14–16
(6)	assertoric	contingent	II	chapters 17–19
(7)	contingent	apodictic	III	chapters 20–22
(8)	apodictic	contingent ⎭		

Theophrastus and Eudemus, as it has already been mentioned, had only the *de dicto* interpretation in mind. On this ground they held the opinion that the conclusion follows the weaker component.

Occam was one of the few in the history of logic who dealt with modal quantified syllogisms in detail. What is most praiseworthy in his work is his endeavour to enrich the list of versions. He proceeded from the

[95] This table was published in the preface of the *Organon* (Budapest 1961) Vol. 1. p. LXXVIII. The table of conclusions drawn from these premises is on page 184/a in the same edition.

following six modalities: necessary, possible, contingent, impossible, actual, and other modalities (known, unknown, presumed, doubtful, apparent). Since these modalities can occur both as major and as minor premises in syllogisms, he ended up with thirty-six versions altogether.[96]

Kant did not deal with modal syllogisms. His view of modal propositions, however, had a great influence on later logicians interested in the subject.

In modern logic it was Theophrastus's approach which had gained ground to the extent that even such a follower of Aristotle as Überweg wrote the following:

> Theophrastus and Eudemus are certainly right here; for even with the syllogisms that pertain to the real relations of possibility, actuality, and necessity, every limitation which lies in one of the premisses is carried over into the conclusion.[97]

Of course, this idea also had an influence on the way Aristotle's modal syllogistics was approached. The majority of modern commentators, as I shall discuss later, misinterpreted Aristotle's arguments concerning this issue.

Among the mathematical logicians it was Łukasiewicz who thoroughly analysed the Aristotelean modal logic. He has summed up the main points of his investigation as follows:

> There are two reasons why Aristotle's modal logic is so little known. The first is due to the author himself: in contrast to the assertoric syllogistic which is perfectly clear and nearly free of errors, Aristotle's modal syllogistic is almost incomprehensible because of its many faults and inconsistencies ...
>
> The second reason is that modern logicians have not as yet been able to construct a universally acceptable system of modal logic which would yield a solid basis for the interpretation and appreciation of Aristotle's work.[98]

I agree with the second reason, but with the first only in so far as the Aristotelean modal logic syllogistic is, indeed, not without errors. Still, to call it "almost incomprehensible" is a gross exaggeration.

[96] Cf.: Prantl, Vol. III. pp. 402–408.
[97] Überweg, p. 320.
[98] Łukasiewicz, p. 133.

Łukasiewicz ignored modal syllogisms of the *de re* type. Thus, he took the same position on this question as Überweg, Maier, and all those who admit *de dicto* only. This approach has contributed a great deal to the fact that he considered Aristotle's modal logic practically incomprehensible.

II

Let me begin the discussion of modal quantified syllogisms with inferences concerning propositions wherein the same modality occurs. Aristotle made, first of all, the following remarks on syllogisms based on necessary propositions:

> There is hardly any difference between syllogisms from necessary premisses and syllogisms from premisses which merely assert. When the terms are put in the same way, then, whether something belongs or necessarily belongs (or does not belong) to something else, a syllogism will or will not result alike in both cases, the only difference being the addition of the expression 'necessarily' to the terms. For the negative statement is convertible alike in both cases, and we should give the same account of the expressions 'to be contained in something as in a whole' and 'to be predicated of all of something'.[99]

He commented on the difference between necessary and categorical syllogisms as follows:

> But in the middle figure [i.e., second figure—Gy. T.] when the universal statement is affirmative, and the particular negative, and again in the third figure when the universal is affirmative and the particular negative, the demonstration will not take the same form, but it is necessary by the 'exposition' of a part of the subject of the particular negative proposition, to which the predicate does not belong, to make the syllogism in reference to this: with terms so chosen the conclusion will necessarily follow. But if the relation is necessary in respect of the part taken, it must hold of some of that term in which this part is included: for the part taken is just some of that. And each of the resulting syllogisms is in the appropriate figure.[100]

That is, the proof of the validity of necessary syllogisms differs from that of categorical syllogisms in two cases: mode *AOO* of the second figure and

[99] Aristotle. 1928. 29b36–30a3.
[100] *Op. cit.*, 30a6–14.

mode OAO of the third figure. Aristotle traced back modes Baroco and Bocardo of categorical syllogisms to the first figure by *reductio ad impossibile*. If he had applied the same method with the above modes of necessary syllogisms, too, he should have had recourse to the contradictory of the conclusion.

Let us take a conclusion of type O: it is necessary that some S is not P. Its contradictory is: it is possible that all S is P. This proposition predicating possibility should have been joined to any of the necessary premises in order to apply *reductio ad impossibile*. While the validity of categorical syllogisms was proved by using purely categorical propositions only, here propositions of mixed character would have occurred in the proof. To avoid this Aristotle suggested a proof with an exposition which provides that both premises could take the form of necessary propositions.[101] Thus, he made a distinction between necessary and categorical syllogisms as far as the proof of the validity of the modes is concerned.

Syllogisms with contingent propositions have hardly been dealt with in the history of logic. Theophrastus had a considerable part in it, as is well known. He identified the contingent with the possible. This mistaken oversimplification came to be widely accepted. Heinrich Maier, for example, wrote the following:

> Thus the whole theory of possibility inferences presents an essentially different picture than in Aristotelian syllogistics. Theophrastus simplified the Aristotelian doctrine, and one can say that he did not worsen it. His critique was directed at all the points in which the master's theory gave rise to doubt, and we cannot here deny that his corrections are real improvements.[102]

Let us see now those syllogisms where *both premises are contingent propositions*.

> Whenever A may possibly belong to all B, and B to all C, there will be a perfect syllogism to prove that A may possibly belong to all C. This is clear

[101] For details see: *op. cit.*, 30a4–33.
[102] Maier, p. 213.

from the definition: for it was in this way that we explained 'to be possible for one term to belong to all of another'.[103]

It is a mode of type Barbara:

> It is contingent that all M is P
> It is contingent that all S is M
> _____
> It is contingent that all S is P

Aristotle discussed the modes including contingent premises at great length. His analysis resulted in the following:

> It is clear that if the terms are universal in possible premises a syllogism always results in the first figure, whether they are affirmative or negative, only a perfect syllogism results in the first case, an imperfect in the second. But possibility must be understood according to the definition laid down, not as covering necessity. This is sometimes forgotten.[104]
>
> In the second figure whenever both premisses are problematic, no syllogism is possible, whether the premisses are affirmative or negative, universal or particular.[105]

As an explanation for this statement he pointed out that the modes of the second figure cannot be traced back to those of the first either by conversion or by *reductio ad impossibile*, therefore their validity cannot be proved.

> In the last figure a syllogism is possible whether both or only one of the premisses is problematic. When the premisses are problematic the conclusion will be problematic. . . [106]

Let us transcribe Barbara in the *de re* interpretation:

> All M is contingently-P
> All S is contingently-M
> _____
> All S is contingently-P

[103] Aristotle. 1928. 32b38–33al.
[104] *Op. cit.*, 33b18–24.
[105] *Op. cit.*, 36b26–27.
[106] *Op. cit.*, 39a3–5.

Since contingently-M and M cross one another, this inference is not valid. The same holds true of the other modes of the first figure.

In the third figure, however, there are valid modes, indeed, e.g.:

> All M is contingently-P
> All M is contingently-S
> _____
> Some contingently-S is contingently-P

Let us now interpret Barbara *de dicto*. We may transcribe this formula as follows:

> It is possible that all M is P and it is possible that some M is not P
> It is possible that all S is M and it is possible that some S is not M
> _____
> It is possible that all S is P and it is possible that some S is not P

Do these premises provide a sufficient reason for drawing the conclusion? I think they do not. For example:

> It is contingent that all representative runners are athletes living at Oxford.
> It is contingent that all athletes living at Cambridge are representative runners.
> _____
> It is contingent that all athletes living at Cambridge are athletes living at Oxford.

The premises are true, still the conclusion is f. It is probably no accident that Aristotle did not come forth with an example to illustrate the contingent Barbara, although he did so in the case of the categorical syllogism. It can be demonstrated by examples to the contrary that no valid conclusion can be drawn from contingent premises in the case of other figures, either.

Syllogisms where both premises are propositions expressing possibility had a strange fate, indeed. Aristotle did not at all consider them.—The schoolmen formulated in theses what conclusions may and must not drawn from given modal premises. The usual specifications, however, fail to include such theses as '*posse ad posse valet consequentia*', or '*non valet consequentia*'. Occam was among the rare exceptions who dealt with this relation. He came to the conclusion that as far as both premises are propositions predicating possibility, no valid conclusion can be arrived at in any of the modes.[107]

To support his statement, he referred to the following example:

> It is possible that every coloured is white.
> It is possible that every black is coloured.
> _____
> It is possible that every black is white.

It can be demonstrated with examples that not only Barbara but all the other modes also fail to lead to valid conclusions.

Now let us try to explain why Aristotle refused to deal with syllogisms expressing possibility. At one point he wrote the following:

> ... it is not necessary that some animal should not be good, since it is possible for every animal to be good. Or if that is not possible, take as the term 'awake' or 'asleep': for every animal can accept these.[108]

Commentators have pointed out that Aristotle, discussing the above syllogism, regarded the proposition 'it is possible for every animal to be good' as true. His remark "if that is not possible" implies that some of his pupils did not share his opinion in this respect. Therefore Aristotle chose another example in the conviction that it was beyond any doubt, namely: 'it is possible for every animal to be awake or asleep'.

He could have analysed the modes one after the other without any greater difficulties. He was, however, confronted with the following dilemma: if a proposition like 'it is possible for every animal to be awake' is *t*, then syllogisms concerning possibility are not valid. In order to arrive at valid syllogisms, he should have described the proposition at issue as *f*. Thus, there was nothing left but to keep silent about these inferences.

---

[107] Occam, *Summa Logicae*. Cf.: Prantl Vol. III. p. 402.
[108] Aristotle. 1928. 31b7–10.

It is necessary at least to refer to other versions of syllogisms constructed of propositions in the same modality. These versions result, on the one hand, from our going beyond the scope of categorical propositions and, on the other, from our considering even such modalities which have hardly, or not at all, been dealt with in logic (e.g., only a fact).

An example of a syllogism beyond the domain of categorical propositions is:

> It is necessary that all M is P
> It is possible that only this S is M
> ___
> It is possible that only this S is P

III

I proceed now to the study of syllogisms for which one premise is a *necessary* and the other is an *actual* proposition. If the major premise is necessary I shall call it version (I) of the mode concerned, and if the minor one is necessary it will be version (II).

Let us begin with Aristotle's following arguments:

> It happens sometimes also that when *one* premiss is necessary the conclusion is necessary, not however when either premiss is necessary, but only when the major is, e.g. if A is taken as necessarily belonging or not belonging to B, but B is taken as simply belonging to C: for if the premisses are taken in this way, A will necessarily belong or not belong to C. For since A necessarily belongs, or does not belong, to every B, and since C is one of the Bs, it is clear that for C also the positive or the negative relation to A will hold necessarily.[109]

It is Barbara and Celarent respectively that he had in mind here. The expounded form of the former is:

> A necessarily belongs to all B
> B belongs to all C
> ___
> A necessarily belongs to all C

---

[109] *Op. cit.*, 30a15–22.

According to D1 and in the formulation of traditional logic, it takes the following form:

$$\frac{\begin{array}{l}\text{It is necessary that all } M \text{ is } P \\ \text{It is a fact that all } S \text{ is } M\end{array}}{\text{It is necessary that all } S \text{ is } P}$$

The formula of D2 is:

$$\frac{\begin{array}{l}\text{All } M \text{ is necessarily } P \\ \text{All } S \text{ is } M\end{array}}{\text{All } S \text{ is necessarily } P}$$

According to D3 we get the following result:

$$\frac{\begin{array}{l}\text{All } M \text{ is } nec\text{-}P \\ \text{All } S \text{ is } M\end{array}}{\text{All } S \text{ is } nec\text{-}P}$$

Theophrastus, bearing D1 in mind, argued that all what follows from the given premises is the conclusion: it is a fact that all S is P. That is, he presupposed that the conclusion should follow the minor premise.

Alexander tried to prove Aristotle's view valid by *reductio ad impossibile*.[110] His reasoning runs as follows: let us take the negation of the conclusion of a syllogism and join it to any one of the premises of the same syllogism. If these two premises lead by valid inference to a conclusion contradictory to the other premise of the given syllogism, then the syllogism concerned is valid.

Let us apply this procedure to the syllogism under discussion. The negation of the proposition 'it is necessary that all S is P' is: 'it is possible that some S is not P'. Let us add it to the minor premise of the given syllogism: 'it is a fact that all S is M'. These premises make up the following syllogism:

$$\frac{\begin{array}{l}\text{It is possible that some } S \text{ is not } P \\ \text{It is a fact that all } S \text{ is } M\end{array}}{\text{It is possible that some } M \text{ is not } P}$$

[110] Alexander, In *Aristotelis I. Commentarium*. pp. 126,9–127, 16.

This conclusion contradicts the major premise of the original syllogism, namely 'it is necessary that all M is P'. However, the above inference is not valid:

> It is possible that some living beings are not animals (t).
> It is a fact that all living beings in this room are dogs (t).
> ___
> It is possible that some dogs are not animals (f).

It was a common belief in the Middle Ages, on the ground of Theophrastus's theory, that if one premise of a syllogism is a necessary and the other an actual proposition, then the conclusion cannot be more than an actual proposition.

Maier, in opposition to the Aristotelean thesis, referred, among others, to the following example: [111]

> Every man is necessarily a living being.
> Every self-moving is actually man.
> ___
> Every self-moving is actually, but not necessarily, man.

The predicate of the conclusion is an obvious misprint: 'living being' should be understood instead of 'man'.

At the same time Maier quoted the two main objections made to the above example in the history of logic. One of them is that the minor premise of this inference is 'merely fiction'. In my opinion it would be more exact to call it false because animals, for instance, are also self-moving. The other objection is that Aristotle defined a proposition as affirmative only if the extension of the subject was included in that of the predicate. In this example, however, the subject is a wider term than the predicate.

Maier rejected these objections on the ground of their being based on experience. The use of *reductio ad impossibile*, however, cannot be characterized so. At the end of the chapter on the syllogism at issue[112] he declares that this problem cannot be solved until the syllogisms concerning possibility have not been analysed. At the place of reference,[113] however, he calls this problem (to which in the meantime

___

[111] Maier, p. 126.
[112] *Op. cit.*, p. 136.
[113] *Op. cit.*, p. 217.

19*

another one has been added) a puzzle which all the same will be solved in passage 3. This passage can be found in the next thick volume on page 254 where this problem is declared to be not so much a puzzle but rather an absurdity. Then, without using the word 'solution', Maier suggests the introduction of a distinction between the necessity of the inference and that of the conclusion. That is to say, and that is already my comment, the familiar form of inference is not adequate to decide whether or not the conclusion is a necessary proposition. So we are left empty-handed after all the great promises.

One of the best commentators on Aristotle, Ross, could not cope with this problem, either. He wrote that Aristotle's doctrines are obviously mistaken for the following reason: Aristotle tries to demonstrate that the premises support not only that all C is A but also that all C is necessarily A exactly the same way as all B is of necessity A (i.e. the major premise), that is, in consequence of the unaltered necessity of their peculiar nature. In reality, all they refer to is that in so far as all C is B, then C is also A, but not in consequence of the unaltered necessity of its peculiar nature, but due to such a temporary necessity which results from its temporary participation in the nature of B.[114]

Accordingly, in the major premise 'all B is of necessity A' (1) B is by its inner nature A while in the conclusion 'all C is actually A' (2) C is A only by temporarily participating in the nature of B. Let us substitute the following propositions into the given formulas:

(1)   All men are of necessity living beings.
(2)   All Greeks are actually living beings.

Are the Greeks living beings only because they are temporarily men?

A. Becker correctly recognized that Aristotle used primarily the interpretation D3 when assessing the syllogism under discussion.[115] After all this, it is difficult to understand why some logicians, e.g. Sándor Szalai, judge the approach of the Stagirite still from the point of view of D2.

> Some modal modes of inference, however, which Aristotle had accepted
> as axiomatically valid and whose invalidity in their original form was

---

[114] Cf.: *Aristotle's Prior and Posterior Analytics.* p. 43.
[115] Becker, p. 39.

recognized as early as by Theophrastus, are indefensible or extremely difficult to defend ...

Let us consider the following (as Theophrastus must have done, albeit with a different example): it may happen that things which belong to a certain class necessarily possess certain properties. For example, according to our present knowledge of theoretical physics, it is necessary that all objects made of metal should be good heat-conductors. Let us take now a group of objects which belong to the class of metal objects not necessarily but only actually. Say: every piece of cutlery is made of metal for nowadays cutlery happens to be made of metal. But it is not necessary at all: there were times when people ate with wooden forks and spoons, and in the future we may have plastic cutlery. But today, at least in Europe, all cutlery is made of metal, with very few exceptions indifferent to our discussion. Thus, can the following syllogism be accepted as valid?

> Every object made of metal is necessarily a good heat-conductor.
> Every piece of cutlery consists of objects made of metal.
> Hence every piece of cutlery is necessarily a good heat-conductor.

No, it is obviously not valid! If the premises are true it is true that every piece of cutlery (today, in Europe) is a good heat-conductor, but it is not true that they are *necessarily* good heat-conductors—it simply does not follow from the premises![116]

If I do not misunderstand Aristotle, I can give the following answer to the above problem: if the term of 'objects made of metal' is included in the extension of the term of 'necessarily good heat-conductors', and the term of 'cutlery' as described above is included in the extension of the term of 'objects made of metal', then the term of 'cutlery' is included in the extension of the term of 'necessarily good heat-conductors'.

Let us sum up the above. On the ground of D3 Aristotle was right in finding this inference valid. The opinion of Theophrastus and his followers is correct from the point of view of a *de dicto* interpretation, but the fact that they considered this aspect only, and criticized Aristotle accordingly, is to be rejected. Although Alexander and his followers (e.g.,

---

[116] Aristotelés, *Organon*. (Budapest 1961) Vol. I. pp. XCI–XCII.

Łukasiewicz)[117] wanted to defend Aristotle's position, nothing has been gained by it since they built their arguments on the validity of the *de dicto* version.

What happens if the major premise is an *actual proposition* and the minor one is a *necessary proposition*?

> But if the major premise is not necessary, but the minor is necessary, the conclusion will not be necessary. For if it were, it would result both through the first figure and through the third that *A* belongs necessarily to some *B*. But this is false; for *B* may be such that it is possible that *A* should belong to none of it. Further, an example also makes it clear that the conclusion will not be necessary, e.g. if *A* were movement, *B* animal, *C* man: man is an animal necessarily, but an animal does not move necessarily, nor does man.[118]

Theophrastus, naturally under the aspect of *de dicto* interpretation, agreed that the conclusion is an actual proposition in this case. Łukasiewicz, however, admitted necessary conclusions from the same point of view. What is the reason for this difference? Both of them proceeded from the fact that the two versions of Barbara do not differ principally, therefore their difference seen with Aristotle is unjustified. Later on, however, they took different turns. Theophrastus held that the conclusion always follows the minor premise. According to Łukasiewicz, however, the (II) should follow the (I) which he had already accepted as valid. In my view the conclusion concerned must be an actual proposition.

The other modes of the first figure (Celarent, Darii, Ferio) need not be discussed in detail for what has been said of Barbara—*mutatis mutandis*—applies also to D1 and D3, respectively. With regard to D2, it must be noted, however, that Celarent and Ferio are not valid, e.g.:

No coloured (*M*) is necessarily white (*P*).
All snow-white (*S*) is coloured (*M*).
_____
No snow-white (*S*) is white (*P*).

[117] Łukasiewicz, § 55.
[118] Aristotle. 1928. 30a23–32.

IV

Let us follow further Aristotle's argumentation.

> In the second figure, if the negative premiss is necessary, then the conclusion will be necessary, but if the affirmative, not necessary. [Aristotle here implied: in case both premises are universal, since otherwise this statement would not hold good.] First let the negative be necessary; let *A* be possible of no *B*, and simply belong to *C*. Since them the negative statement is convertible, *B* is possible of no *A*. But *A* belongs to all *C*; consequently *B* is possible of no *C*. For *C* falls under *A*.[119]

Here Aristotle has Cesare in mind. Let us see first version (I), according to D3:

> No *P* is necessarily-*M*
> All *S* is *M*
> _____
> No *S* is necessarily-*P*

Why does a necessary *P* occur in the conclusion? Aristotle thought that by the conversion of the major premise this mode can be traced back to Celarent (I) in the conclusion of which there is a necessarily-*P*:

> No *M* is necessarily-*P*
> All *S* is *M*
> _____
> No *S* is necessarily-*P*

But: (1) the conversion of the major premise is an error, e.g., no horse is necessarily-race-horse; no race-horse is necessarily-horse. (2) the formula to be traced back is not valid:

> No coloured is necessarily-white.
> All snow-white is white.
> _____
> No snow-white is necessarily-coloured.

This example is suitable to demonstrate—with the hyphens omitted— also that the mode under discussion is not valid according to D2, either.— Cesare (I), however, is valid in any respect, e.g.:

[119] *Op. cit.*, 30b7–13.

No *P* is *M*

All *S* is necessarily-*M*

No *S* is *P*

Aristotle wrote the following about Camestres:

> The same result would be obtained if the minor premiss were negative: for
> if *A* is possible of no *C*, is possible of no *A*: but *A* belongs to all *B*,
> consequently *C* is possible of none of the *B*s: for again we have obtained the
> first figure. Neither then is *B* possible of *C*: for conversion is possible
> without modifying the relation.[120]

This is Camestres (II):

All *P* is *M*

No *S* is necessarily-*M*

No *S* is necessarily-*P*

All one can say here is that neither the conversion, nor the formula is
valid.

Let us proceed to the next combination:

> But if the affirmative premiss is necessary, the conclusion will not be
> necessary. Let *A* belong to all *B* necessarily, but to no *C* simply. If then the
> negative premiss is converted, the first figure results. But it has been proved
> in the case of the first figure that if the negative major premiss is not
> necessary the conclusion will not be necessary either. Therefore the same
> result will obtain here.[121]

It is Camestres (I):

All *P* is necessarily-*M*

No *S* is *M*

No *S* is *P*

The validity of this formula can be proved by tracing it back to Celarent
(II).

---

[120] *Op. cit.*, 30b13–18.
[121] *Op. cit.*, 30b19–24.

The rest of the modes of the second figure need not be dealt with, for Festino behaves like Cesare, and Baroco like Camestres. I could not say anything essentially new about the third figure, either.

Aristotle summed up the above said as follows:

> It is clear then that a simple conclusion is not reached unless both premisses are simple assertions, but a necessary conclusion is possible although one only of the premisses is necessary. But in both cases, whether the syllogisms are affirmative or negative, it is necessary that one premiss should be similar to the conclusion. I mean by 'similar', if the conclusion is a simple assertion, the premiss must be simple; if the conclusion is necessary, the premiss must be necessary. Consequently this also is clear, that the conclusion will be neither necessary nor simple unless a necessary or simple premiss is assumed.[122]

I would like to make the following remark here: the first sentence of the quotation is obviously loosely formulated since in the foregoing, among others, such inferences have been discussed, where one premise is a necessary proposition while the conclusion can be a simple assertion. It would not have been worth mentioning it merely for this reason. However, it turns our attention to the fact that Aristotle failed to reckon with the combination wherein the premises are necessary propositions and the conclusion is still a simple assertion, for example:

All M is necessarily-P
All S is necessarily-M
_____
All S is P

In this case we have acted similarly to the case when Barbari was concluded instead of Barbara in categorical syllogisms. Thus, the following statement does not hold true, either: "... if the conclusion is a simple assertion, the premiss must be simple".

Medieval logicians had a saying referring to the difficulties inherent in our theme: no ass would try the modalities (*de modalibus non gustabit asinus*). It does not follow, of course, that he who takes the risk and tries modalities cannot act as foolishly as an ass. I leave it to the reader to judge my attempts.

[122] *Op. cit.*, 32a6–14.

12. THE LOGIC OF TIME AND SPACE

I

While there is rich historical material at our disposal on certain subjects of logic, the sources concerning the problem of time and space are rare and partly extralogical. Aristotle influenced posterity on this matter mainly with two statements. One of them was formulated in the *Metaphysics*:

> But while the usual exists, can nothing be said to be always, or are there eternal things? This must be considered later, but that there is no science of the accidental is obvious; for all science is either of that which is always or of that which is for the most part.[123]

The other can be read in the *Organon*:

> By a noun we mean a sound significant by convention which has no reference to time, and of which no part is significant apart from the rest...
>
> A verb is that which, in addition to its proper meaning, carries with it the notion of time. No part of it has any independent meaning, and it is a sign of something said of something else. I will explain what I mean by saying that it carries with it the notion of time. 'Health' is a noun, but 'is healthy' is a verb; for besides its proper meaning it indicates the present existence of the state in question. Moreover, a verb is always a sign of something said of something else, i.e. of something either predicable of or present in some other thing. Such expressions as 'is not-healthy', 'is not-ill', I do not describe as verbs; for though they carry the additional note of time, and always form a predicate, there is no specified name for this variety; but let them be called indefinite verbs, since they apply equally well to that which exists and to that which does not. Similarly 'he was healthy', 'he will be healthy', are not verbs, but tenses of a verb; the difference lies in the fact that the verb indicates present time, while tenses of the verb indicate those times which lie outside the present.[124]

It would be a task for linguists to analyse to what extent the contents of the second quotation have been determined by the specifics of the ancient Greek language. Let me note the following from a logical point of view: (1)

[123] Aristotle. *Metaphysica*. 1027a17–21.
[124] Aristotle. 1928. 16a19–20, 16b6–19.

there are words, nouns and adjectives, that, besides their main meaning express time as well, e.g., dawn, birthday, the thirty-years war. (2) The emphasis on the present tense is just as much unjustified as the distinction between the verb and the verbal form on this ground. (3) It is not only the indefinite (in fact, negative) verb that can refer both to the existent and the nonexistent, e.g., mermaids swim.

In the Middle Ages three kinds of conditional proposition were distinguished: *propositio causalis, temporalis, localis* (propositions predicating causality, time, and place, respectively).

Kant did not find it necessary to fit the propositions under discussion into his table of propositions. He analysed time from a philosophical point of view only and tried to demonstrate it, first of all, as a subjective category:

> I perceive that appearances follow one another, that is, that there is a state of things at one time the opposite of which was in the preceding time. Thus I am really connecting two perceptions in time. Now connection is not the work of mere sense and intuition, but is here the product of a synthetic faculty of imagination, which determines inner sense in respect of the time-relation. But imagination can connect these two states in two ways, so that either the one or the other precedes in time. For time cannot be perceived in itself, and what precedes and what follows cannot, therefore, by relation to it, be empirically determined in the object. I am conscious only that my imagination sets the one state before and other after, not that the one state precedes the other in the object. In other words, the *objective relation* of appearances that follow upon one another is not to be determined through mere perception.[125]

Lenin was against this approach:

> Recognising the existence of objective reality, i.e. matter in motion, independently of our mind, materialism must also inevitably recognise the objective reality of time and space, in contrast above all to Kantianism, which in this question sides with idealism and regards time and space not as objective realities but as forms of human understanding. The basic difference between the two fundamental philosophical lines on this question

[125] Kant, *Kritik*. pp. 283–284; Kemp Smith, pp. 218–219.

is also quite clearly recognised by writers of the most diverse trends who are in any way consistent thinkers.[126]

Let us have a closer look at the following figure as a summary of time operators:

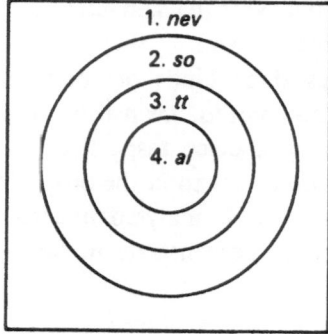

Fig. 29.

The designations should be understood as: *nev* = never (never ever, no time), *so* = sometimes (occasionally, some time, at times), *tt* = this time (now, at present, at a certain time, then), *al* = always (at every time, any time, for ever).

Ring 2 represents 'only sometimes' and ring 3 'only this time'. Rings 2 and 4 together express 'sometimes but not only this time' which equally excludes 'this time' and 'never'. Rings 2 and 3 together imply 'sometimes but not always' and in my interpretation equally exclude 'always' and 'never'. The possible combinations are:

0	— void	2 *o* 3	— *so* but not *al*
1	— *nev*	2 *o* 4	— *so* but not only *tt*
2	— only *so*	3 *o* 4	— *tt*
3	— only *tt*	1 *o* 2 *o* 3	— not *al*
4	— *al*	1 *o* 2 *o* 4	— not only-*tt*
1 *o* 2	— not *tt*	1 *o* 3 *o* 4	— not only-*so*
1 *o* 3	— only *tt* or *nev*	2 *o* 3 *o* 4	— *so*
1 *o* 4	— *al* or *nev*	1 *o* 2 *o* 3 *o* 4	— anything

---

[126] Lenin, *Materialism and Empirio-Criticism.* p. 176.

I find the following interpretations possible as well: 2: sometimes but not now, 2 *o* 3: at times, now and then, 2 *o* 4: frequently, 1 *o* 3 *o* 4: now or never.

Figure 29 can be used also as a survey of space operators. The following answers are given to the question 'where?': (1) nowhere, (2) somewhere, (3) at this place (here, there), (4) everywhere, anywhere. The answers given to the question 'from where?' are: (1) from nowhere, (2) from somewhere, (3) from here, from there, from this or that place, (4) from everywhere, from anywhere. The answers to the question 'where to?' are: (1) to nowhere, (2) to somewhere, (3) hereto, thereto, (4) to everywhere, to anywhere. This survey has revealed that space operators are analogous to time operators. Therefore, for simplicity's sake, I shall dwell upon the latter only since their analysis holds true—*mutatis mutandis*—of the former, too.

Time operators may refer to terms (*de re*) as well as to propositions as a whole (*de dicto*). The former version will be indicated by joining the temporal to the term through a hyphen. Examples of the two versions are:

> *De re*:   *S* is always-*P*
> *De dicto*:   *S* is sometimes *P*

II

And now let us see the evaluation of temporal propositions. I start with the following propositions: (1) there is always *S*, (2) there is always no *S* (there is never *S*), (3) there is not always *S*, (4) there is not always no *S* (there is sometimes *S*). I shall evaluate them on the ground of Figure 2 (see Table 46).

Six kinds of relations can be distinguished with regard to these formulas: subordinated, superordinated, contrary, subcontrary, contradictory, and identical. The fourth is subaltern to the first, and the third to the second. Consequently, the first is subaltern to the fourth, and the second to the third. The first and the second are contrary to one another. The third and the fourth are in a subcontrary relation. The first and the third, and the second and the fourth respectively are contradictory. All formulas are identical with themselves.

If we want to avoid ambiguous values the conditions must be modified.
The shade-lines mean: there is never, and the dot means: there is always.
Then the values of the propositions discussed above will be as in Table 47:

TABLE 46   TABLE 47

	(1)	(2)
(1)	f	t/f
(2)	t/f	f
(3)	t	t/f
(4)	t/f	t

	(1)	(2)
(1)	f	t
(2)	t	f
(3)	t	f
(4)	f	t

The relations between the propositions are reduced to equivalence and
contradiction in this case. Accordingly, we shall obtain only equivalent
values in the table which sums up the inference possibilities and which I
omit for simplicity's sake.

Let us take now the following propositions: (1) S is always P, (2) S is
always not P (S is never P), (3) S is not always P, (4) S is not always not P
(S is sometimes P). The diagrams in Figure 12 represent temporarily
indefinite relations according to their original interpretation (e.g., S is
identical with P, S is subordinated to P). Just as propositions can differ
from one another in temporal aspect so can also the relations of terms. It is
reasonable to build the evaluation of the propositions under discussion on
such term relations which always hold true (e.g., S is always identical with
P). Under this condition the following values are obtained:

TABLE 48

	(1)	(2)	(3)	(4)	(5)	(6)	(7)	(8)	(9)	(10)	(11)	(12)
(1)	f	f	t	t	f	f	f	f	t	t	f	f
(2)	f	f	f	f	t	t	t	t	f	f	f	f
(3)	t	t	f	f	t	t	t	t	f	f	f	f
(4)	t	t	t	t	f	f	f	f	t	t	f	t

The relations of the propositions can be found out from this table.

Let us proceed to the following versions: (1) there is sometimes S, (2)
there is sometimes no S, (3) it does not hold true that there is sometimes S

(there is never S), (4) it does not hold true that there is sometimes no S. They need not be discussed at length because they are successively equivalent to the following propositions: (1) there is not always no S, (2) there is not always S, (3) there is always no S, (4) there is always S.

It is just as easy to cope with the following formulas: (1) S is sometimes P, (2) S is sometimes not P, (3) it does not hold true that S is sometimes P (S is never P), (4) it does not hold true that S is sometimes not P. Their equivalents are: (1) S is not always not P (2) S is not always P, (3) S is always not P, (4) S is always P.

Let us go further: (1) now there is S, (2) now there is no S, (3) not now there is S, (4) not now there is no S. Their evaluation is based on Figure 2. The meaning of shade-lines is: 'there is never', and that of the dot: 'there is always'. Their values agree successively with the values in Table 47.

Let us examine this time in relation to propositions 'S is P': (1) S is P this time, (2) S is not P this time, (3) it does not hold true that S is P this time, (4) it does not hold true that S is not P this time. Their evaluation is based on Figure 12 and the diagrams represent such term relations which are always valid.

TABLE 49

	(1)	(2)	(3)	(4)	(5)	(6)	(7)	(8)	(9)	(10)	(11)	(12)
(1)	f/t	f/t	t	t	f	f	f	f	t	t	f/t	f/t
(2)	t/f	t/f	f	f	t	t	t	t	f	f	t/f	t/f
(3)	t/f	t/f	f	f	t	t	t	t	f	f	t/f	t/f
(4)	f/t	f/t	t	t	f	f	f	f	t	t	f/t	f/t

Let us examine, finally, the following time: (1) sometimes there is S, (2) sometimes there is no S, (3) not sometimes there is S, (4) not sometimes there is no S. The shade-lines in Figure 12 mean 'sometimes there is', and the dot means: 'sometimes there is not'.

TABLE 50

	(1)	(2)
(1)	t	t
(2)	t	t
(3)	f	f
(4)	f	f

The same time occurs in the following versions: (1) S is sometimes P, (2) S is sometimes not P, (3) it does not hold true that S is sometimes P, (4) it does not hold true that S is sometimes not P. One can analyse them more easily by considering the successively equivalent versions:

(1) S is sometimes P and S is sometimes not P

(2) S is sometimes not P and S is sometimes P

(3)   it does not hold true that S is sometimes P, or it does not hold true that S is sometimes not $P \leftrightarrow S$ is never P or S is always P

(4)   it does not hold true that S is sometimes not P or it does not hold true that S is sometimes $P \leftrightarrow S$ is always P or S is never P

Their evaluation is based on Figure 12, and the diagrams represent term relations which always hold true:

TABLE 51

	(1)	(2)	(3)	(4)	(5)	(6)	(7)	(8)	(9)	(10)	(11)	(12)
(1)	t	t	f	f	f	f	f	f	f	f	t	t
(2)	t	t	f	f	f	f	f	f	f	f	t	t
(3)	f	f	t	t	t	t	t	t	t	t	f	f
(4)	f	f	t	t	t	t	t	t	t	t	f	f

III

All I wanted to demonstrate in the previous evaluation of temporal existential propositions is that these propositions can be evaluated if adequate criteria are chosen. This procedure has at least the following shortcomings: (1) the selection of the criteria might have seemed arbitrary, (2) the question as to what results are obtained with other criteria in mind has not been answered, (3) the comparison of the values of different temporal propositions has not been permitted.

The best way to overcome these shortcomings is in principle to take all the kinds of time indicated in Figure 29 as a basis of evaluation. It would not be useful, however, because (1) such a table would be difficult to survey, (2) how many kinds of time are considered has nothing to do with the interrelations of the propositions concerned. Therefore, as a compromise, I shall select here some of the typical kinds:

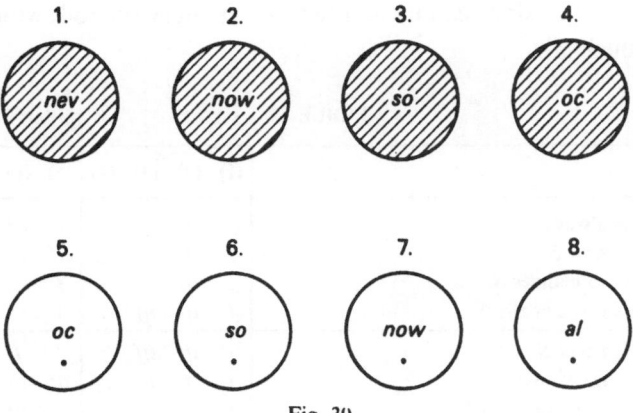

Fig. 30.

The above circles are to be understood like this:

(1)   there is never
(2)   there is not now
(3)   there is not sometimes
(4)   there is not occasionally
(5)   there is occasionally
(6)   there is sometimes
(7)   there is now
(8)   there is always

In order to arrive at existential propositions, the above time operators have to be complemented by any subject (*S*). Let us see now the values of 'there is always *S*'. The result will be obviously *t* in case (8) e.g., there is always movement. If *S* is substituted by something which does not exist (1)–(4), then the value of the proposition will be *f*, e.g., there is always resurrection. In case (5) the result is *f* for that which occasionally exists does not exist sometimes by definition and is contradictory to 'there is always'. In order to define the value of cases (6)–(7) let us keep in mind that there are always and not always existing things among those which exist sometimes and now, respectively. Therefore, the value of the proposition in the above cases can be *t* but can be *f* as well. I think the detailed analysis

20

of the further versions can be omitted. In summary the following values are obtained:

TABLE 52

	(1)	(2)	(3)	(4)	(5)	(6)	(7)	(8)
(1) There is always S	f	f	f	f	f	t/f	t/f	t
(2) There is now S	f	f	t/f	t/f	t/f	t/f	t	t
(3) There is sometimes S	f	t/f	t/f	t	t	t	t	t
(4) There is occasionally S	f	t/f	t/f	t	t	t/f	t/f	f
(5) There is never S	t	t/f	t/f	f	f	f	f	f
(6) There is no S now	t	t	t/f	t/f	t/f	t/f	f	f
(7) There is sometimes no S	t	t	t	t	t	t/f	t/f	t/f
(8) There is occasionally no S	f	t/f	t/f	t	t	t/f	t/f	f
(9) There is not always S	t	t	t	t	t	t/f	t/f	f
(10) There is not now S	t	t	t/f	t/f	t/f	t/f	f	f
(11) There is not sometimes S	t	t/f	t/f	f	f	f	f	f
(12) There is not occasionally S	f	t/f	t/f	f	f	t/f	t/f	f
(13) There is not always no S	f	t/f	t/f	t	t	t	t	t
(14) There is not now no S	f	f	t/f	t/f	t/f	t/f	t	t
(15) There is not sometimes no S	f	f	f	f	f	t/f	t/f	t
(16) There is not occasionally no S	f	t/f	t/f	f	f	t/f	t/f	f

Table 52 reveals the following: (1) if several criteria are used at the same time, no proposition can be evaluated unambiguously. However, one can find for each proposition a criterion which makes the solution possible. (2) It is easy to recognize the equivalence, negations, etc. among propositions, and this additional knowledge contributes to the survey of temporal immediate inferences, on the one hand, and to the operation of tracing them back, on the other.

Syllogisms can be composed of temporal propositions, e.g.:

M is always P
S is always M
_____
S is always P

Any categorical syllogism can be correlated to a temporal one.

All that has been said so far allows us to state that there is an analogy between the categorical and the corresponding temporal propositions. Aristotle wrote the following on this issue:

> ... if *A* is predicated universally of *B* and *B* of *C*, *A* too must be predicated always and in every instance of *C*, since to hold in every instance and always is of the nature of the universal.[127]

Let us compare the above said to the following statement of Quine:

> In putting statements of ordinary language over into the forms *A*, *E*, *I*, and *O* we must be on the alert for irregularities of idiom, and look beneath them to the intended sense. One such irregularity is omission of '-ever', as in 'Who hesitates is lost', 'I want to go where you go', 'When it rains it pours', 'She gets what she goes after'. Another irregularity is the nontemporal use of 'always', 'whenever', 'sometimes', 'never'. E.g., the statement:
> The sum of the angles of a triangle is always equal to two right angles really means:
> The sum of the angles of any triangle is equal to two right angles, and may be rendered 'All *F* are *G*' where '*F*' represents 'sums of angles of triangles' and '*G*' represents 'equal to two right angles'.[128]

Klaus, too, was of a similar opinion:

> The universal operator can appear in everyday language in a quite hidden manner... It also plays a role, for instance in such expression as 'always', 'all the time', etc. These expressions are furthermore extremely ambiguous. Compare, for instance, the statements: 'This week it rained all the time', and 'Equilateral Euclidian triangles are always equiangular'.
> The first determination of 'always' or 'all the time' is a temporal one. The logical structure of the statement becomes visible if we formulate: "For every point in time, *x*, it is the case that if *x* belongs to this week, then it rained at *x*."
> The second example contains no temporal element. This statement has the same content as: "For all *x*, it is the case that, if *x* is an equilateral Euclidian triangle, then it is also equiangular.[129]

---

[127] Aristotle. 1928. 96a14–17.
[128] Quine, p. 67.
[129] Klaus, p. 236.

20*

According to Klaus there is no temporal moment in the triangle example. He presumably regards this assertion as evident since he does not explain it. It is, however, not evident and, in fact, not even acceptable for me. The thesis concerned is not *a priori* valid, but an abstraction of a relation which exists in reality. Nevertheless,

> ... the basic forms of all being are space and time, and being out of time is just as gross an absurdity as being out of space.[130]

In Quine's opinion 'always' is not used in the sense of time in the example under discussion. It is possible that some do not want to use it in this way. But is that sufficient reason to generalize? I find the thesis at issue also meaningful in a temporal sense. In logic one must proceed from what someone has actually said and not from what he intended to say. Accordingly, the example under discussion is a temporal proposition. If it is taken as such, both ambiguity and linguistic anomaly are out of the question.

IV

With regard to propositions concerning the future, Aristotle raised the following question in Chapter 9 of the '*Hermeneutics*':

> In the case of that which is or which has taken place, propositions, whether positive or negative, must be true or false. Again, in the case of a pair of contradictories, either when the subject is universal and the propositions are of a universal character, or when it is individual, as has been said, one of the two must be true and the other false; whereas when the subject is universal, but the propositions are not of a universal character, there is no such necessity. We have discussed this type also in a previous chapter.
>
> When the subject, however, is individual, and that which is predicated of it relates to the future, the case is altered.[131]

As to the relations of the above propositions, Aristotle quoted the following examples:

---

[130] Engels, *Anti-Dühring*. p. 61; Eng. tr. (Moscow 1954) p. 76.
[131] Aristotle. 1928. 18a28–33.

Every man is white	— No man is white
Socrates is white	— Socrates is not white
Man is white	— Man is not white
A sea-fight will take place on the next day	— A sea-fight will not take place on the next day

As far as the relation of the universal affirmative and the universal negative propositions is concerned, Aristotle maintained that "one of the two must be true and the other false". This assertion, however, does not hold good. An example to the contrary is: all books are interesting—no book is interesting.

Aristotle analysed the following suppositions about affirmative and negative propositions relating to the future: (1) both of them can be t, (2) both of them can be f, (3) one of them is definitely t, the other is definitely f, (4) one of them is indefinitely t, the other is indefinitely f.

He rejected the first supposition upon the following consideration:

> For if all propositions whether positive or negative are either true or false, then any given predicate must either belong to the subject or not, so that if one man affirms that an event of a given character will take place and another denies it, it is plain that the statement of the one will correspond with reality and that of the other will not. For the predicate cannot both belong and not belong to the subject at one and the same time with regard to the future. Thus, if it is true to say that a thing is white, it must necessarily be white; if the reverse proposition is true, it will of necessity not be white. Again, if it is white, the proposition stating that it is white was true; if it is not white, the proposition to the opposite effect was true.[132]

Aristotle's opinion can be shared in so far as the propositions concerned cannot be t at one and the same time. I have, however, two objections to the above arguments. (1) It does not follow from the thesis that every affirmation or negation is either t or f that everything is or is not of necessity. The antecedent does not exclude the possibility that, e.g., something actually is or is not.

(2) If an affirmation and a negation are in a relation of contradiction, then one of them being t, the other is f. Aristotle, however, has not

[132] *Op. cit.*, 18a34–18b2.

restricted the relation of an affirmation and a negation to contradiction only, as has been seen in case of universal propositions. And as soon as the propositions concerned are contrary to one another, it does not hold true any more that one of them is *t* and the other *f*.

How can one explain the error Aristotle has made here? On the one hand, it is due to the fact that he failed to distinguish between a negative proposition and a negation. The formula 'no *S* is *P*', for example, is a negative proposition but not a negation of the formula 'all *S* is *P*'. The negation of the latter is: 'some *S* is not *P*'. Furthermore, it is a well-known fact that a negative proposition is not the only way of negation, e.g., no *S* is *P*—some *S* is *P*.

On the other hand, he followed the principle that all affirmations or negations are either *t* or *f*. This statement is true only when taken separately: (a) some affirmation is *t* or *f*, (b) some negation is *t* or *f*. As a whole, however, it does not hold good that out of two propositions confronted as affirmation and negation (as an affirmative and a negative proposition in the Aristotelean interpretation) one is *t* and the other *f*.

He disproved the second supposition as follows:

> Again, to say that neither the affirmation nor the denial is true, maintaining, let us say, that an event neither will take place nor will not take place, is to take up a position impossible to defend. In the first place, though facts should prove the one proposition false, the opposite would still be untrue. Secondly, if it was true to say that a thing was both white and large, both these qualities must necessarily belong to it; and if they will belong to it the next day, they must necessarily belong to it the next day. But if an event is neither to take place nor not to take place the next day, the element of chance will be eliminated. For example, it would be necessary that a sea-fight should neither take place nor fail to take place on the next day.[133]

The law of the excluded middle holds true of the propositions under discussion only because they are in a relation of contradiction.

The third supposition is nothing else but the position of the fatalists. In their opinion whatever exists does so necessarily and what is *t* is necessarily true. I do not cite here in detail Aristotle's arguments against this view

---

[133] *Op. cit.*, 18b17–25.

because the main points are summed up in the proof of the fourth supposition which he found correct:

> Everything must either be or not be, whether in the present or in the future, but it is not always possible to distinguish and state determinately which of these alternatives must necessarily come about. Let me illustrate. A sea-fight must either take place to-morrow or not, but it is not necessary that it should take place to-morrow, neither is it necessary that it should not take place, yet it is necessary that it either should or should not take place to-morrow. Since propositions correspond with facts, it is evident that when in future events there is a real alternative, and a potentiality in contrary directions, the corresponding affirmation and denial have the same character. This is the case with regard to that which is not always existent or not always not-existent. One of the two propositions in such instances must be true and the other false, but we cannot say determinately that this or that is false, but must leave the alternative undecided. One may indeed be more likely to be true than the other, but it cannot be either actually true or actually false. It is therefore plain that it is not necessary that of an affirmation and a denial one should be true and the other false. For in the case of that which exists potentially, but not actually, the rule which applies to that which exists actually does not hold good. The case is rather as we have indicated.[134]

Aristotle was right in stating that any of the propositions at issue can be t and f, respectively. He thought, however, that these propositions ("when the subject . . . is individual, and that which is predicated of it relates to the future") differ from the rest in this very respect. I do not think they do. (1) The same applies to the past and the present respectively, e.g., Anaximander liked music—Anaximander did not like music. It is not necessary that one of the propositions should be t or f if it is an actual proposition. (2) I find the emphasis on singular propositions unjustified. Let us compare these assertions: all my books have a leather binding — some of my books have no leather binding. Which of these propositions should be t?

---

[134] *Op. cit.*, 19a27–19b4.

In my view Schaff has introduced a correct distinction.

Let us first analyse the various groups of propositions concerning the future. Let us see, e.g., the following two statements: 'John will be in the Soviet Union within a year, at the same time', or 'Within a year, since July, it will be summer again'...

Propositions relating to the future are difficult to evaluate from the point of view of truth because, according to the theory of objective truth, truth is understood as a definite relation of the proposition to the actual state of affairs (to reality); the propositions relating to the future (i.e. foresight in the most general sense of the word) lack the other part of this relation, at least in the form usually associated with the truth of propositions and in the way propositions are compared to some—already given—state of affairs (e.g., when we say: 'This flag is red', and the flag is indeed red).

This deficiency is obvious and unavoidable in such propositions as the one in our first example which concerns John's trip to the Soviet Union in the future: this deficiency is closely interlinked with the uncertainty typical of such propositions. Therefore, these propositions cannot be described as true until afterwards (*ex post*), that is, when a certain state of affairs has actually occurred. It is only then that the proposition can be compared to the real state of affairs and that, as a result, its truth can be accurately evaluated. Then, however, the very character of the proposition has already changed.

It does not hold true of the second type of propositions where foresight is based on a permanent and necessary succession of certain recurrent causes and effects, that is, on the functioning of certain *laws* of nature and society. If a certain concrete state of affairs of the reality under examination is exactly known, then, knowing the *laws* of evolution of reality, one can foresee a definite future event not with a certain degree of probability but *with certainty* since such events take place of necessity. Therefore, propositions which express such a foresight can be, indeed, described as true or false.[135]

[135] Schaff, pp. 90–92.

V

Aristotle's statement about simultaneity will be our point of departure as regards the relations of tenses.

> The term 'simultaneous' is primarily and most appropriately applied to those things the genesis of the one of which is simultaneous with that of the other; for in such cases neither is prior or posterior to the other. Such things are said to be simultaneous in point of time. Those things, again, are 'simultaneous' in point of nature, the being of each of which involves that of the other, while at the same time neither is the cause of the other's being. This is the case with regard to the double and the half, for these are reciprocally dependent, since, if there is a double, there is also a half, and if there is a half, there is also a double, while at the same time neither is the cause of the being of the other.[136]

Elsewhere he commented on anteriority and posteriority:

> The words 'prior' and 'posterior' are applied ... to some things (on the assumption that there is a first, i.e. a beginning, in each class) because they are nearer some beginning determined either absolutely and by nature, or by reference to something or in some place or by certain people; e.g. things are prior in place because they are nearer either to some place determined by nature (e.g. the middle or the last place), or to some chance object; and that which is farther is posterior.—Other things are prior in time; some by being farther from the present, i.e. in the case of past events (for the Trojan war is prior to the Persian, because it is farther from the present), others by being nearer the present, i.e. in the case of future events (for the Nemean games are prior to the Pythian, if we treat the present as beginning and first point, because they are nearer the present).[137]

Several problems arise in connection with anteriority and posteriority. Let me mention among them the fallacy of *post hoc, ergo propter hoc*, i.e., when the temporal succession of two phenomena is regarded as a causal relation and, as such, also as a ground for inference. For example: yesterday he ate fish, and therefore he has a bad stomach today. The fact

---

[136] Aristotle. 1928. 14b24–32.
[137] Aristotle, *Metaphysica*. 1018b8–19.

that somebody ate fish yesterday does not necessarily lead to the conclusion that this is why he has a bad stomach today.

Having once noticed the temporal succession of two phenomena, some establish causality between them, that is, they think the appearance of one of them necessarily entails that of the other. For example, when the coincidence of sunspots and economic crises in capitalist countries had been observed, certain bourgeois economists developed a whole theory about the appearance of sunspots being the cause of economic crises.

Such a way of thinking has nothing to do with logic. Temporal succession is but one of the necessary characteristics of causality and, therefore, does not by itself provide sufficient reason to conclude that there is causal relation.

Hume represented the other extreme:

> When we look about us towards external objects, and consider the operation of causes, we are never able, in a single instance, to discover any power or necessary connexion; any quality, which binds the effect to the cause, and renders the one an infallible consequence of the other. We only find, that the one does actually, in fact, follow the other. The impulse of one billiard-ball is attended with motion in the second. This is the whole that appears to the *outward* senses. The mind feels no sentiment or *inward* impression from this succession of objects: Consequently, there is not, in any single, particular instance of cause and effect, anything which can suggest the idea of power or necessary connexion.[138]

That is, he reduced the interrelations of events to temporal relations only.

Simultaneity, anteriority, and posteriority have been discussed in the foregoing. Other temporal relations are:

Subaltern,	e.g.: today—this week.
Crossing,	e.g.: between January 1 and 10—between January 5 and 15.
Contradictory,	e.g.: this week—not this week.
Contrary,	e.g.: beginning—end.

---

[138] Hume, *Enquiries.* p. 63.

Figures 29, 18, and 25 and their interpretations make possible a comparison of temporal, quantified, and modal propositions. As we have already seen, there is an analogy between temporal and quantified propositions. The same applies to the relation of modal and temporal and quantified propositions respectively. The following comparison may satisfactorily illustrate the analogy without lengthy argument:

(1)   It is necessary that S is P → It is possible
                                                      that S is P
(2)   All S is P → Some S is P
(3) S is always P → S is sometimes P

There is a relation of subordination between the propositions in the same line. That is, if the premise is t, then the conclusion is necessarily t. Having selected the corresponding pairs of propositions with the help of the above figures, we obtain the same connections in all cases.

What are the inference possibilities of modal and temporal propositions? The figure combining the two kinds of operators makes the answer easier:

As a reminder, let me repeat the designations: (1) ip=impossible, (2) po=possible, (3) so=sometimes, (4) ft=fact, (5) tt=this time, (6) al=always, (7) nec=necessary.

Ring 2 represents 'only possible', ring 3 'only sometimes', ring 4 'only fact', ring 5 'only once', and ring 6 'only always'.

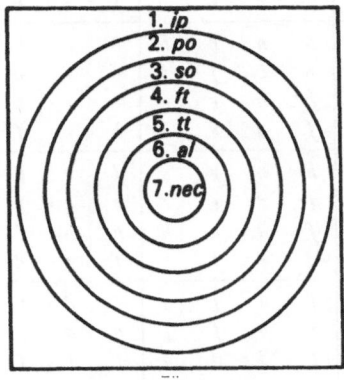

Fig. 31.

'Never' (not sometimes) is implied by rings 1 and 2 together. It means that something is either impossible (e.g., the absolute zero degree will never be reached), or possible (e.g., man has never been to Mars yet).

The subaltern relations can be directly read from Figure 31. So, e.g., what exists necessarily always exists, but it does not hold true the other way round. The same relation applies to rings 2 and 3, 4 and 5, 2 and 4, 3 and 7, etc. Perhaps the relation between rings 4 and 5 requires some explanation because one could suppose that they are equivalent. An example to the contrary is: if one knows of a Greek philosopher when he was born one can take it for granted as a fact that he was born. The fact that he was born, however, does not imply when he was born.

No further explanation is needed for stating that any modal proposition can be compared to any temporal one. Table 53 contains their relations. The numbers arranged vertically refer to the modal propositions successively, and those horizontally to the temporal ones.

It is, of course, also possible to construct a table such that it would demonstrate the relations of temporal propositions to modal ones, proceeding from the former propositions.

TABLE 53

	(1)	(2)	(3)	(4)	(5)	(6)	(7)	(8)	(9)	(10)	(11)	(12)	(13)	(14)	(15)	(16)
(1)	c	e	e	e	e	e	e	e	e	e	e	e	e	e	e	g
(2)	i	b	h	h	h	b	b	b	h	h	h	b	b	b	h	a
(3)	i	b	h	h	h	b	h	h	b	b	h	b	b	b	h	a
(4)	i	h	h	h	h	b	h	b	h	b	h	b	b	b	b	a
(5)	i	h	h	h	b	h	h	b	h	b	b	h	b	b	b	a
(6)	i	f	f	h	h	b	f	f	d	d	h	b	b	d	d	a
(7)	i	e	h	h	h	b	e	e	e	e	h	b	b	b	e	a
(8)	i	e	h	h	f	d	e	b	h	e	f	d	b	b	e	a
(9)	i	e	f	f	h	b	e	d	f	e	h	b	d	d	e	a
(10)	i	e	f	f	f	d	e	e	e	e	f	d	d	d	e	a
(11)	i	h	h	f	f	d	h	h	h	b	f	d	d	b	b	a
(12)	i	f	f	f	d	f	f	d	f	d	d	f	d	d	d	a
(13)	i	f	f	f	f	d	f	d	f	d	f	d	d	d	d	a
(14)	i	d	f	f	f	d	f	f	d	d	f	d	d	d	f	a
(15)	i	d	f	f	f	d	d	d	f	f	f	d	d	d	f	a
(16)	g	e	e	e	e	e	e	e	e	e	e	e	e	e	e	c

CHAPTER FIVE

13. RELATIONS OF TERMS AND PROPOSITIONS

I

Aristotle defined relative terms (*relativa, termini relativi*) as follows:

> Those things are called relative, which, being either said to be *of* something else or *related to* something else, are explained by reference to that other thing. For instance, the word 'superior' is explained by reference to something else, for it is superiority *over something else* that is meant. Similarly, the expression 'double' has this external reference, for it is the double *of something else* that is meant.[1]

Then he added:

> ... if a man definitely apprehends a relative thing, he will also definitely apprehend that to which it is relative. Indeed this is self-evident: for if a man knows that some particular thing is relative, assuming that we call that a relative in the case of which relation to something is a necessary condition of existence, he knows that also to which it is related. For if he does not know at all that to which it is related, he will not know whether or not it is relative. This is clear, moreover, in particular instances. If a man knows definitely that such and such a thing is 'double', he will also forthwith know definitely that of which it is the double. For if there is nothing definite of which he knows it to be the double, he does not know at all that it is double. Again, if he knows that a thing is more beautiful, it follows necessarily that he will forthwith definitely know that also than which it is more beautiful. He will not merely know indefinitely that it is more beautiful than something which is less beautiful, for this would be supposition, not knowledge. For if he does not know definitely that than which it is more beautiful, he can no longer claim to know definitely that it is more beautiful than something else

---

[1] Aristotle. 1928. 6a36–6b2.

303

which is less beautiful: for it might be that nothing was less beautiful. It is, therefore, evident that if a man apprehends some relative thing definitely, he necessarily knows that also definitely to which it is related.[2]

The Stagirite considered examples where the recognition of relativity would have no sense without knowledge of those related. Let us take, however, the relative term 'effect' as an example. If I see that I have fever I know that this symptom has some cause. It is not all sure, however, that I also know the cause of fever. Even if I know that John is somebody's son it is not at all sure that I know his father. That is, one can know for sure that something is relative even if the individual factors of the relation are indefinite.

Further comes the following argumentation:

> The term 'to have' is used in various senses. In the first place it is used with reference to habit or disposition or any other quality, for we are said to 'have' a piece of knowledge or a virtue. Then, again, it has reference to quantity, as, for instance, in the case of a man's height; for he is said to 'have' a height of three cubits or four cubits ... Or it refers to that which has been acquired; we are said to 'have' a house or a field.[3]

Let us take an example: John's house. This term implies a possessive relation. What is the difference between this term and those discussed before? In my opinion it differs from them by being a compound term which cannot be broken down to simple ones.

Medieval logic primarily studied the role of relative terms in propositions and in inferences and demonstrated that, e.g., inferences composed of existential propositions with relative terms are valid: there is a master, hence there is a servant. Yet, one must not draw any conclusion from propositions like 'S is P' if they contain a relative term, e.g., the father is running, hence his son is running. Their main concern was the interpretation of propositions containing a compound relative term (*terminus obliquus*). Duns Scotus, for example, interpreted the proposition 'the donkey of man runs' as follows: 'there is a man whose donkey runs'.

---

[2] *Op. cit.*, 8a37–8b15.
[3] *Op. cit.*, 15b17–21, 27.

In traditional logic, Jevons's approach deserves attention:

> It is further usual to divide terms according as they are *relative* or *absolute*, that is, non-relative...
>
> The fact, however, is that everything must really have relations to something else, the water to the elements of which it is composed, the gas to the coal from which it is manufactured, the tree to the soil in which it is rooted. By the very laws of thought, again, no thing or class of things can be thought of but by separating them from other existing things from which they differ. I cannot use the term mortal without at once separating all existing or conceivable things into the two groups *mortal* and *immortal*; metal, element, organic substance, and every other term that could be mentioned, would necessarily imply the existence of a correlative negative term, non-metallic, compound, inorganic substance, and in this respect therefore every term is undoubtedly relative. Logicians, however, have been content to consider as relative terms those only which imply some peculiar and striking kind of relation arising from position in time or space, from connexion of cause and effect, etc.; and it is in this special sense therefore the student must use the distinction.[4]

Jevons recognized that the usual examples are but special cases of relative terms. But how should one understand his assertion that all terms are relative? It could mean that all terms reflect the objects approximately only and not completely. In so far the author is right, but the relativity of terms and the definition of relative terms are two different things. The former belongs to the general characterization of terms while the latter to the study of a certain kind of terms.

The above context allows us even to interpret all terms as relative. The fact that a considerable part of our terms do not express any relation contradicts this view. The term 'metal', e.g., potentially includes, but does not express, explicitly, its relation to non-metal. Furthermore, a term like 'more beautiful picture' is incomplete in itself while 'metal' is complete, independently of 'non-metal'. Jevons shows his uncertainty by suggesting in the last part of the above quotation that the traditional approach should be followed.

[4] Jevons, 1872. pp. 25–26.

In my opinion relative terms are those which express some kind of relation. They may be terms which are conditional upon one another (e.g. bigger — smaller), but also which express compatibility (e.g. big house) or even contradiction (e.g. quadrangular circle). Relative terms have as many versions as relations are distinguished. They will be surveyed later.

II

Under the name of class calculus the following relations and operations are discussed:

Terms to which the same individuals belong are identical. All elements of S are also elements of P, and the other way round.

Terms where all elements of S are also elements of P are in a relation of inclusion. This relation, as opposed to identity, is not subject to the condition that all elements of P should be elements of S as well, although this possibility is not excluded.

S is a real part of P if all elements of S are also elements of P but it cannot hold true the other way round.

The union of S and P includes the elements which belong to one of the two classes at least.

The cross section of S and P includes the elements which belong to both classes.

Two classes are in a relation of complementaries if one of them contains all the elements which do not belong to the extension of the other.

Two classes are disjunct if they have not a single element in common.

All the above operations and relations can be negated, e.g., not-complementary.

To make the connections between the above term relations clear let us have a closer look at Figure 32.

The designations mean the following: dj = disjunct (what is in the universe, outside circle 2), un = union (circle 2), i = inclusion (circle 3), id = identity (circle 4).

Disjunct excludes union, and union excludes disjunct, but allows for inclusion and identity. Inclusion allows for identity. Identity at the same time implies inclusion and union as well. Inclusion at the same time implies union as well.

Ring 2 represents cross section, i.e., only union (only *un*), and ring 3 the real part, i.e., only inclusion (only *i*).

The expression "partial coincidence" (*pc*) will be used to designate the relation represented by rings 2 and 3 together. This relation differs from union by excluding not only the disjunct but also identity. The relation

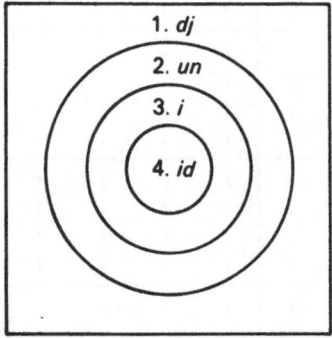

Fig. 32.

implied by ring 2 and circle 4 together is called: union but not only inclusion. It typically excludes both the disjunct and the real part. Accordingly, we get the following summary:

0	— indefinite (absolute)	2 *o* 3	— *pc*
1	— *dj*	2 *o* 4	— *un* but not only-*i*
2	— only *un*	3 *o* 4	— *i*
3	— only *i*	1 *o* 2 *o* 3	— not *id*
4	— *id*	1 *o* 2 *o* 4	— not only-*i*
1 *o* 2	— not *in*	1 *o* 3 *o* 4	— not only-*un*
1 *o* 3	— only *in o dj*	2 *o* 3 *o* 4	— *un*
1 *o* 4	— not *pc*	1 *o* 2 *o* 3 *o* 4	— *u*

Let us now consider propositions wherein the above relations are present. Here, first of all, the relation of logical and linguistic expressions should be explained. They coincide in certain cases, e.g., '*S* is identical with

21

TABLE 54

	(1)	(2)	(3)	(4)	(5)	(6)	(7)	(8)	(9)	(10)	(11)	(12)
(1)	f	f	f	f	f	f	f	f	f	f	f	f
(2)	f	f	f	f		t	t	t	f	f	f	f
(3)	f	f	f	f	f	f	f	f	f	f	f	f
(4)	f	f	t	f	f	f	f	f	f	t	f	f
(5)	f	f	f	t	f	f	f	f	t	f	f	f
(6)	t	t	f	f	t	t	t	t	f	f	t	t
(7)	f	f	t	f	t	t	t	t	f	t	f	f
(8)	f	f	f	t	t	t	t	t	t	f	f	f
(9)	t	t	t	f	f	f	f	f	f	t	t	t
(10)	t	t	f	t	f	f	f	f	t	f	t	t
(11)	f	f	t	t	f	f	f	f	t	t	f	f
(12)	t	t	t	f	t	t	t	t	f	t	t	t
(13)	t	t	f	t	t	t	t	t	t	f	t	t'
(14)	f	t	t	t	t	t	t	t	t	t	t	f
(15)	t	t	t	t	f	f	f	f	t	t	t	t
(16)	t	t	t	t	t	t	t	t	t	t	t	t

P'—'Iran is identical with Persia'. Let us take, however, the formula 'S includes P'. An example 'fruit includes apple' would sound senseless. It can be said, however, that 'apple belongs to fruits'. It is always a question of interpretation to state what relation is expressed by a sentence. The relation of two terms will be described usually by the formula $R(S, P)$ where R = relation.

On the ground of Figure 12 the above relative propositions produce the various values (see Table 54 above).

Table 54 reveals, among others, the relations between relative propositions and the immediate inferences based on them. E.g., the proposition 'S is disjunct to P' is contrary to the proposition 'S cross section P'.

In order to define the convertibility of a relative proposition it is sufficient to consider its values according to diagrams 10 and 11 respectively. If these values are identical, then the proposition concerned can be simply converted.

By comparing the values of a relative proposition to those of any quantified ones, we can reveal the relation of the two propositions. Let us take, e.g., the following proposition as first: *S* is not a real part of *P*, and this one as second: some *S* is *P*. There is a subcontrary relation between these two propositions.

If we are interested only in the relation, and not in the values, of two propositions Figure 33 will do, where the relations of relative and quantified propositions can be seen:

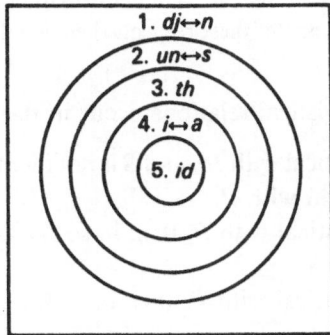

Fig. 33.

To avoid misunderstandings let me repeat the meaning of these designations: *id* = identical, *i* = inclusion, *a* = all, *th* = this, *un* = union, *s* = some, *dj* = disjunct, *n* = nothing. Let us check the above stated subcontrary relation. If the proposition '*S* is not a real part of *P*' is *t*, then 'some *S* is *P*' is *t*/*f*. If, however, the premise of *f* the conclusion is *t*.

The relations of relative and modal propositions can also be stated. All we have to do is to find and compare the values of two propositions in the corresponding tables.

Let me note also that there are quantified, modal, and temporal relative propositions as well, e.g.:

Some *S* is identical with *P*
It is necessary that *S* should not only-include *P*
S is always disjunct to *P*

21*

Relative propositions can constitute syllogisms. Let us first take identity as a basic relation:

> ... look and see if, supposing the one to be the same as something, the other also is the same as it: for if they be not both the same as the same thing, clearly neither are they the same as one another.
>
> Moreover, examine them in the light of their accidents or of the things of which they are accidents: for any accident belonging to the one must belong also to the other, and if the one belong to anything as an accident, so must the other also. If in any of these respects there is a discrepancy, clearly they are not the same. [5]

The following syllogistical relations occur in the above text:

(1)   If P is identical with M, and S is not identical with M, then S is not identical with P.

(2)   If S is identical with P, then in so far as M belongs to P, M belongs to S.

(3)   If S is identical with P, then in so far as P belongs to M, S belongs to M.

The following relations are only implied:

(4)   If M belongs to S but does not belong to P, then S is not identical with P.

(5)   If S belongs to M but P does not belong to M, then S is not identical with P.

Finally, we can suppose that Aristotle implicitly accepted the following inference as well: if M is identical with P and S is identical with M, then S is identical with P.

Since no system of relative propositions has been developed in traditional logic the inference discussed above can be met but rarely:

> ... if two things are identical with a third common thing they are identical with each other. This is a law of thought of a very simple and obvious character, and we may observe concerning it:

---

[5] Aristotle. 1928. 152a32–38.

(1)   That all people think in accordance with it, and agree that they do so as soon as they understand its meaning.

(2)   That they think in accordance with it whatever may be the subject about which they are thinking.

Thus if the things considered are:

London
The Metropolis,
The most populous city in Great Britain,

since 'the Metropolis is identical with London', and 'London is identical with the most populous city in Great Britain', it follows necessarily in all minds that 'the metropolis is identical with the most populous city in Great Britain'.[6]

The relation concerned has been discussed in mathematical logic as the law of the transitivity of identity. One of the ways to formulate it is: $w = x \cdot x = y \cdot \supset \cdot w = y$.

The syllogisms based on inclusion coincide with those based on universal propositions, as Aristotle sees them:

Whenever three terms are so related to one another that the last is contained in the middle as in a whole, and the middle is either contained in, or excluded from, the first as in or from a whole, the extremes must be related by a perfect syllogism... If A is predicated of all B, and B of all C, A must be predicated of all C ...[7]

Similarly, he treated also the following propositions as identical:

S is not at all included in P ↔ No S is P
S is partially included in P ↔ Some S is P

Figure 33 reveals that in these three cases there is, indeed, equivalence between the quantified and the relative propositions. Therefore, it might seem justified that traditional logic dealt with one of them only. And it is easy to answer why exactly the former was dealt with since usually nobody says that glass is not at all included in heat-conductors. The sentence 'no glass is a heat-conductor' is, however, clearly understandable by all.

[6] Jevons, 1872. pp. 2–3.
[7] Aristotle. 1928. 25b31–35, 38–39.

What should be done, then, with the relation of identity? It was not too problematic. Syllogistic was restricted to propositions *A*, *E*, *I*, *O* which contain the quantifiers all, none, and some. However, it is by using these very quantifiers that apparently equivalent propositions can be constructed. Therefore, inferences based on identity were treated but incidentally, and even then, outside the syllogistic. The investigation into inferences based on relative propositions by and large merged into the study of categorical syllogisms. I think what has been said so far gives us sufficient reason to state that quantified and relative propositions are in the relation of partial coincidence. Consequently, none of them makes the study of the other unnecessary.

IV

Propositions can be divided into simple and compound ones. This distinction has been introduced by Aristotle. " ... of propositions one kind is simple, i.e. that which asserts or denies something of something, the other composite, i.e. that which is compounded of simple propositions."[8] Aristotle stated the following of compound propositions:

> There is no unity about an affirmation or denial which, either positively or negatively, predicates one thing of many subjects, or many things of the same subject, unless that which is indicated by the many is really some one thing. I do not apply this word 'one' to those things which, though they have a single recognized name, yet do not combine to form a unity. Thus, man may be an animal, and biped, and domesticated, but these three predicates combine to form a unity. On the other hand, the predicates 'white', 'man', and 'walking' do not thus combine. Neither, therefore, if these three form the subject of an affirmation, nor if they form its predicate, is there any unity about that affirmation.[9]

Accordingly, if the subject or the predicate of a proposition consists of several words there are two possibilities: in so far as these several words refer to the essence of a thing the proposition is simple, e.g. man is a biped, domesticated animal. If, however, these several words express additional

[8] *Op. cit.*, 17a20–22.
[9] *Op. cit.*, 20b13–21.

properties the proposition concerned is compound, e.g., 'the white man walks'—since this proposition should be understood as follows: 'the man is white and walks'.—There are only simple propositions in the Aristotelean syllogistic.

Compound propositions were studied first by the peripatetics, who distinguished two kinds of them, namely, hypothetical and disjunctive propositions.—According to Stoic terminology propositions may be simple and not-simple respectively. Apart from those mentioned above, conjunctive and causal propositions were also counted among the latter.

Avicenna held that hypothetical and disjunctive propositions are units of thought while conjunctive ones are but aggregates. Some medieval logicians, e.g., Occam, ranked even temporal and local propositions (expressing time and place respectively) among the compound ones.

According to another view, simple propositions have only one subject and one predicate while compound ones have several subjects and one predicate, or one subject but several predicates, or several subjects and several predicates.

Traditional logic offers a rather varied picture so far as our theme is concerned. For orientation's sake, let me introduce a few typical positions.

The distinction of simple and compound propositions played only a secondary role in the Kantian classification. Kant dealt with this question only with regard to the classification of propositions according to their relations as he put it. He made the following remark:

> All relations of thought in judgments are: (a) of the predicate to the subject, (b) of the ground to its consequence, (c) of the divided knowledge and of the members of the division, taken together, to each other. In the first kind of judgments we consider only two concepts, in the second two judgments, in the third several judgments in their relation to each other. The hypothetical proposition 'If there is a perfect justice, the obstinately wicked are punished', really contains the relation of two propositions, namely, 'There is a perfect justice', and 'The obstinately wicked are punished'. Whether both these propositions are in themselves true, here remains undetermined. It is only the logical sequence which is thought by this judgment. Finally, the disjunctive judgment contains a relation of two or more propositions to each other, a relation not, however, of logical sequence, but of logical opposition...[10]

[10] Kant, *Kritik* p. 144; Kemp Smith, pp. 108–109.

As far as our theme is concerned the above approach raises the following questions or even more: Does the classification into categorical, hypothetical, and disjunctive propositions cover all the relations possible in propositions? What justifies the omission of conjunctive propositions, for example? Why can one make a distinction between simple and compound propositions only in the classification of propositions according to their relations? After all, propositions may be simple and compound also according to their quality, extension, and mode, e.g.:

> Simple: It is necessary that S is P
> Compound: It is necessary that if S is P, then S_1 is P_1

Mill introduced the following consideration:

> For brevity . . . and to avoid repetition, the propositions are often blended together: as in this, 'Peter and James preached at Jerusalem and in Galilee', which contains four propositions: Peter preached at Jerusalem, Peter preached at Galilee, James preached at Jerusalem, James preached in Galilee.
>
> We have seen that when the two or more propositions comprised in what is called a complex proposition are stated absolutely, and not under any condition or proviso, it is not a proposition at all, but a plurality of propositions; since what it expresses is not a single assertion, but several assertions, which, if true when joined, are true also when separated. But there is a kind of proposition which, though it contains a plurality of subjects and of predicates, and may be said in one sense of the word to consist of several propositions contains but one assertion; and its truth does not at all imply that of the simple propositions which compose it. An example of this is, when the simple propositions are connected by the particle *or*; as, either A is B or C is D; or by the particle *if*; as, A is B if C is D.[11]

This argumentation is but a somewhat modified version of Avicenna's view.

While Kant and Mill considered the distinction of simple and compound propositions as a secondary aspect, others attached primary importance to it. In modern logic, it was, first of all, Überweg who represented the view that the classification of propositions should proceed

[11] Mill, p. 82.

from this distinction. This opinion was shared by Tavanets in his monograph on propositions. As he saw it, simple propositions are composed of two terms, and compound ones of several simple propositions.

Some, however, challenge this theory:

> Disjunctive propositions, especially of the antecedent–consequent type, are often erroneously considered as *compound propositions*. In reality they are simple propositions even if numerous terms (e.g., in a proposition of the antecedent-consequent type: *A*, *B*, *C*, *D*) occur in them.
>
> Simple propositions are those where *one* assertion (or denial) is made. *Compound propositions, however, are those where several interrelated assertions (or denials) are made.*
>
> It is a mistake to define compound propositions as having several subjects or several predicates. The criterion for a proposition to be simple or compound is how many assertions (or denials) are made in it.
>
> On the other hand, a compound proposition is neither a mere combination of two or more propositions but a *qualitatively* new proposition where the simple propositions it is based upon are related according to their *meaning*.
>
> A compound categorical proposition is, for example, 'Kant, Fichte, and Hegel are the main representatives of classical German philosophy'. This proposition has three objects and three assertions are made in it.
>
> Affirmative exceptive–disjunctive propositions may be both simple and compound, depending on the character of disjunction. Negative exceptive-disjunctive propositions are always compound...
>
> The propositions of the antecedent–consequent type may also be both simple and compound (in the former case having a simple or a compound antecedent, on the one hand, and a simple or a compound consequent, on the other).[12]

I do agree that a simple proposition may also have more than two terms. But this is why I regard the categorical proposition quoted above as a simple proposition with a compound subject rather than as a compound proposition since it cannot be broken down into further propositions.— At the beginning of the above quotation disjunctive propositions and those of the antecedent–consequent type are said to be simple while at the

[12] *Logika.* (Budapest 1956) p. 166.

end they are described as potentially both simple and compound. I think the propositions at issue are compound because they can be broken down into simple ones.

V

It is a known fact that propositions have an extension just as much as terms. The extension of the latter, e.g. man, includes all the individuals that belong to this term. The extension of a proposition, e.g. 'John goes for an excursion', includes all the occurrences of this relation. It is, of course, possible that some proposition refers to one case only, e.g. 'John went for an excursion in the vicinity today'. It corresponds to the singular term.

As early as in traditional logic, circles were found suitable to represent the extension of propositions as well as of terms.[13] This recognition, however, has not been applied consistently. Let us take a proposition p. The only restriction for this designation is that it can be replaced by propositions only. The circles to be used later on represent all the possible instances of some relation p.

Existentially indefinite propositions will be designated by an empty circle. For example, the proposition 'Martians like music' is indefinite since at the level of our present knowledge we cannot decide whether or not such a connection exists. There is but one relation which can be asserted of the above proposition, namely: p is identical with p.

If the existence of a connection is to be considered, then it must be indicated. Accordingly, first of all, the following two versions are possible: 1. nep, 2. ep (ne = non-existent connection, e = existent connection). They can be illustrated by circles where the shade-lines imply non-existence, and the dot existence. The basic relations of existentially indefinite propositions are: (1) nep is identical with nep, (2) nep contradicts ep. (3) ep contradicts nep, (4) ep is identical with ep. These four relations can be reduced to two: (a) the propositions are compatible in cases (1) and (4), (b) they are incompatible in cases (2) and (3).

Circles are equally suited to represent the relations of propositions and terms as well. Überweg illustrated the subaltern relation, or subor-

---

[13] Drobisch, pp. 56–57.

dination, between the antecedent (A) and the consequent (B) of a hypothetical proposition as follows:[14]

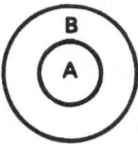

Fig. 34.

The use of this designation has the following justification: those instances (situations) in which connection *A* occurs are included in the extension of those instances in which connection *B* occurs. E.g., all instances when it rains are subordinated to those instances when the pavement becomes wet. It follows from the nature of subordination that there are instances when the pavement becomes wet without rain.

It is more reasonable to use the expression 'formal implication' instead of 'subordination', on the one hand, because the former has come into general use for denoting the above-mentioned relation of propositions, and on the other, because the expression '*p* implies *q*' requires less explanation than '*p* is subordinated to *q*'. They are, however, essentially the same: if there is *p* there is also *q*, but if there is *q* it is not sure that there is also *p*.

Another expression should also be modified on this occasion. I shall speak of admittance instead of crossing. The formula '*p* admits of *q*' expresses the relation where the propositions are compatible (not-disjunct) but exclude implication, e.g., 'if today is Tuesday, then we are in Belgium'.

The relation of two propositions will usually be indicated by the formula *R(p, q)*. Individual relations will be illustrated in Figure 12. According to the above, the relations represented there are to be understood as follows:

(1)   *nep* admits of *neq*
(2)   *neq* implies *neq*

[14] Überweg, p. 214.

(3)   *neq* implies *nep*
(4)   *nep* is identical with *neq*
(5)   *nep* is disjunct to *neq*
(6)   *nep* is disjunct to *eq*
(7)   *ep* is disjunct to *neq*
(8)   *ep* is disjunct to *eq*
(9)   *ep* is identical with *eq*
(10)   *ep* implies *eq*
(11)   *eq* implies *ep*
(12)   *ep* admits of *eq*

Let us examine also the formulae: p, and not p. There is a relation of contradiction between these formulae: if one of them is t, then the other is f. Let us now take a universe which contains all possible propositions. The above formulae can be illustrated like this in this universe:

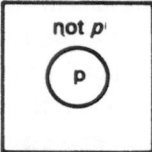

Fig. 35.

14. CONDITIONAL PROPOSITIONS
AND INFERENCES

I

It is Theophrastus (and, to a lesser degree, Eudemus) who has usually been credited with the discovery of conditional propositions[15] and syllogisms based on them. Such propositions, in fact, can be found as early as with Aristotle:

... if no pleasure is good, then no good will be pleasure.[16]

[15] I prefer the expression 'conditional' to 'hypothetical' because, while the latter refers to the subjective nature of the relation between propositions, the former insists on the objectivity of this connection.
[16] Aristotle. 1928. 25a7.

Even the thought of conditional inference had occurred to him:

> ... if a man should suppose that unless there is one faculty of contraries, there cannot be one science, and should then argue that not every faculty is of contraries, e.g. of what is healthy and what is sickly: for the same thing will then be at the same time healthy and sickly. He has shown that there is not one faculty of all contraries, but he has not proved that there is not a science. And yet one must agree. But the agreement does not come from a syllogism, but from an hypothesis... Many other arguments are brought to a conclusion by the help of an hypothesis; these we ought to consider and mark out clearly. We shall describe in the sequel their differences, and the various ways in which hypothetical arguments are formed...[17]

The realization of this goal, however, cannot be found in his still extant works, and the task to elaborate the details has fallen to his posterity.

Conditional propositions, as interpreted by the peripatetics, express a conditionally existing connection. This view appears in various formulations right through the history of logic. Let us see one of its many versions:

> A conditional proposition may be further described as one which makes a statement under a certain condition or qualification restricting its application. In the hypothetical form this condition is introduced by the conjunction *if*, or some other word equivalent to it. Thus:
>
> 'If iron is impure, it is brittle'
>
> is a hypothetical proposition consisting of two distinct categorical propositions, the first of which, 'Iron is impure', is called the *antecedent*; the second, 'It is brittle', the *consequent*. In this case 'impurity' is the condition or qualification which limits the application of the predicate brittle to iron.[18]

Does the above interpretation hold true of all conditional propositions? In order to answer this question the following should be taken into consideration: some logical works mention two kinds of conditional propositions. One of them is the consecutive proposition, e.g., 'some writers are women, hence some women are writers'. The other is the causal

[17] *Op. cit.*, 50a20–26, 39–50b2.
[18] Jevons, 1872. pp. 160–161.

proposition, e.g., 'the glass has broken because it fell off the table'. None of these propositions deals with a conditionally existing connection but rather with established facts.

The components of a conditional proposition were regarded as variables as early as by Aristotle who designated them by letters (if A, then B). The Stoics, contrary to the peripatetics, introduced ordinary numbers, and therefore the conditional proposition took the following form: if the first, then the second.

There is a debate in logical literature whether conditional propositions are simple or compound propositions. The definition of compound propositions has been discussed at length in our previous chapter. Referring back to it, I regard conditional propositions as compound because they can be broken down into simple propositions.

Aristotle defined conditional propositions as a connection of theses where one inevitably entails the other.—The Stoics, on the contrary, allowed for the replacement of variables by any proposition.

It was generally Aristotle's view which had been followed in traditional logic.

> If there is no objective connection between its two components the proposition said to be conditional is a conditional sentence only grammatically, but not logically. Thus, 'if the sun rises, then the tree is in the garden' is a conditional sentence grammatically but not a conditional proposition logically because there is no objective connection between the two component propositions. Erroneous thinking operates with many such conditional propositions, e.g., 'if the sun rises, I shall be with my friend'. It is a completely false statement because what I wanted to express with my sentence is not a condition but a date. Correctly I should say: 'by the time the sun rises, I shall be with my friend'...
>
> In the same way a clear distinction should be made between conditional propositions and those grammatically conditional sentences which express volitional facts, for propositions always reflect the facts of thought and never those of the will. Thus: 'if it does not rain tomorrow, then I shall go for a walk' is a conditional sentence grammatically but not a conditional proposition logically because it conveys a volitional fact and there is no objective connection between the rain and my going for a walk, the latter being an essential requirement of conditional propositions.[19]

[19] Huszár, p. 110.

Mathematical logicians, on the contrary, have followed the line of the Stoics:

> The only aspect of content that we still hold onto is the property of statements of being true or false. Aside from this prerequisite we attach no further limitations on statements and their compounding, so that we also admit the following notorious compound statements, which in terms of content are obviously completely senseless:
>
> (a)   If $1 + 1 = 2$, then copper is a metal.
> (b)   If $1 + 1 = 2$, then copper is not a metal.
> (c)   If $1 + 1 = 3$, then copper is a metal.
> (d)   If $1 + 1 = 3$, then copper is not a metal.[20]

I agree that the variables may be replaced by any proposition. Accordingly, the example 'if it does not rain tomorrow, then I shall go for a walk' is not only a conditional sentence but also a conditional proposition. What the value of this proposition depends on is, of course, another question.

II

The Megarian Stoic school raised the question, *what is the value of a conditional proposition on the whole*? They disagreed among themselves with regard to the answer. The Megarian Philo distinguished the following instances on the ground of two simple propositions:

	Antecedent	Consequent	If the antecedent, then the consequent
(1)	*t*	*t*	*t*
(2)	*t*	*f*	*f*
(3)	*f*	*t*	*t*
(4)	*f*	*f*	*t*

[20] Klaus, p. 73.

He quoted the following examples to illustrate them:

(1)   If it is day-time, then there is light.
(2)   If it is day-time, then it is night.
(3)   If the Earth flies, then the Earth exists.
(4)   If the Earth flies, then the Earth has wings.

He gives no explanation, however, why these very values are arrived at.[21]

Diodorus maintained that Philo's evaluation does not hold true as soon as the implication refers to a definite time, e.g., the present. This view will be explained later in detail when the relation of implication and time operators will be analysed.

Let us quote, finally, a third opinion:

> Those who judge (implication) by what is implicit (ἐμφάσει κρίνοντες) say that the connected (proposition) is true when its consequent is potentially (δυνάμει) contained in the antecedent. According to them the (proposition) 'if it is day, it is day' and every repetitive connected (proposition) is probably false, since nothing can be contained in itself.[22]

As the above quotations and other sources that have come down to us reveal, a hot controversy was going on among the Stoics on this question. The remark of Callimachus, a librarian in Alexandria in the 2nd century B.C., had become a common saying: even the crows on the roofs keep on cawing which are the true implications.

Traditional logic adopted mainly the following approach to the evaluation of conditional propositions as a whole: either no opinion was pronounced on the subject or some of the views of the Stoics were taken over. The word 'mainly' has been used because there were, however rare, exceptions.

> Since conditional propositions express actual statements they are either true or false. *A conditional proposition is true only if the proposition as a whole is true on the ground of the objective connection it expresses* and not if only the individual proposition in it are true since a conditional proposition

---

[21] Cf.: Sextus Empiricus: *Adversus Mathematicos* in Prantl, Vol. I. p. 454.
[22] Quoted in: Bochenski, p. 136; Ivo Thomas tr., p. 119.

conveys not only the truth of its component propositions but rather the objective connection of these two components as the second is but a consequent of the antecedent, this being the only sense of a conditional proposition. E.g., 'if the lamp is burning, then it is evening'. Here both components may be true, i.e., the lamp may be burning and it may be evening as well, still the conditional proposition as a whole is false because there is no objective connection between the burning of the lamp and the evening. That is, the lamp may be burning and still it is not evening, and it may be evening and still the lamp is not burning. But if one says: 'if the sun is set, then it is evening' this conditional proposition is true because there is an objective connection between the sunset and the evening.[23]

Here the value of conditional propositions is related to the connection, and not to the values, of their components. By objective connection, as the second example has revealed, the necessary relation of the antecedent and the consequent is meant. Further on I shall demonstrate the limitations of this conception.

Another author, Tavanets, writes as follows:

... a conditional proposition is true if the supposition that the phenomenon referred to in the antecedent of the proposition necessarily exists in reality implies the assertion that the phenomenon referred to in the consequent of the proposition exists as well. Similarly, a conditional proposition is true also if the supposition that the phenomenon referred to in the consequent does not exist in reality necessarily entails the assertion that the phenomenon referred to in the antecedent does not exist, either. A conditional proposition is, however, false if the supposition that the phenomenon referred to in the antecedent exists does not imply the assertion that the phenomenon referred to in the consequent also exists. Similarly, a conditional proposition is false also if the supposition that the phenomenon predicated in the consequent does not exist does not entail the assertion that the phenomenon predicated in the antecedent does not exist, either.[24]

Here, too, conditional propositions are restricted to the necessary relation of antecedent and consequent. And the question what value a conditional proposition will obtain by supposing the nonexistence of the

[23] Huszár, p. 112.
[24] Tavanets, pp. 126–127.

22

antecedent and the existence of the consequent has not been answered, either.

Mathematical logic has developed, among others, the following view:

> An affirmation of the form 'if p then q' is commonly felt less as an affirmation of a conditional than as a conditional affirmation of the consequent. If, after we have made such an affirmation, the antecedent turns out true, then we consider ourselves committed to the consequent, and are ready to acknowledge error if it proves false. If on the other hand the antecedent turns out to have been false, our conditional affirmation is as if it had never been made.
>
> Departing from this usual attitude, however, let us think of conditionals simply as compound statements which, like conjunctions and alternations, admit as wholes of truth and falsity. Under what circumstances, then, should a conditional as a whole be regarded as true, and under what circumstances false? Where the antecedent is true, the above account of common attributes suggests equating the truth value of the conditional with that of the consequent; thus a conditional with true antecedent and true consequent will count as true, and a conditional with true antecedent and false consequent will count as false. Where the antecedent is false, on the other hand, the adoption of a truth value for the conditional becomes rather more arbitrary; but the decision which proves most convenient is to regard all conditionals with false antecedents as true.[25]

As a first step towards the solution of the problem, let us accept the suggestion of certain logicians that material implication and proposition of the form 'if . . . then' must be distinguished. Let us accept also that the value of a material implication depends on its components only. However, let me choose another procedure for the evaluation of conditional propositions.

My analysis will be based on the diagrams of Figure 12, but S and P are replaced now by p and q respectively. For simplicity's sake let us begin with cases (5)–(8). In these cases p and q are disjunct, so the proposition concerned is f, e.g., 'if the sun shines, then iron is metal' (diagram 8).

It will be t, however, when the relation between them is identity (9) and implication (10) respectively, e.g., 'if it rains, then the pavement will be wet'.

[25] Quine, pp. 12–13.

In case (11) '*q* implies *p*, therefore *p* implies *q*' is *f*. In case (12) '*p* admits of *q*' is *f* because this relation excludes implication. On the ground of the above said the following table of values can be constructed:

TABLE 55

	(1)	(2)	(3)	(4)	(5)	(6)	(7)	(8)	(9)	(10)	(11)	(12)
If *p*, then *q*	*f*	*f*	*t*	*t*	*f*	*f*	*f*	*f*	*t*	*t*	*f*	*f*

As far as the circular schemes are interpreted as the designations of value sets (the dot means *t*, and the shade-lines mean *f*), the above table reveals the following relations: if both components are *t* (8)–(12), then the proposition at issue is *t/f*. If the antecedent is *t* and the consequent is *f* (7), then it is *f*. The result will be the same if the antecedent is *f* and the consequent is *t* (6). Finally, if both components are *f* (1)–(5), then the result is *t/f*. To make it clearer let me sum it up in a table:

TABLE 56

p	*q*	if *p*, then *q*
t	*t*	*t/f*
t	*f*	*f*
f	*t*	*f*
f	*f*	*t/f*

III

Wolff and his followers were of the opinion that conditional propositions can also be classified according to quality, extension, modality, and time. Drobisch, e.g., lists the following versions: [26]

in all cases (or always) if there is *S*, then there is (not) *P*
in some cases (sometimes) if there is *S*, then there is (not) *P*
in one case if there is *S*, then there is (not) *P*

[26] Drobisch, p. 56.

Others held that this classification does not hold true for conditional propositions.

> Whether the propositions which are represented as antecedent and consequent are affirmative or negative, general or singular, descriptive or explanatory has of course no influence at all on the essence of the assertion; and all attempts to introduce distinction of quantity etc. into hypothetical propositions rest upon a 'confusion of hypothetical propositions with statements on temporal relations'.[27]

In order to disprove this view it is reasonable to give a differentiated analysis of the above problems. The components of a conditional proposition may express existent and non-existent connections respectively:

(1)   if there is p, then there is q
(2)   if there is p, then there is no q
(3)   if there is no p, then there is q
(4)   if there is no p, then there is no q

E.g., if there is an instance when space is curved, then there is an instance when space deviates. The question arises whether the words 'there is an instance' are not unnecessary in this proposition (1). Let me refer to the fact that the designations p and q respectively do not assert anything of the existence of the connection at issue, though, at most, might imply it. (2) I have explained in the foregoing why I find it necessary to make a distinction between the formulae 'p' and 'there is p'.

Conditional propositions may include both affirmative and negative components:

(1)   if p, then q
(2)   if p, then not-q
(3)   if not-p, then q
(4)   if not-p, then not-q

It is the formula 'if not-p, then p' which may arouse interest among the propositions with a negative component. If p is t, then the value of this formula is t according to mathematical logic. This instance was studied

[27] Sigwart, p. 244.

first by a Jesuit scholar of the 16th century (in Łukasiewicz' words) after whom the above evaluation process was named the Clavius rule, or *consequentia mirabilis*.

In so far as the above formula is regarded as a material implication its evaluation is perfect. If, however, it is interpreted as a conditional proposition, then the result is miraculous indeed. That is, if, for example, the proposition 'it rains' is t, then the proposition 'if it does not rain, then it rains' is also t. Whenever there is contradiction between the antecedent and the consequent I find the conditional proposition f.

Let us substitute indefinite propositions for p and q respectively, e.g., if S is P, then S_1 is P_1. It is this formula of the conditional proposition which became most common in traditional logic. The values of this formula, as stated with the help of Figure 12, successively agree with the values of the formula 'if p, then q'.

Let us try now to replace p and q by quantified propositions:

If all S is P, then some S is P
If all S is P, then no S is P

The first proposition can be illustrated by diagram 10 and the second by Diagram 8, and their values are t and f, respectively.—The fact that the consequent is also quantified in the above formulae should be noted as well, since it is still a question under debate whether or not the predicate of simple propositions can be quantified. If, however, the analogy between simple and compound propositions is accepted and the quantification of the second component of compound propositions is regarded as meaningful, then there is no reason to protest against the quantification of the predicate.

There are conditional propositions whose components are modal propositions. Here also, let me mention but two out of the numerous versions:

If it is necessary that S is P, then it is a fact that S is P
If it is impossible that S is P, then it is not a fact that S is P

It is easy to see that immediate inferences have occurred in the previous examples. The relation between the validity of such propositions and the value of conditional propositions will be discussed later.

Sigwart rightly recognized that whether or not a proposition is conditional does not depend on the kind of components involved. However, there may be various relations in a conditional proposition which depend on their components, and knowledge of the structural differences between the components is an essential requirement for the evaluation of conditional propositions.

The following problem arose with regard to the time operators involved in the components of an implication:

> Diodorus says that the connected (proposition) is true when it begins with true and neither could nor can end with false. This runs counter to the Philonian position. For the connected (proposition) 'if it is day, I converse' is true according to Philo, in case it is day and I converse, since it begins with the true (proposition) 'it is day' and ends with the true (proposition) 'I converse'. But according to Diodorus (it is) false. For at a given time it can begin with the true (proposition) 'it is day' and end with the false (proposition) 'I converse', suppose I should fall silent ... (and) before I began to converse it began with a true (proposition) and ended with the false one 'I converse'. Further, the (proposition) 'if it is night, I converse' is true according to Philo in case it is day and I am silent; for it (then) begins with false and ends with false. But according to Diodorus (it is) false; for it can begin with true and end with false, in case the night is past and I am not conversing. And also the (proposition) 'if it is night, it is day' is according to Philo true in case it is day, because, while it begins with the false (proposition) 'it is night', it ends with the true (proposition) 'it is day'. But according to Diodorus it is false because, while it can begin—when night is come—with the true (proposition) 'it is night', it can end with the false (proposition) 'it is day'.[28]

The controversy continued in medieval logic. Pseudo-Scotus made an attempt to consider both aspects.

> Die Konsequenz (*consequentia*) wird so eingeteilt: Die eine ist material, die andere formal. Die formale Konsequenz ist jene, welche mit (in) allen Termini hält, wenn (nur) ähnliche Ordnung (*dispositio*) und Form der Termini bestehen... Die materiale Konsequenz ist jene, welche nicht mit

---

[28] Sextus Empiricus, *Adversus Mathematicos*. in Bochenski, p. 135; Ivo Thomas tr., pp. 117—118.

(in) allen Termini hält, wenn (nur) ähnliche Ordnung und Form (dieser Termini) festgehalten werden, so dass eine Änderung nur der Termini vor sich geht. Und diese Konsequenz ist zweifach: die eine ist schlechthin wahr, die andere ist wahr für jetzt (*ut nunc*). Die schlechthin wahre Konsequenz ist jene, welche auf eine formale durch die Hinzunahme einer einzigen notwendigen Aussage zurückgeführt werden kann. Die materiale für jetzt richtige Konsequenz ist jene, welche auf eine formale durch die Hinzunahme einer wahren kontingenten Aussage zurückgeführt werden kann.[29]

Buridan held the following view:

Of material consequences some are said to be consequences simply, since they are consequences without qualification, it being impossible for their antecedents to be true without their consequents. ... And it is to be known that to this kind of consequences *ut nunc* belong permissive consequences, e. g. 'Plato says to Socrates: if you come to me I will give you a horse'. The proposition may be a genuine consequence, or it may be a false proposition and no consequence, since (I) if the antecedent is impossible, viz. because Socrates cannot come to Plato, then the consequence is simply speaking a genuine consequence, because from the impossible anything follows as will be said below. But if (2) the antecedent is false but not impossible, then the consequence is valid *ut nunc*, because from whatever is false anything follows, as will be said later, provided, however, that we restrict the name 'consequence *ut nunc*' to consequences *ut tunc*, whether concerning the past, future, or any other determinate time. But if (3) the antecedent is true, so that Socrates will come to Plato, then perhaps we should say that it is still a genuine consequence because it can be made formal by the apposition of true (propositions), when one knows whatever Plato wills to do in the future, that his wish will persist and that he will be able to carry it out; and when all circumstances are taken account of according to which he wills it and he suffers no hindrance, so that he will be able to and will do what and when he wills; if you then modify this proposition so that is true according to the ninth book of the *Metaphysics*, i. e. 'Plato wills to give Socrates a horse when he comes to him; therefore Plato will give Socrates a horse'. If then these propositions about Plato's will and power are true, then Plato uttered a genuine consequence *ut nunc* to Socrates, but if they are not true he told Socrates a lie.[30]

[29] *Op. cit.*, p. 222.
[30] *Op. cit.*, pp. 224–226; Ivo Thomas tr. 193–195.

To make the comparison of the above opinions easier let us take the case when both components of the implication are *t*. According to Philo the value of such an implication is *t* whatever time operator is involved in the individual components. Diodorus, on the contrary, believed that if the components are in the present tense the result will be *f*. According to Pseudo-Scotus material implications have two versions: one with a general reference, and one with reference to the present. Both implications may be *t* if the conditions he prescribed are fulfilled. Buridan's approach was essentially the same.

I have the following suggestion for the solution of this problem. The starting point should be the nature of the relation between the components of an implication. If all we know is that both components are *t*, then the value of the implication is *t*/*f*. Thus, both Philo and Diodorus were wrong. What does it depend on whether or not an implication has a definite value? Some think the decisive factor is whether the assertion has a general reference or to the present only. But what happens if both kinds of reference are present? Then the implication can be *t*, e. g., 'if always *q*, then now *p*', but also *f*, e. g., 'if now *p*, then never *p*'. Consequently, the difference between the time operators is not an adequate criterion. There can be no such problem with the relations of propositions since there can be but one relation between two definite components of an implication. Since in the proposition 'if it is day-time, then I debate' *p* only admits of *q* but does not imply it, its value is *f*.

IV

I proceed to those operators which belong to conditional propositions as a whole. Let us examine, first of all, whether it is justified to speak of negative conditional propositions. Some deny it:

> ... a proposition expressing the relation of the antecedent and the consequent involves both affirmation and negation as a component: *S* is *P*, or *S* is not *P*. But it does not play the same role as in categorical propositions any more. Here merely the consequence of some occurrence is expressed. Therefore, propositions which have essentially an antecedent-consequent structure have but one, namely *affirmative*, quality because always some connection is asserted in them.[31]

[31] *Logika*. (Budapest 1956) p. 165.

If the assertion that all propositions can be both affirmative and negative is accepted—and there is no reason to doubt it—, then to ignore negative conditional propositions is unjustified in my view. The values of such propositions, under the aspects applied in the previous chapter, are:

TABLE 57

	(1)	(2)	(3)	(4)	(5)	(6)	(7)	(8)	(9)	(10)	(11)	(12)
If *p*, then not *q*	*f*	*f*	*f*	*f*	*t*	*t*	*t*	*t*	*f*	*f*	*f*	*f*

Conditional propositions can be quantified, e. g., in some cases if *p* then *q*. Naturally, any proposition can be substituted for *p* and *q* respectively. Überweg distinguished the following versions: [32]

(1) In all cases if there is *A*, then there is *B*
(2) In some cases if there is *A*, then there is *B*
(3) In no case if there is *A*, then there is *B*
(4) In some cases if there is *A*, then there is no *B*

One need not, of course, restrict oneself to 'all' and 'some', respectively. Any quantifier can be added to a conditional proposition.

Let me select the following types of modal conditional propositions:

(1) It is necessary that if *p*, then *q*
(2) It is a fact that if *p*, then *q*
(3) It is possible that if *p*, then *q*
(4) It is contingent that if *p*, then *q*

Their values are:

TABLE 58

	(1)	(2)	(3)	(4)	(5)	(6)	(7)	(8)	(9)	(10)	(11)	(12)	(13)
(1)	*f*	*f*	*t*	*t*	*f*	*f*	*f*	*f*	*t*	*t*	*f*	*f*	*f*
(2)	*f*	*f*	*t*	*t*	*f*	*f*	*f*	*f*	*t*	*t*	*f*	*f*	*f*
(3)	*f*	*f*	*t*	*t*	*f*	*f*	*f*	*f*	*-t*	*t*	*f*	*f*	*t*
(4)	*f*	*f*	*f*	*f*	*f*	*f*	*f*	*f*	*f*	*f*	*f*	*f*	*t*

[32] Überweg, pp. 213, 218, 220, 223.

Traditional logic generally interpreted the formula 'if p, then q' according to (1) rather than (2). It has led to an unjustified restriction since (2) may be t even in such cases when (1) is f, e. g., it is a fact that if productivity improves production costs decrease.

In mathematical logic the following approach has developed with regard to this problem: it has turned out as early as the beginning of this century that material implication leads to paradoxes in interpretation. To avoid them C. I. Lewis introduced, in his work published in 1918 what he called strict implication, that is, 'if p, then necessarily q'. He believed that strict implication corresponds to the common use of the formula 'if, then' and thereby the above-mentioned paradoxes can be avoided. Since mathematical logic has proved his hope to be unfounded it is unnecessary to dwell upon his theory.

Finally, let me mention the temporal conditional propositions, for example:

(1)   If p, then always q
(2)   If p, then never q
(3)   If p, then sometimes q
(4)   If p, then sometimes not q

v

Now let us compare conditional and categorical propositions. Basically two trends have developed in the history of logic. One of them insisted on their difference, the other on their similarity.

The former school of thought proceeded from the supposition that these two kinds of propositions are related to one another as *conditional to unconditional*.

> Propositions are distinguished into two kinds, according as they make a statement conditionally or unconditionally. Thus the proposition, 'If metals are heated they are softened', is conditional, since it does not make an assertion concerning metals generally, but only in the circumstances when they become heated. Any circumstance which must be granted or supposed before the assertion becomes applicable is a *condition*.[33]

[33] Jevons, p. 62.

To identify categorical and unconditional propositions is just as big a mistake as to identify conditional and hypothetical ones. Let us take the following categorical proposition as an example: 'the pavement is wet'. What makes this proposition unconditional? Perhaps the fact that the necessary conditions are not known or stated? This is simply a factual proposition.

Herbart was the first to object to the view that categorical propositions—by themselves—make a statement on existence. Drobisch argued as follows: It is together with the subject only that the predicate is postulated, but even the subject is not so much postulated (*gesetzt*) but presumed only (*vorausgesetzt*).[34] If we admit that, e. g., the formula 'all *S* is *P*' does not assert anything of existence, then the distinction that categorical propositions refer to existence *by being unconditional propositions* while conditional ones leave this question open *by being conditional propositions* becomes unfounded, too.

The other trend tried to find what the above kinds of propositions have in common. Wolff went further than anybody else when stating that there is only a linguistic difference between conditional and categorical propositions. This view has gained numerous adherents (Herbart, Beneke, etc.).

The idea that the propositions concerned can be mutually obverted into one another comes close to the previous approach. Accordingly, the formula '*all S is P*' is equivalent to the formula '*if S, then P*' (Drobisch, *op. cit.*, p. 56).

Jevons quoted the following example: 'if iron is impure, it is brittle', and obverted it into: 'impure iron is brittle'. The fact that the subject of the antecedent and the consequent is the same term (iron) made the appearance that they, indeed, can be obverted even more convincing. According to Jevons this obversion, though a bit more complicated, is still possible even with different subjects: 'if the barometer is falling, bad weather is coming'—'the circumstances of the barometer falling are the circumstances of bad weather coming'.[35]

---

[34] Drobisch, p. 60.
[35] Jevons, pp. 163–164.

Kant disagreed with this opinion based on the following consideration:

> Some people believe that it is easy to turn a hypothetical statement into a
> categorical statement. But this is not the case, since both are of completely
> different natures. In categorical judgement nothing is problematic,
> everything is assertoric; in hypothetical judgements on the contrary, only
> the consequence is assertoric. In the latter type I can thus join two false
> judgements together; for only the correctness of their coupling is relevant—
> *the form of the consequence,* upon which the logical truth of these
> judgements rests.—There is an essential difference between the two
> statements: All bodies are divisible, and If all bodies are compounded, then
> they are divisible. In the first statement I assert the thing straightforwardly;
> in the latter only under a problematically expressed condition.[36]

Überweg took a middle position between the two extremes. He thought
obversion possible under certain conditions. In his opinion, e.g., the
proposition 'all men are mortal' can be converted into 'if somebody is a
man, then he is mortal'. That is, transformation is possible if there is
inherence (inclusion, subordination) in the categorical proposition and
dependence (implication) in the conditional one.[37]

Überweg was aware of the role relations play in propositions. These
relations have, indeed, something in common and this has created the
illusion that categorical propositions can be converted into conditional
ones. Yet the fact that categorical propositions are simple, and conditional
ones are compound contradicts obversion.

Attempt were made to demonstrate the *congruence of the components* of
these two kinds of propositions. Kant argued that the consequent of a
conditional proposition corresponds to the copula of a categorical one.[38]

Mill wrote as follows:

> What, then, is the subject, and what the predicate of the hypothetical
> proposition? [If the Koran comes from God, Mahomet is the prophet of
> God—this is the proposition under discussion—Gy. T.] 'The Koran' is not
> the subject of it, nor is 'Mahomet'; for nothing is affirmed or denied either
> of the Koran or of Mahomet. The real subject of the predication is the entire

---

[36] Kant, *Schriften.* p. 536.
[37] Überweg, p. 234.
[38] Cf.: Kant, *Schriften* p. 536.

proposition, 'Mahomet is the prophet of God'; and the affirmation is, that this is a legitimate inference from the proposition, 'The Koran comés from God'. The subject and predicate, therefore, of an hypothetical proposition are names of propositions. The subject is some one proposition. The predicate is a general relative name applicable to propositions; of this form—'an inference from so and so' . . .

The distinction, therefore, between hypothetical and categorical propositions, is not so great as it at first appears. In the conditional, as well as in the categorical form, one predicate is affirmed of one subject, and no more: but a conditional proposition is a proposition concerning a proposition; the subject of the assertion is itself an assertion.[39]

The goal of these endeavours was to trace the conditional proposition back to the categorical one, on the one hand, and to demonstrate the congruence of the two kinds of propositions, on the other. To support their theses, both parties misinterpreted the original function of the subject and the predicate, respectively.

As a summary: the attempts at the reconciliation of these two kinds of propositions have proved to be unfounded and arbitrary. Their difference results from one being a simple proposition and the other a compound one rather than from the confrontation of the conditional and unconditional. Their relation is characterized by an analogy to be illustrated by the Venn diagrams. This analogy becomes obvious if we regard the components (terms and antecedent-consequent respectively) as sets and correlate the relations of these very components (e. g., subordination—implication). Then we shall obtain the same values in the corresponding cases.

VI

Traditional logic discussed immediate inferences mainly with regard to categorical propositions only. Why were these operations not analysed also in relation to conditional propositions? As is generally known, mostly the unquantified versions of the latter were investigated, and these versions allow for an immediate conclusion in very few cases only. Therefore let us try to go beyond this very limited field.

[39] Mill, p. 83.

Let us extend our analysis to the following quantified conditional propositions:

(1)   In all cases if p, then q
(2)   In no case if p, then q
(3)   In some cases if p, then q
(4)   In some cases if p, then not q

It requires no further explanation that the same inferences based on the logical square are possible with these propositions as with the corresponding categorical ones.

The operation of conversion is usually treated at length with regard to categorical propositions only, yet it is hardly mentioned in relation to conditional ones. That is presumably due to the fact that quantified conditional propositions cannot be converted. For example, the proposition 'if it is raining, then the pavement is wet' does not imply that 'if the pavement is wet, then it is raining'.

The conversion of quantified conditional propositions was interpreted by Überweg as follows:

(1)   In all cases if there is A, there is B—At least in a part of cases where there is B, there is A
(2)   In some cases if there is A, there is B—In some cases if there is B, there is A
(3)   It never holds true that if there is A, there is B—It never holds true that if there is B, there is A
(4)   In some cases if there is A, there is no B—must not be converted

Accordingly, the same rules of conversion apply to conditional propositions as to categorical ones.[40]

Obversion can be performed even on unquantified conditional propositions: if p, then q, hence if p, then not not-q. Neither is problematic the obversion of quantified propositions, e.g., in all cases if p, then q, hence in no case if p, then not q.

[40] Cf.: Überweg, pp. 214, 218, 220, 224.

The contraposition of the formula 'if p, then q' leads to the proposition 'if not q, then not p'. In the case of quantified propositions the following relations are arrived at:

In all cases if p, then q	→ In all cases if not q, then not p
In no case if p, then q	must not be converted
In some cases if p, then q	must not be converted
In some cases if p, then not q	→ In some cases if not q, then not not-p

Naturally, many more versions are possible. There is no need, however, to dwell upon them since they can be established relatively easily under the above conditions.

The first formulation of the conditional syllogism can be met in Theophrastus (to make it more clear I shall put it in three lines):

> If there is A, then there is B
> If there is B, then there is C
> ―――――――――――――――――
> If there is A, then there is C

Further versions can be constructed by using quantified modal and temporal conditional propositions respectively, e.g.:

> It is necessary that if r, then q
> It is possible that if p, then r
> ―――――――――――――――――
> It is possible that if p, then q

Theophrastus established the following inferences from the combination of conditional and indefinite propositions:

(1)
> If there is A, then there is B
> There is A
> ―――――――――――
> Hence there is B

(2)
> If there is A, then there is B
> There is no B
> ―――――――――――
> Hence there is no A

These versions were called later *modus ponens* (1) and *modus tollens* (2), respectively.

Boethius dealt with conditional syllogisms in two books, and he did it so thoroughly that he has for a long time been considered to be the inventor of these inferences.[41] He increased the two instances introduced by Theophrastus to eight.

If there is *A*, then there is *B*	If there is *A*, then there is no *B*
There is *A*	There is *A*
———	———
There is *B*	There is no *B*
If there is no *A*, then there is *B*	If there is no *A*, then there is no *B*
There is no *A*	There is no *A*
———	———
There is *B*	There is no *B*
If there is *A*, then there is *B*	If there is *A*, then there is no *B*
There is no *B*	There is *B*
———	———
There is no *A*	There is no *A*
If there is no *A*, then there is *B*	If there is no *A*, then there is no *B*
There is no *B*	There is *B*
———	———
There is *A*	There is *A*

In traditional logic the following position has developed: the typical feature of the syllogism concerned is that if two occurrences are related as antecedent and consequent, then in case the antecedent exists the consequent also occurs and when the consequent is not present the antecedent is also absent. The rule according to which one may conclude with the assertion of the consequent from that of the antecedent and from the negation of the antecedent to that of the consequent is built on this very relation.

[41] Cf.: Prantl, Vol. I. p. 700.

Let us see, finally, an inference whose major premise is a modal conditional proposition. This version can be met as early as in Aristotle:

> If it is necessary that *B* should be when *A* is, it is necessary that *A* should not be when *B* is not. If then *A* is true, *B* must be true: otherwise it will turn out that the same thing both is and is not at the same time. But this is impossible.[42]

Naturally, not all modalities provide for a valid conclusion, e.g.:

$$\text{It is possible that if } p, \text{ then } q$$
$$\underline{p }$$
$$\text{Hence } q$$

Traditional logic, implicitly or explicitly, regarded conditional syllogisms as inferences concerning probability upon the consideration that their conclusion is a conditional proposition. It explains the fact that conditional-categorical syllogisms were favoured at the cost of conditional ones. However, what makes an inference syllogistic or probable is not the modality of its conclusion but whether or not the premises unequivocally define the conclusion.

15. DISJUNCTION AND CONJUNCTION

I

Disjunctive propositions and inferences based on them are met first with the Peripatetics. They regarded the disjunctive proposition as one of the conditional kind:

Conditional proposition

Hypothetical proposition   Disjunctive proposition

They thought that the components of a disjunctive proposition must exclude one another. If the major premise is such a proposition and the

---

[42] Aristotle. 1928. 53b12–15.

23

minor premise is a categorical proposition, then the result will be a disjunctive syllogism, e.g., A or B or C; A is B; A is not C.

Aristotle's following statement has presumably contributed to the development of the above interpretation of disjunctive propositions:

> The term 'other in species' is applied to things which being of the same genus are not subordinate the one to the other, or which being in the same genus have a difference, or which have a contrariety in their substance; and contraries are other than one another in species (either all contraries or those which are so called in the primary sense), and so are those things whose definitions differ in the *infima species* of the genus (e.g. man and horse are indivisible in genus, but their definitions are different), and those which being in the same substance have a difference. 'The same in species' has the various meanings opposite to these.[43]

As the Stoics suggested, not only exclusive disjunctions are possible but also what they called permissive disjunction. The sources that have come down to us, however, do not reveal how they interpreted the latter.[44]— Medieval evaluations of disjunctive propositions will be discussed in the next chapter.

Kant was of the following opinion:

> ... the disjunctive judgment contains a relation of two or more propositions to each other, a relation not, however, of logical sequence, but of logical opposition in so far as the sphere of the one excludes the sphere of the other, and yet at the same time of community, in so far as the propositions taken together occupy the whole sphere of the knowledge in question. The disjunctive judgment expresses, therefore, a relation of the parts of the sphere of such knowledge, since the sphere of each part is a complement of the sphere of the others, yielding together the sum-total of the divided knowledge. Take, for instance, the judgment, 'The world exists either through blind chance, or through inner necessity, or through an external cause'. Each of these propositions occupies a part of the sphere of the possible knowledge concerning the existence of a world in general; all of them together occupy the whole sphere. To take the knowledge out of one of these spheres means placing it in one of the other spheres, and to place it

---

[43] Aristotle, *Metaphysica*. 1018a38–1018b8.
[44] Cf.: Bochenski, p. 138.

in one sphere means taking it out of the others. There is, therefore, in a disjunctive judgment a certain community of the known constituents, such that they mutually exclude each other, and yet thereby determine *in their totality* the true knowledge. For, when taken together, they constitute the whole content of one given knowledge.[45]

Accordingly, the components of a disjunctive proposition have to exclude one another, and, if taken together, have to cover all the possibilities in the given respect.

Others, like the Stoics, held that the components of disjunction do not necessarily have to exclude one another.

Thus if we say that 'a good book is valued either for the usefulness of its contents or the excellence of its style', it does not by any means follow because the contents of a book are useful that its style is not excellent. We generally choose alternatives which are inconsistent with each other; but this is not logically necessary.[46]

Mathematical logicians deny the uniqueness of the exclusive disjunction, among others, upon the following consideration:

If we want to establish indisputable instances of the exclusive use of 'or', we must imagine circumstances in which the person who uses 'or' has a positive purpose of denying, explicitly within the given statement, the joint truth of the components. Such examples are rare, but they exist. In an example given by Tarski it is supposed that a child asks his father to take him to the beach and afterwards to the movie. The father replies, in a tone of refusal, 'We will either go either to the beach or to the movie.' Here the exclusive use is clear; the father means simultaneously to promise and to refuse. But it is much easier to find cases in which the nonexclusive interpretation is obligatory. For example, when it is decreed that passports will be issued only to persons who were born in the country or who are married to natives of the country, this does not mean that passports will be refused to persons who were born in the country and are married to natives.[47]

---

[45] Kant, *Kritik* pp. 144 145; Kemp Smith, p. 109.
[46] Jevons, 1872. p. 166.
[47] Quine, pp. 4 5.

Let us sum up what has been said so far. We have seen that the use of the copula 'or' allows for two kinds of interpretation. In one of them disjunction expresses only exclusion, which implies that only one of the components can be true. The other one understands disjunctive propositions primarily as having at least one true component and allows for compatibility of the components (permissive disjunction).

I share the view that both components of a disjunction can be true. Therefore, it is a mistake to restrict it to exclusion. It does not follow, however, from acceptance of both components being true that they must be compatible. I myself mean by the formula 'p or q' that the components, of whatever value, are incompatible, disjunct. For distinction's sake I shall denote the permissive disjunction by the formula 'p and q respectively'.

II

The Stoics made the first attempt at an evaluation of the disjunctive proposition.

> The disjunctive (proposition) consists of (contradictorily) opposed (propositions), e.g. of those to the effect that there are proofs and that there are not proofs. . . . For as every disjunctive is true if (and only if) it contains a true (proposition) and since one of (two contradictorily) opposed (propositions) is evidently always true, it must certainly be said that the (proposition) so formed is true.[48]

It was widely debated in the Middle Ages whether the exclusive or the permissive aspect should be primary in defining the value of disjunctive propositions. Petrus Hispanus wrote the following on this issue: the condition for a disjunctive proposition to be true is that one of its two components must be true, e.g. man is a living being or raven is a stone; or we admit of both components to be true, but it is not peculiar (*proprie*), e.g., man is a living being or a horse is capable of neighing. The falsity of such a proposition requires that both components should be false, e.g., man is not a living being or horse is a stone.[49]

---

[48] Sextus Empiricus, *Adversus Mathematicos*. in Bochenski, *Formale Logik*. p. 137: Ivo Thomas tr., p. 119.
[49] Cf.: Petrus Hispanus, *Summulae Logicales*. Vol. I. (Turin 1947) p. 23.

Burleigh, however, took a contrary position:

> Some say that for the truth of a disjunctive it is always required that one part be false, because if both parts were false it would not be a true disjunctive; for disjunction does not allow those things which it disjoins to be together, as Boethius says. But I do not like that. Indeed I say that if both parts of a disjunctive are true, the whole disjunctive is true. And I prove it thus: If both parts of a disjunctive are true, one part is true; and if one part is true, the disjunctive is true. Therefore (arguing) from the first to the last: if both parts of a disjunctive are true, the disjunctive is true. ...
>
> I say therefore, that for the truth of a disjunctive it is not required that one part be false.[50]

All I want to note here is that the above proof goes in a circle.

Our subject is commented in mathematical logic, among others, in this way:

> ... it is inadmissible to restrict the composition of true or false propositions using the logical particle 'or' by demanding that some kind of inner relationship exist between the propositions compounded; we would in any case find it difficult to say precisely what such a relationship is. This means that we must also consider as logically sensible such statements as, 'It is raining outside or I am giving a lecture.' In fact it gets even worse. On the basis of what has been said, no matter how psychologically repugnant it may seem, we are compelled to consider the first three of the following four statements not only to be sensible but also to be *true*, and the fourth statement to be at least *sensible* even if it is false!
>
> (a) $2 \times 2 = 4$ or Berlin is the capital of Germany.
> (b) $2 \times 2 = 4$ or Berlin is the capital of China.
> (c) $2 \times 2 = 5$ or Berlin is the capital of Germany.
> (d) $2 \times 2 = 5$ or Berlin is the capital of China.[51]

The values of these propositions—following the above sequence—were defined in mathematical logic like this:

---

[50] Burleigh, *De puritate artis logicae*. in Bochenski, *Formale Logik*. pp. 228–229; Ivo Thomas tr. p. 197.
[51] Klaus, pp. 61–62.

TABLE 59

p	q	(1)	(2)	(3)
t	t	t	f	f
t	f	t	t	t
f	t	t	t	t
f	f	f	f	t

Let us see now another interpretation of disjunctive propositions according to which they express the incompatibility of their components. The evaluation should be based on the same relations as in the case of conditional propositions. In this way the following values are obtained:

TABLE 60

	(1)	(2)	(3)	(4)	(5)	(6)	(7)	(8)	(9)	(10)	(11)	(12)
p or q	f	f	f	f	t	t	t	t	f	f	f	f

Let me note that disjunctive propositions, like conditional ones, can also be combined with quantifiers, modalities, etc., for example, it is necessary that p or q. But I shall not dwell upon these combinations.

Despite the lack of direct sources one can presume that the equivalence of the formula 'p or q' to the formula 'if not p, then q' was established by the Stoics (cf.: Bochenski: *op. cit.*, p. 138).—Boethius traced disjunctive propositions back to conditional ones in the following way: [52]

Either there is A, or there is B	$\rightarrow$	If there is no A, then there is B
Either there is no A, or there is no B	$\rightarrow$	If there is A, then there is no B
Either there is A, or there is no B	$\rightarrow$	If there is no A, then there is no B
Either there is no A, or there is B	$\rightarrow$	If there is A, then there is B

[52] Prantl, Vol. I. p. 719.

This equivalence is beyond doubt as far as the relation of permissive disjunction and material implication is concerned. But the relation between disjunctive and conditional propositions is supposed to be the same. Let us demonstrate on a concrete example where this supposition can lead to.

(1)   Either the pavement is wet, or it is raining (p o q).

(2)   If the pavement is not wet, then it is raining ($p{\rightarrow}q$).

I do not know how others think of it, but I myself find the second proposition rather strange, especially if, relying on the truth of the components of disjunction, this conditional proposition must be accepted as true.

I see here the following equivalence: p o q $\leftrightarrow$ if p, then not q. If the antecedent is an alternative proposition the equivalence is obvious, e. g., glass in not a heat-conductor, or glass is a heat-conductor $\leftrightarrow$ if glass is not a heat-conductor, then glass is not a heat-conductor. There is no problem with exclusive disjunction, either. For example: all chairs are pieces of furniture, or no chair is a piece of furniture $\leftrightarrow$ if all chairs are pieces of furniture, then some chairs are pieces of furniture.

Now the antecedent should be a disjunctive proposition by which I mean that its components are incompatible, disjunct, e. g., a horse has four legs, or snow is white. It would be an obvious mistake to draw from this the conclusion that if a horse has four legs, then snow is not white. If, however, not q is interpreted as a complementary of q, then the conditional proposition would sound like this: if a horse has four legs, then we do not assert that snow is white. The result is, of course, not very edifying, though not false, either. I had to consider it in order to demonstrate the general validity of the above equivalence.

III

Disjunctive propositions can be obverted and converted respectively:

Obversion: p or q $\leftrightarrow$ p or not not-q
Conversion: p or q $\leftrightarrow$ q or p

No contraposition can be performed, however; that is, the formula 'p or q' does not imply 'not-q or not-p'. E. g., this book is interesting, or no book is interesting (*t*)—some books are interesting, or this book is not interesting (*f*).

A premise composed of disjunctive propositions does not lead to a correct conclusion, e. g.:

> No *S* is *P*, or some *S* is *P*
> All *S* is *P*, or no *S* is *P*
> ———————————————————
> All *S* is *P*, or some *S* is *P*

Disjunctive syllogism can be found first with Theophrastus who distinguished their two modes.[53]

(1)	*A* or *B* or *C*		*A* or *B* or *C* or *D*
	A is *B*	respectively	*A* is *B*
	A is not *C*		*A* is neither *C*, nor *D*
(2)	*A* or *B* or *C*		*A* or *B* or *C* or *D*
	A is not *B*	respectively	*A* is neither *B*, nor *C*
	A is *C*		*A* is *D*

Formula (1) was called affirmatively negative mode (*modus ponendo tollens*) in traditional logic. The validity of this inference is based on the following connection: if two propositions are incompatible, then the assertion of one of them entails the negation of the other.

> *p* or *q*
> *p*
> ———————
> not *q*

Formula (2) was called negatively affirmative mode (*modus tollendo ponens*). If the major premise is a disjunctive proposition in this mode, then the latter is false, as has been pointed out by many, e. g.:

> All books are interesting, or no book is interesting.
> Not all books are interesting.
> ———————————————————————
> Hence no book is interesting.

---

[53] *Op. cit.*, p. 388.

However, this mode becomes correct as soon as the major premise is a permissive disjunction, since it implies that at least one of the components is true. Thus, if the minor premise asserts the falsity of one of the two components, then the other component must be true.

p and q respectively

not p
_____

q

If, however, the major premise is a permissive disjunction, then no valid inference is arrived at in the affirmatively negative mode.

An inference is called conditional disjunctive syllogism when one of the premises consists of two or more conditional propositions and either the other premise, or the conclusion, or both of them, are disjunctive propositions. Depending upon the number of the conditional propositions, they are distinguished into dilemma, trilemma, and polylemma.

Dilemmas were probably known already to the Stoics, but their logical analysis was carried out only by the Roman rhetors.[54] The standpoint of traditional logic can be summed up like this: there are two main kinds of conditional disjunctive syllogism, namely the simple and the compound modes. In the simple mode the conclusions of conditional propositions are identical:

If S is P, then S_1 is P_1, if S_2 is P_2, then S_1 is P_1

Either S is P, or S_2 is P_2
_____

Hence S_1 is by all means P_1

The compound form of conditional disjunctive syllogisms is characterized by a difference between the conclusion of the conditional propositions involved. In this kind of syllogisms both the minor premise and the conclusion are disjunctive propositions. For example:

If the student studies the night before the examination, then he will be tired at the examination, but if he does not study, he will sit for it unprepared.

_____

[54] Cf.: *op. cit.*, p. 510.

The student either will study the night before the examination, or will not study.

Consequently, he will either be tired at the examination, or will sit for it unprepared.

Its formula is:

If S is P, then S_1 is P_1, if S_2 is P_2, then S_3 is P_3
Either S is P, or S_2 is P_2

Consequently, either S_1 is P_1, or S_3 is P_3

Let us see, first of all, the distinction between the two modes. So far I have called compound modes those which can be broken down into simple ones. Here we have seen that the conclusions of the conditional propositions are identical in one of the versions and different in the other. In my view this is not a question of relation of the simple to the compound. It is rather that the former is a specific case of the latter. If those who applied these terms had been consistent they should have called 'if S is P, then S_1 is P' a simple conditional proposition and the formula 'if S is P, then S_1 is P_1,' a compound one. I think the modes concerned do not require special designations.

IV

The so-called reversible dilemmas raised many problems already in ancient times. Let us see one of them. Protagoras took up the education of a young man named Euathlus under the condition that the pupil would pay half of the tuition when the teaching was finished, and the other half when he would have won his first lawsuit. Having finished his education, Euathlus paid half of the tuition. He refused to pay the other half, saying that he had lost interest in the legal career, that he would not take up any lawsuit, and that according to their agreement he had to pay the other half of the tuition only after having won his first lawsuit.

Protagoras then sued his pupil upon the following consideration: if he wins the lawsuit, then the court will adjudge the required sum to him, and if he loses it his pupil will have won his first lawsuit, so he is compelled to pay according to their agreement.

The pupil argued as follows: if I win the suit it means that the court rejects Protagoras' demand, and if I lose it I am not obliged to pay according to our agreement.

Gellius, having related this story, concludes that the judges considered the arguments of both parties unaccountable and unsolvable, and in order not to bring in a sentence which—whichever party was favoured—by itself would cause further debate, they did not pronounce on the matter but postponed the suit for the unforeseeable future.

The explanations suggested during the history of logic try, first of all, to detect a logical fallacy in this dilemma. They do so in the hope that by avoiding this fallacy the dilemma would be solved in a way that the teacher, himself a lawyer, would get the other half of the tuition fee. Some think, for example, that the mistake consists in the fact that the original agreement was not a complete disjunctive proposition: there were but two alternatives involved, namely that Euathlus either wins or loses his first lawsuit, and the possibility that Euathlus might not want to be a lawyer and consequently would not take up any lawsuit was left out.

However, completeness is not reached even in this way since the young man, having finished the course, might have sued his teacher for not having taught him properly, thus claiming to be absolved from his obligation to pay for the tuition. But, after all, with the requirement of completeness fulfilled, can one really solve this problem?

Before answering this question let me repeat that traditional logic interpreted disjunction as an exclusion. It means that a disjunction is valid only if either one or the other component is true. A conditional disjunctive syllogism, however, as we have in the example with the lawyer and his pupil, may lead also to a disjunctive conclusion where both of the components are true. It should have been a warning that the traditional interpretation of disjunction is too restricted. But instead of correcting the interpretation itself there was an attempt to apply the principle of exclusion in this case as well—in the endeavour to answer the question as to whether or not Euathlus was compelled to pay his tuition.

Let us transcribe the dilemma into an explicit form and then examine this way of putting the question:

> If the lawyer wins the lawsuit, then he will get the other half of the tuition fee, but if he does not win it, then he will also get the other half of the tuition fee.

> The lawyer either wins the lawsuit, or does not win the lawsuit.

> The lawyer gets the other half of the tuition fee.

It is obvious that the antecedents of these conditional propositions negate one another (he wins it—he does not win it). But mutually contradictory antecedents cannot lead to the same conclusion with the same degree of necessity. If the winning of the case necessarily entails the payment of the tuition fee, then losing it cannot necessarily entail the same, and the other way round. The possibility exists that the lawyer gets his money regardless of whether he wins the case or not, but it is excluded that he would get it necessarily in both cases.

To make these considerations easier to understand let us choose a simpler example:

> If it is raining, then the pavement becomes wet, and if it is not raining, then the pavement becomes wet.

> It is either raining, or not raining.

> The pavement becomes wet.

The pavement may be wet if it is raining and also if it is not raining. In the first case the antecedent necessarily entails the consequent. In the latter, however, the pavement becomes wet not because it was raining but for some other reason (e. g., the street has been sprinkled).

Let us return to our original example: if the lawyer loses the case it does not follow from it that he will get his money. It might follow from the agreement only according to which the young man, having won the case, is obliged to pay. If the case is seen from Euathlus's point of view, the result will be the same:

> If the young man wins the case, then he does not have to pay, and if he loses the case he does not have to pay.

> The young man either wins the case, or loses the case.

> The young man does not have to pay.

In the foregoing I have tried to demonstrate that if the antecedents of conditional propositions are contradictory they cannot lead to the same conclusion with equal necessity. Thereby I have implicitly presumed a necessary relation between the antecedent and the consequent at least in one of the cases, e. g., if the lawyer wins the case, then he gets the other half of the tuition fee. But does this presumption prove to be true indeed? The answer should be "no" for it might happen that the pupil dies after the trial and the lawyer has won the case in vain as he will not get his money anyway.

In the example of light being refracted if it passes through a prism, there is a necessary connection between the antecedent and the consequent. The winning of the case, however, does not inevitably guarantee that the lawyer can claim his due. Consequently, it may occur that he will not get any money whether he wins the case or not. The answer to the question whether or not the young man has to pay the other half of the tuition fee is: it does hold true in any of these cases that he *has to*. Under the given conditions it cannot be decided unequivocally what consequence will follow. It is possible in both cases that the young man will pay, but it is just as possible that he will not.

The conclusion of a dilemma is unconditionally true if at least: (1) there is a necessary connection between the antecedent and the consequent in the conditional proposition, (2) if the consequents are in a relation of incompatibility, e. g.:

> If a river flows into an inland sea, then it has a river delta, and if it does not flow into an inland sea, then it has an estuary.

> The Danube flows into an inland sea, or it does not flow into an inland sea.

> Thus the Danube has either a river delta, or an estuary.

Now let us return to the formula of conditional disjunctive syllogisms:

If p then q, if r then s

p or r

q or s

Since any proposition can be substituted for these variables, this formula is not universally valid. Its conclusion can be equally true and false. Now it is easy to see that the validity of the syllogism concerned could not have been decided on the ground of this formula. Jevons commented on this issue as follows:

> Dilemmatic arguments are however more often fallacious than not, because it is seldom possible to find instances where two alternatives exhaust all the possible cases, unless indeed one of them be the simple negative of the other in accordance with the law of the excluded middle. Thus if we were to argue that 'if a pupil is fond of learning he needs no stimulus, and that if he dislikes learning no stimulus will be of any avail, but as he is either fond of learning or dislikes it, a stimulus is either needless or of no avail', we evidently assume improperly the disjunctive minor premise. Fondness and dislike are not the only two possible alternatives, for there may be some who are neither fond of learning nor dislike it, and to these a stimulus in the shape of rewards may be desirable. Almost anything can be proved if we are allowed thus to pick out two of the possible alternatives which are in our favour, and argue from these alone.[55]

But why is this inference fallacious? Because logicians accept as correct only those inferences whose conclusion is necessarily true. In my view this criterion is too narrow. An inference is correct if, proceeding from the given premises, we can state the value of the conclusion. Let us take the following old dilemma as an example: The crocodile has kidnapped the child of a mother and is ready to return it under the condition that the mother gives a true answer to the question whether it will return the child. The mother answers: you will not return the child. Then the crocodile says: if you said the truth, then I do not return your child for if I returned it it would not be true what you have said. But if you did not say the truth I do not return your child because you have not fulfilled the condition of giving a true answer. Consequently, I cannot return your child in any case.

The mother gives the following answer: if I said the truth, then you have to return my child according to our agreement. If I did not say the truth, then you have to return the child for if you did not give it back it would be true what I have said.

[55] Jevons, 1872. pp. 168–169.

From the point of view whether or not its conclusion is necessarily true this inference cannot be described as correct. If, however, the value of the conclusion (the crocodile does or does not return the child) is said to be f/t. then it has correctly stated what follows from the given premises.

When *fallacies* are discussed in logical literature, often dilemmas are quoted as examples because these inferences only *seem* to be correct— according to the usual criterion. Such cases as the 'lawyer', the 'crocodile', etc., are, indeed, fallacies when presented as if conditions (1) and (2) described above were provided for. If they are not presented in this way, but the question what conclusion follows from the given premises is to be answered, the above examples cannot be regarded as fallacies any more.

v

The Stoics were the first to deal with conjunction. They regarded the conjunctive (or copulative) proposition as a compound proposition which consists of simple theses. In their opinion a conjunctive proposition is true if each of its components is true.

For the Stoics, conjunction and implication were of equal rank. Later on, however, conjunction was pushed into the background so that Boethius, the systematizer of ancient logic, did not even mention it. According to Avicenna and Algazali hypothetical and disjunctive propositions convey a uniform thought while conjunctive propositions are merely aggregates.

Mill has put this view into the following words:

> ... what is called a complex (or compound) proposition is often not a proposition at all, but several propositions, held together by a conjunction. Such, for example, is this: Caesar is dead, and Brutus is alive: or even this, Caesar is dead, *but* Brutus is alive. There are here two distinct assertions; and we might as well call a street a complex house, as these two propositions a complex proposition. It is true that the syncategorematic words *and* and *but* have a meaning; but that meaning is so far from making the two propositions one, that it adds a third proposition to them. All particles are abbreviations, and generally abbreviations of propositions; a kind of short-hand, whereby something which, to be expressed fully, would have required a proposition or a series of propositions, is suggested to the mind at once.

Thus the words, Caesar is dead and Brutus is alive, are equivalent to these: Caesar is dead; Brutus is alive; it is desired that the two preceding propositions should be thought of together.[56]

For my part I share the opinion of those who regard conjunction, similarly to implication and disjunction, as a compound proposition.

Modern logic (first, the Port-Royal logic) has introduced the following distinctions: propositions which have several subjects and one predicate (S_1 and S_2 are P) were called copulative and those which have one subject and several predicates (S is P_1 and P_2) conjunctive.

If this distinction has any significance at all, it must be completed:

(1) S is P and S is P (identical subjects and identical predicates)
(2) S is P and S_1 is P (different subjects and identical predicates)
(3) S is P and S is P_1 (identical subjects and different predicates)
(4) S is P and S_1 is P_1 (different subjects and different predicates)

Mathematical logicians, following the Stoics, make the value of a conjunctive proposition depend on the values of its components. Accordingly, the following table of values is obtained:

TABLE 61

p	q	$p \cdot q$
t	t	t
t	f	f
f	t	f
f	f	f

Let us now see argumentation with proof included:

If both p and q are true, that is, if what the propositions assert obtains in reality, then the truth of the proposition $p \cdot q$ means that the states of affairs, which are reflected by both propositions *exist together*. Co-existence in reality corresponds to logical conjunction. If on the other hand one of the propositions is false, then the correspondent in reality to the fact that the compound proposition is false is the fact that the state of affairs p (assuming

---

[56] Mill, p. 82.

p is the true proposition and q the false) does not co-exist with q but only coexists with $\bar{q}$. It should however be remarked that while the logical co-existence of propositions is abstracted from the real co-existence of states of affairs, the two are not identical.[57]

One of the consequences of the above considerations is that such propositions as 'I am listening to the radio *and* Berlin is situated on the banks of the river Spree' are regarded as true. Upon what grounds?

> But just such an example — as absurd as this may seem (I am listening to the radio, *and* Berlin is on the Spree) since it is obviously of no use for science nor for everyday practice — is along with others particularly suited to clarify the objective-real basis of logical conjunction. *Conjunction is the logical reflection of the co-existence of states of affairs in objective reality.* In logic we abstract from the concrete kind of co-existence, such as causal relation or temporal order, that is, from the necessary inner relationships, which are the subject matter of dialectics, and we consider merely the co-existence of states of affairs as such. Finally, propositional logic is only interested in the question of when a conjunction yields a true or a false compound proposition — independently of the concrete content of the propositions which are coupled and independently of the particular kind of relationship between them.[58]

Let us call the above operation material conjunction to tell it apart from others. The above interpretation makes the evaluation of this operation obvious.

Let us proceed now to what is called a conjunctive proposition in traditional logic. By this I mean a statement whose components are compatible. A conjunctive proposition—on the ground of the relation of its components—will be obviously t, for example, in the following cases:

(9)   The sun is shining and the sun is shining.

(10)   It is raining and the pavement is wet.

(11)   The pavement is wet and it is raining.

(12)   It is good weather and we go for an excursion.

[57] Klaus, p. 48.
[58] Händel-Kneist, pp. 39–40.

24

If, however, the components are disjunct, then the result will be f, e.g., I am listening to the radio and Berlin is situated on the banks of the river Spree. Let us sum it up:

TABLE 62

	(1)	(2)	(3)	(4)	(5)	(6)	(7)	(8)	(9)	(10)	(11)	(12)
p and q	t	t	t	t	f	f	f	f	t	t	t	t

The above values are the negations of the values of the formula 'p or q'. This result evidently follows from my defining conjunction as a relation of compatible propositions and disjunction as that of incompatible ones. (The negation of conjunction can be expressed in mathematical logic as well by using the conjunctive particle 'or'. There, however, this 'or' is called not disjunction but incongruity and is not interpreted like incompatible.)

Conjunctive propositions can be combined with quantifiers, modalities, etc., just as much as the conditional ones, e.g., it is necessary that p and q. I find their detailed analysis unnecessary.

The proposition concerned can be obverted and converted respectively:

Obversion: p and $q \leftrightarrow p$ and not not-q
Conversion: p and $q \leftrightarrow q$ and p

Contraposition, however, cannot be performed on it, e.g., the pavement is wet and it is not raining (t)—it is raining and the pavement is not wet (f).

16. OTHER RELATIONS OF PROPOSITIONS

A

I

Let us examine the following argumentation

If we want to say that the existence of the relation 'A is B' is *necessary* and *sufficient* to conclude to the validity of the relation 'C is D', then the proposition stated is of the following type: if A is B, then, and only then, S is D. This type of proposition is called *exclusive* and consists of two simple propositions: if A is B, then C is D, and if A is not B, then C is not D.[59]

In traditional logic such propositions were dealt with mainly in relation to exclusion but even there only incidentally. Tavanets was one of the few who found this type of proposition important:

> The exclusive conditional proposition plays an important role in the theory of inference. If the major premise of the conditional-categorical inference is an exclusive conditional proposition, then by conditional-categorical inference we can conclude with certainty not only (1) from the presence of the condition to the presence of the conditioned and (2) from the absence of the conditioned to the absence of the condition, but also (3) from the absence of the condition to the absence of the conditioned, and (4) from the presence of the conditioned to the presence of the condition. An example of the inference from the absence of the condition to the absence of the conditioned is: 'If the opposite sides of a quadrangle are parallel, then, and only then, the diagonals of this quadrangle bisect one another. The opposite sides of this quadrangle are not parallel. Hence the diagonals of this quadrangle do not bisect one another.[60]

The formula 'if p, then, and only then, q' is called equivalence in mathematical logic and is held to be true under the condition that both of its components are t or f, respectively.

TABLE 63

p	q	$p \leftrightarrow q$
t	t	t
t	f	f
f	t	f
f	f	t

Let us examine the above formula from the point of view of the relation expressed in it. It allows for the interpretation that its components are conditional upon one another which means that, with an expression of

[59] *Logika.* (Budapest 1956) p. 170.
[60] Tavanets, p. 131.

24*

Latin origin, it is a biconditional proposition. This is its evaluation on the ground of Figure 12:

TABLE 64

	(1)	(2)	(3)	(4)	(5)	(6)	(7)	(8)	(9)	(10)	(11)	(12)
If p, then, and only then, q	f	f	f	t	f	f	f	f	t	f	f	f

Having considered the above said, I find the following distinction necessary: the interpretation and evaluation that mathematical logic has given correspond to the proposition 'p is equivalent to q' but not to the formula concerned. It even leads to paradoxes, at least in the interpretation. Accordingly one should accept as true, e.g., the following: if iron is metal, then, and only then, peach is fruit. I suggest that we interpret and evaluate the formula at issue as described above.

The following figure will be instrumental in revealing the connections between the propositions under discussion:

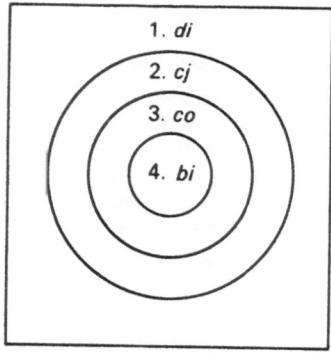

Fig. 36.

The designations are to be understood as follows: di = disjunction, cj = conjunction, co = conditional, bi = biconditional. Ring 2 represents 'only conjunction', ring 3 'only conditional'. Let us sum up these interpretations:

0 — indefinite	2 *o* 3 — *cj*, but not *bi*
1 — *di*	2 *o* 4 — *cj*, but not only-*co*
2 — only *cj*	3 *o* 4 — *co*
3 — only *co*	1 *o* 2 *o* 3 — not *bi*
4 — *bi*	1 *o* 2 *o* 4 — not only-*co*
1 *o* 2 — not *co*	1 *o* 3 *o* 4 — not only-*cj*
1 *o* 3 — *di o* only *co*	2 *o* 3 *o* 4 — *cj*
1 *o* 4 — *di o bi*	1 *o* 2 *o* 3 *o* 4 — *u*

The above relations of propositions assume the following values successively:

TABLE 65

	(1)	(2)	(3)	(4)	(5)	(6)	(7)	(8)	(9)	(10)	(11)	(12)
(1)	f	f	f	f	t	t	t	t	f	f	f	f
(2)	f	f	f	f	f	f	f	f	f	f	f	f
(3)	t	f	f	f	f	f	f	f	f	f	f	t
(4)	f	f	t	f	f	f	f	f	f	t	f	f
(5)	f	f	f	t	f	f	f	t	t	f	f	f
(6)	t	t	f	f	t	t	t	t	f	f	t	t
(7)	f	f	t	f	t	t	t	t	f	t	f	f
(8)	f	f	f	t	t	t	t	t	t	f	f	f
(9)	t	t	t	f	f	f	f	f	f	t	t	t
(10)	t	t	f	t	f	f	f	f	t	f	t	t
(11)	f	f	t	t	f	f	f	f	t	t	f	f
(12)	t	t	t	f	t	t	t	t	f	t	t	t
(13)	t	t	f	t	t	t	t	t	t	f	t	t
(14)	f	t	t	t	t	t	t	t	t	t	t	f
(15)	t	t	t	t	f	f	f	f	t	t	t	t
(16)	t	t	t	t	t	t	t	t	t	t	t	t

This table of values agrees with Table 54. Accordingly, there is an analogy between the compound and the simple relative propositions. Therefore, as far as the inferences based on compound relative propositions are concerned, let me refer to what has been said about the simple ones. However, it should be made clear here if, and to what degree, the value of the propositions discussed depends upon the value of their components. Let us, first of all, agree that in the diagrams the dot represents the set of true propositions and the shade-lines that of the false ones. Diagrams 6 and 7 represent propositions of different values (*t* in one set, and *f* in the other) while the other diagrams contain propositions of the same value.

As Table 65 reveals, versions (3)–(5), (9)–(11), and (15) are t only if their components are equivalent. The other versions (with the exception of the first one) can be *t* even if their components are of different values (but they are *f* only if the components are equivalent). Thus, the value of compound relative propositions depends upon the value of their components, but it cannot be decided on this ground only.

Accordingly, conditional propositions, for example, cannot be *t* if their components are of different values. Consequently, the thesis 'from falsehood, anything may follow' does not hold good in this system.

II

Let us examine the operations of the proposition calculus in a sequence chosen with a later comparison in mind:

In order to judge the position of mathematical logic let us give a trivial interpretation of Table 66 (*comp* = component):

(1)   identically *f*
(2)   both *comp* are *f*; neither one, nor the other *comp* is *t*
(3)   only *q* is *t*
(4)   only *p* is *t*
(5)   both *comp* are *t*; neither one, nor the other *comp* is *f*
(6)   *p* is *f*; at most *q* is *t*
(7)   *q* is *f*; at most *p* is *t*
(8)   both *comp* are *t*, and both *comp* are *f* respectively
(9)   one *comp* is *t*, the other *comp* is *f*
(10)   *q* is *t*; at most *p* is *f*
(11)   *p* is *t*; at most *q* is *f*
(12)   at most one *comp* is *t*; at least one *comp* is *f*
(13)   not only *p* is *t*

[continued p. 361 after Table 66—Ed]

TABLE 66

		pt, qt	*pt, qf*	*pf, qt*	*pf, qf*
(1)	Identically false	*f*	*f*	*f*	*f*
(2)	Negated disjunction	*f*	*f*	*f*	*t*
(3)	Negated converted implication	*f*	*f*	*t*	*f*
(4)	Negated implication	*f*	*t*	*f*	*f*
(5)	Conjunction	*t*	*f*	*f*	*f*
(6)	Negation of *p*	*f*	*f*	*t*	*t*
(7)	Negation of *q*	*f*	*t*	*f*	*t*
(8)	Equivalence	*t*	*f*	*f*	*t*
(9)	Negated equivalence	*f*	*t*	*t·*	*f*
(10)	Affirmation of *p*	*t*	*f*	*t*	*f*
(11)	Affirmation of *q*	*t*	*t*	*f*	*f*
(12)	Negated conjunction	*f*	*t*	*t*	*t*
(13)	Implication	*t*	*f*	*t*	*t*
(14)	Converted implication	*t*	*t*	*f*	*t*
(15)	Disjunction	*t*	*t*	*t*	*f*
(16)	Identically true	*t*	*t*	*t*	*t*

(14)   not only *q* is *t*

(15)   at most one *comp* is *f*; at least one *comp* is *t*

(16)   identically *t*

Their interrelations are illustrated in the following figure (the Arabic numerals refer to the corresponding versions):

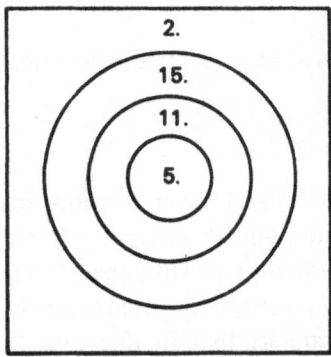

Fig. 37.

Now let us see the same versions expressed by quantifiers (*this* = *p*):

(1)   in all cases *f*
(2)   no *comp* is *t*
(3)   some *comp*, but not *this*, is *t*
(4)   only *this* is *t*
(5)   every *comp* is *t*
(6)   not *this* is *t*
(7)   only *this*, or no *comp*, is *t* respectively
(8)   all or no *comp* is *t* respectively
(9)   some, but not every, *comp* is *t*
(10)   some *comp*, but not only *this*, is *t*
(11)   *this* is *t*
(12)   not every *comp* is *t*
(13)   not only *this* is *t*
(14)   *this* or no *comp* is *t* respectively
(15)   some *comp* is *t*
(16)   in all cases *t*

This interpretation brings to light the analogy between the operations of the propositional calculus and the quantifiers.

Now the interrelations of the above operations will be analysed. Mathematical logic shows great interest in the identities like, e.g., the rules of tautology (the sign of equality implies identity):

$$p \cdot p = p$$
$$p \ o \ p = p$$

Implications, too, have an important role here, e.g.:

$$p \cdot p \rightarrow p \ o \ q$$
$$p \leftrightarrow q \rightarrow p \leftarrow q$$

Nevertheless, the analysis of other relations can be found with few authors only. By carrying out the analysis we arrive at the very same 9 kinds of relations that are contained in Table 15. And there is no reason to give their comprehensive survey as it would coincide with Table 16 (it must be kept in mind, of course, that in this case the table contains the operations of the propositional calculus successively).

Thus, the interrelations of these operations are analogous, among others, to those of the quantified propositions. These interrelations make it possible to perform, e.g., the following inferences:

if $p \cdot q$ is t, then $p \leftrightarrow q$ is t (inference concerning the subaltern) subcontrary)
if $p \cdot q$ is t, then $p \leftrightarrow q$ is t (inference concerning the subaltern)
if p is t, then $p \; o \; q$ is f (inference concerning the contrary)

I think a few examples here is enough, for their overall survey, valid by analogy, has already been given in the first chapter of our part on immediate inferences.

The rules of commutation are treated as further identities:

$$p \cdot q = q \cdot p$$
$$p \; o \; q = q \; o \; p$$
$$p \leftrightarrow q = q \leftrightarrow p$$

When the immediate inferences based on categorical propositions have been discussed we have already seen the following conversion: all S is P, hence some P is S. Can it be correlated to some conversion in the propositional calculus? In my opinion the following: $p \cdot q$, hence $q \; o \; p$. Since these two propositions are in a relation of subordination (if the antecedent is t, then the conclusion is also t, but it does not hold true the other way round) the expression 'conversion into subaltern' can be extended also to the operations concerned. In principle any operation in Table 15 can be combined with conversion. Consequently, the identities discussed as the rules of commutation are specific cases of conversion.

The scope of identities can be widened even further by taking into consideration the negations of the components as well. The operation based on such relations was called obversion in traditional logic, e.g., all S is P, hence no S is not-P. There is a relation of a similar nature in the proposition calculus, e.g., $p \leftrightarrow q$, hence $\sim (p \leftrightarrow \sim q)$.

Contrary to traditional logic which confined itself to the versions with negative minor premise, mathematical logic has thoroughly investigated also the relations based on the negation of the major premise and of all components respectively. The De Morgan identities, for example, belong to them.

Let us compare the operations of the propositional calculus (1) and the compound relative propositions (2). In so far as (1) is interpreted according to the commentary of Figure 37, it is justifiable to disregard the content of propositions, that is, what concrete assertions occur in the components. The definition that the logical value of the result of these operations depends only on the logical value of its components holds also true. Under the given condition there will be no paradox in the interpretation.

Let us try to apply the above aspects in assessing (2). The content of these propositions, however, comprises not only the concrete assertions in their components but also the interrelations between them. To ignore the content of (2), therefore, would imply that the concrete interrelations are also disregarded and the investigation is restricted to a mere registration of the existence of a relation between the components.

But, then, why can we efficiently apply to (2) such operation of (1) as, for example, the decision of the validity of inferences with the help of normal formulas? First of all, because by formulating (2) we switch over to (1) and continue to act according to the latter. Therefore no problem arises in the calculation except when the result is converted back into the language of (2).

But does the fact that there are differences in interpretation, and consequently in evaluation, between (1) and (2) prevent the application of the methods of the former for the latter? In certain cases it does by all means. Let us take the following inference as an example:

> If the pressure on gas decreases, then gas expands.
> If gas expands, then it cools down.
> _____
> If the pressure on gas decreases, then gas cools down.

This inference is formulated according to (1) as $[(p \rightarrow q) \cdot (q \rightarrow r)] \rightarrow (p \rightarrow r)$, and then its validity is proved, among others, with the help of a matrix.

Let us, however, substitute the following propositions for this formula:

If a chair is not a piece of furniture, then snow is white.
If snow is white, then a chair is a piece of furniture.

If a chair is not a piece of furniture, then a chair is a piece of furniture.

This inference is valid according to (1) but not valid according to (2). In consequence the methods of the former cannot be applied to the latter without modification.

Can such restrictions be found which would limitate the specifications of (1) in a way acceptable for (2)? The introduction of strict implication and entailment theory, for example, can be seen as attempts at this goal. It has been pointed out, however, that they failed to solve the above problem.

These attempts were based, among others, on the supposition that (1) and (2) differ in certain cases only (as, first of all, the relation of 'if . . . then' and implication), but otherwise coincide (e.g., the relation of conjunction to propositions with the conjunctive particle 'and'). That is, what they tried to overcome were these differences.

I have tried to demonstrate in the foregoing that propositions of different types rank among (1) and (2) respectively and that, therefore, I find the endeavour to overcome their differences hopeless.

Nevertheless, I see here the following solution: there are already partial achievements in the field of (2) as, for example, the conditional and disjunctive inferences, and they should be completed by taking over from (1) all those aspects which can be used here without the danger of creating paradoxes (e.g., the exploration of identities). However, to make the calculus of (2) complete it is also indispensable to give a thorough analysis of the propositions concerned as regards their specifics.

Let us have a closer look at the following sentences:

The ball is going on, everybody is in high spirits.
The victim was at home, his radio was switched on.

The structure of such linguistic expressions is either not dealt with in logic, or is defined as conjunction.[61] This approach recalls the interpre-

[61] Cf.: Mill, p. 82.

tation of propositions of the form 'S is P' which are described either as
extralogical or as universal propositions.

In my view (1) the above statements are compound propositions as they
can be broken down to simple propositions, and (2) it is unjustified to
regard them as conjunction only because they can be connected also by
other logical conjunctive particles than 'and', e.g.:

> If the victim was at home, then his radio was switched on.
> The victim was either at home, or his radio was switched on.

Since simple propositions can be in any relation to one another I shall
regard the above assertions as indefinite compound propositions and
denote them by the formula 'p; q'. As far as the components of these
propositions are concerned, various statements can be made, e.g.:

> at least one from among p and q is true
> neither p nor q is true
> only q is true from among the two propositions
> either only p, or only q is true
> both p and q are true

All that is to be considered in order to evaluate these propositions is the
values of their components. E.g., the proposition 'neither p nor q is true' is
t if both of its components are f.

A definite compound proposition consisting of two propositions has, on
the contrary, three factors: (a) one proposition (p), (b) the other
proposition (q), (c) the relation of the two propositions (e.g., if ... then).
In the consideration of the values of their components in itself a sufficient
basis for evaluating such propositions? As is well-known, the Venn
diagrams which represent three domains only are not sufficient for
evaluating, for example, the proposition 'all not-S is P', therefore they
should be complemented by the universe as well (i.e., four domains are
needed). It is also well-known that the solution of an equation with three
unknown quantities requires not two but three equations. From these
considerations I think that three criteria are necessary in order to evaluate
a definite compound proposition.

B

I

Causality is one of the the categories which aroused the interest of philosophers as early as in ancient times. Democritus's statement that for him it was of greater value to find the explanation of causal connection than to become the king of Persia is usually quoted as a starting point.

Aristotle made an attempt to define the fundamental types of causes:

> Evidently we have to acquire knowledge of the original causes (for we say we know each thing only when we think we recognize its first cause), and causes are spoken of in four senses. In one of these we mean the substance, i.e. the essence (for the 'why' is reducible finally to the definition, and the ultimate 'why' is a cause and principle); in another the matter or substratum, in a third the source of the change, and in a fourth the cause opposed to this, the purpose and the good (for this is the end of all generation and change).[62]

This view prevailed as long as to the Renaissance. The schoolmen gave the following names to the four types of causes successively: (1) *causa formalis* (formal cause), (2) *causa materialis* (material cause), (3) *causa efficiens* (efficient cause), (4) *causa finalis* (final cause). In addition, the following theses were established:

Causa praecedit effectum (cause is prior to effect).

Causa posita ponitur causatum (the postulation of the cause entails the postulation of the effect).

Causa cessante cessat effectus (if the cause ceases to exist, then also the effect ceases to exist).

Causa causae est etiam causa causati (the cause of the cause is also a cause of the effect).

From the Renaissance onward it was, first of all, the efficient cause among the four Aristotelean types which attracted attention.

> Some of the grounds for the Renaissance reduction of causes to the *causa efficiens* were the following: (a) it was, of all the four, the sole clearly conceived one; (b) hence it was mathematically expressible; (c) it could be

---

[62] Aristotle, *Metaphysica*. 983a24–34.

assigned an empirical correlate, namely, an event (usually a motion) producing another event (usually another motion) in accordance with fixed rules; the remaining causes, on the other hand, were not definable in empirical terms, hence they were not empirically testable; (d) as a consequence, the efficient cause was controllable; moreover, its control was regarded as leading to the harnessing of nature, which was the sole aim of the instrumental (pragmatic) conception of science advocated by Bacon and his followers.[63]

To avoid misunderstandings let me note that it would be a mistake to rank Bacon among one-sided empiricists.

... my course and method, as I have often clearly stated and would wish to state again, is this—not to extract works from works or experiments from experiments (as an empiric), but from works and experiments to extract causes and axioms, and again from those causes and axioms new works and experiments, as a legitimate interpreter of nature.[64]

In the modern age, the mechanistic approach to cause began to widely gain ground. This theory reduced causes to forces which bring about a change of place. Qualitative changes were either ignored, or traced back to mechanical changes as far as possible. Especially Galilei, Newton, and Laplace should be mentioned among the adherents of mechanistic causality.

The adherents of skepticism made use of the shortcomings of this vulgar approach to causality.

The generality of mankind never find any difficulty in accounting for the more common and familiar operations of nature—such as the descent of heavy bodies, the growth of plants, the generation of animals, or the nourishment of bodies by food: But suppose that, in all these cases, they perceive the very force or energy of the cause, by which it is connected with its effect, and is for ever infallible in its operation. They acquire, by long habit, such a turn of mind, that, upon the appearance of the cause, they immediately expect with assurance its usual attendant, and hardly conceive it possible that any other event could result from it. It is only on the discovery of extraordinary phenomena, such as earthquakes, pestilence,

---

[63] Bunge, pp. 32–33.
[64] Francis Bacon, *Works*. Vol. I. p. 212 (*Novum Organum*, 1/CXVII); Anderson tr., p. 107.

and prodigies of any kind, that they find themselves at a loss to assign a proper cause, and to explain the manner in which the effect is produced by it. It is usual for men, in such difficulties, to have recourse to some invisible intelligent principle as the immediate cause of that event which surprises them, and which, they think, cannot be accounted for from the common powers of nature. But philosophers, who carry their scrutiny a little farther, immediately perceive that, even in the most familiar events, the energy of the cause is as unintelligible as in the most unusual, and that we only learn by experience the frequent *Conjunction* of objects, without being ever able to comprehend anything like *Connexion* between them.[65]

Kant, too, assessed causality from the position of subjective idealism, but did not regard it as apparent. "All alterations take place in conformity with the law of the connection of cause and effect."[66] Let me mention also that not only causality but even interaction have been included in his table of categories.

Contrary to this mechanistic interpretation of causality describing cause and effect as factors external to one another, Hegel took the following point of view:

Consequently, *effect contains nothing whatever that cause does not contain.* Conversely, *cause contains nothing which is not in its effect.* Cause is cause only in so far as it produces an effect, and *cause is nothing but this determination, to have an effect, and effect is nothing but this, to have a cause.* Cause as such implies its effect, and effect implies cause; in so far as cause has not yet acted, or if it has ceased to act, then it is not cause, and effect in so far as its cause has vanished, is no longer effect but an indifferent actuality.[67]

Hegel paid especially great attention to the category of interaction. He was a severe critic of the metaphysical approach which studied occurrences by isolating them and was, therefore, one-sided.

When one reads Hegel on causality, it appears strange at first glance that he dwells so relatively lightly on this theme, beloved of the Kantians. Why? Because, indeed, for him causality is only *one* of the determinations of

---

[65] Hume, *Enquiries* pp. 69–70.
[66] Kant, *Kritik.* p. 283; Kemp Smith, p. 218.
[67] Hegel, *Wissenschaft,* p. 704. A. V. Miller tr., p. 559.

universal connection, which he had already covered earlier, in his *entire* exposition, much more deeply and all-sidedly; *always* and from the very outset emphasizing this connection, the reciprocal transitions, etc., etc.[68]

From the second half of the 19th century onward, partly under the influence of Hume's ideas, the view that causality is an outdated category to be replaced by the concept of functional relation has kept spreading. As justification it was argued that function is a broader concept than causality, on the one hand, and is free of the metaphysical confusion attached to causality, on the other.

> It is, of course, possible to represent the relation between cause and effect in the form of functional dependence, namely: effect is the function of cause. In this way, however, the most important element of causality—i.e., that cause as a real occurrence brings about and determines effect, another occurrence—remains obscure. The most various, also external and less important and even arbitrary, dependences can be described in the form of functional relation. Idealists dissolve causality into functional dependence upon the consideration that—as they put it—the question of how occurrences come into being, whether or not their existence has a cause, is of no importance for science. What is important is but the existence of some dependence between the occurrences which can be expressed by a certain formula.[69]

The interrelations of the categories under discussion are illustrated in the following figure:

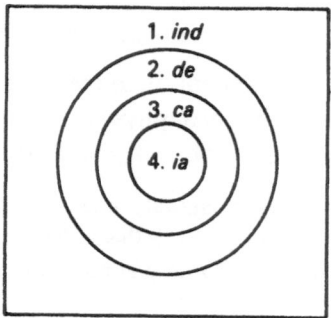

1. *ind*
2. *de*
3. *ca*
4. *ia*

Fig. 38.

[68] Lenin, *Philosophical Notebooks.*, p. 162.
[69] *A marxista filozófia alapjai.* (Budapest 1959) p. 227.

The designations are to be understood as follows: *ia* = interaction, *ca* = causality, *de* = dependence, *ind* = independence. Ring 2 represents 'dependence but not causality', and ring 3 'causality but not interaction'.

As is well known, the ancient atomists were fatalists, and consequently, saw a necessity in all causal relations. As Leucippus put it: nothing comes into being without a reason; everything comes into being from certain causes and of necessity.

Aristotle disagreed with this one-sided view:

> ... since not all things either are or come to be of necessity and always, but the majority of things are *for the most part*, the accidental must exist; for instance a pale man is not always nor for the most part musical, but since this sometimes happens, it must be accidental (if not, everything will be of necessity). The matter, therefore, which is capable of being otherwise than as it usually is, must be the cause of the accidental. And we must take as our starting-point the question whether there is nothing that is neither always nor for the most part. Surely this is impossible. There is, then, besides these something which is fortuitous and accidental.[70]

Elsewhere he wrote the following:

> Evidently there are not causes and principles of the accidental, of the same kind as there are of the essential; for if there were, everything would be of necessity. If *A* is when *B* is, and *B* is when *C* is, and if *C* exists not by chance but of necessity, that also of which *C* was cause will exist of necessity, down to the last *causatum* as it is called (but this was supposed to be accidental). Therefore all things will be of necessity, and chance and the possibility of a thing's either occurring or not occurring are removed entirely from the range of events.[71]

Thomas Aquinas pronounced a similar view:

> It is not true that, given any cause whatever, the effect must follow of necessity. For some causes are so ordered to their effects, as to produce them, not of necessity, but in the majority of cases, and in the minority to fail in producing them. But that such causes do fail in the minority of cases is due to some hindering cause.[72]

---

[70] Aristotle, *Metaphysica*. 1027a9–29.
[71] *Op. cit.*, 1065a6–13.
[72] Quoted in Bunge, p. 103.

25

The mechanical concept of nature which has developed in modern times has contributed to fatalism's gathering strength again. Of its adherents, let me mention Spinoza and Holbach who regarded the world as a huge mechanism. The functioning of this mechanism is a predetermined and necessary chain of causes and effects. Accident is what has no cause, or whose cause is unknown.

This one-sided approach brought about as a reaction a similarly one-sided opposite trend:

> ... so [arbitrary—Gy. T.] must we also esteem the supposed tie or connexion between the cause and effect, which binds them together, and renders it impossible that any other effect could result from the operation of that cause. When I see, for instance, a Billiard-ball moving in a straight line towards another, even suppose motion in the second ball should by accident be suggested to me, as the result of their contact or impulse, may I not conceive, that a hundred different events might as well follow from that cause? May not both these balls remain at absolute rest? May not the first ball return in a straight line, or leap off from the second in any line or direction?[73]

Dialectical materialism rejects both the fatalistic and the skeptical interpretation of causality.

> If we bring together in a rifle the priming, the explosive charge, and the bullet and then fire it, we count upon the effect known in advance from previous experience, because we can follow in all its details the whole process of ignition, combustion, explosion by the sudden conversion into gas and pressure of the gas on the bullet. And here the sceptic cannot even say that because of previous experience it does not follow that it will be the same next time. For, as a matter of fact, it does sometimes happen that it is *not* the same, that the priming or the gunpowder fails to work, that the barrel bursts, etc. But it is precisely this which *proves* causality instead of refuting it, because we can find out the cause of each such deviation from the rule by appropriate investigation: chemical decomposition of the priming, dampness, etc., of the gunpowder, defect in the barrel, etc., etc., so that here the test of causality is so to say a *double* one.[74]

---

[73] Hume, *Enquiries* pp. 29–30
[74] Engels, *Dialektik der Natur*. p. 245; Clements Dutt tr. pp. 305–306.

Let me sum up what has been said so far:

(1) A causal relation can be necessary just as much as accidental.

(2) Necessary and accidental causal relations are contrary, but at the same time they interact.

(3) Of the two relations the necessary is the decisive factor.

II

Causal propositions can be met first with Ramon Lull who regarded them as a kind of conditional proposition: the antecedent of the proposition expresses the cause and its consequent the effect, e.g., the water is boiling because its temperature has been raised to a hundred degrees. This view came to be widely accepted in traditional logic.

In mathematical logic, among others the following standpoint has been established:

Not all propositional compounds are extensional. Let us consider the following compound propositions:

(a) 'The planets move in ellipses around the sun *because* they are attracted by the sun';

and compare it to the compound:

(b) 'The planets move in ellipses around the sun, and they are attracted by the sun.'

If we denote in (a) the clause before 'because' with p, the subsequent clause with q, then (a) and (b) acquire the form:

(a) p because q, (b) p and q.

Both compounds are true since p and q are both true. There is, however, a fundamental difference. If in (b) we replace the statements p or q with *other* arbitrary true statements, then (b) remains true, but (a) does not! . . .

If for instance, we replace in 'p because q' p by 'In 1956 the danger of a new war existed for a time', and q by 'In 1956 the planet Mars reached its maximum proximity to the earth', then (b) remains true, but (a) becomes false and is transformed into a nonsensical astrological assertion; 'p because q' is thus an intensional propositional compound. The characterization of such compounds consists in the fact that they do not retain their truth values when their component propositions are arbitrarily replaced by propositions with the same truth value.

25*

The consideration of the concrete content of individual statements takes us out of the field of formal logic in the narrower sense and into that of the individual sciences. If we want to ascertain, for *which p* or *q* the compound '*p* because *q*' is true or false, we would have to collate huge tables of data, which would have to be taken from the concrete material of the individual sciences.[75]

The distinction between the formulas '*p* and *q*' and '*p* because *q*' is based on the consideration that the value of the former depends on the value of its components only while that of the latter does not. Let us suppose, however, that even conjunction cannot be evaluated on the ground of the value of its components only. Accordingly, if in (b) components *p* and *q* are replaced by any other true proposition, (b) does not inevitably stay true. Consequently, the above distinction becomes pointless.

There is another standpoint worth of attention, namely that the study of the relation '*p* because *q*' leads to the domain of special sciences. Is it really necessary to know whether there is *really* a relation of cause and effect between two given facts in order to define the value of the proposition '*p* because *q*'? I think it is not. That is, all we have to know in order to evaluate the proposition '*p* · *q*' in mathematical logic is that *if* both components are *t*, then their conjunction will be *t*. Accordingly, in order to evaluate the proposition '*p* because *q*' it is sufficient to know under what conditions it is *t*, and it is not the task of logic to state whether or not the given condition exists in some concrete case. Consequently, the study of the relation '*p* because *q*' does not end up in the field of the individual special sciences.

The proposition '*p* because *q*' is of the type of converted conditional propositions. If the proposition '*p* because *q*' is *t*, then the proposition 'if *q*, then *p*' is also *t*, but it does not hold true the other way round. The values of the converted conditional proposition can be seen in the following table:

TABLE 67

	(1)	(2)	(3)	(4)	(5)	(6)	(7)	(8)	(9)	(10)	(11)	(12)
If *q*, then *p*	*f*	*t*	*f*	*t*	*f*	*f*	*f*	*f*	*t*	*f*	*t*	*f*

[75] Klaus, pp. 86–87.

These very values apply to the proposition '*p* because *q*' if the original condition is modified in a way that in cases (2) and (11) respectively the relation '*q* only implies *p*' is replaced by '*q* is the cause of *p*'. Without the modification of the original condition the result will be t/f in the above cases.

For simplicity's sake only the conjunctive particle 'because' has been used in the foregoing. Yet, a causal proposition can be expressed in other linguistic terms as well, e.g., 'due to the diffusion of some parasite the vines have died out'. When evaluating 'due to $p\,q$' (*p* is the cause of *q*), we should rely on the values of conditional propositions with the above modifications.

The simple form of causal inference is:

> *p* is the cause of *r*
> *r* is the cause of *q*
> —————————————
> *p* is the cause of *q*

I share the following opinion about the above consideration:

> ... causal propositions include a decisive *intensive* element, that is: their operational value depends on what kind of *intensive* (i.e. content) relations there are between the elementary propositions joined in them to such an extent that their formal logical definition is *not* possible merely on the ground of the truth on the falsity of the propositions joined in them (that is: without more exact knowledge of content, or in the special terminology of mathematical logic, in the form of a truth function).
>
> It implies, however, that the means, or at least the *present* means, of mathematical logic, *even if* they are usually of a wider scope than those of formal logic, can*not* provide a valid mode of inference (i.e. which would follow of necessity as soon as the premises are accepted) as a strict proof of propositions of the form '*A* because *B*'.
>
> It does *not* mean, however, that the connection 'because' between two propositions is *senseless* and *impossible to interpret* (as some mathematical logicians would assert—on the ground of their 'private' individual philosophy which is not proved by mathematical logic, although this philosophy might have been developed in reference to this very logic!). *Neither* does it mean that 'because' is *beyond logic* (there is no reason to declare it extralogical by raising the question of intensive relations in general and thereby to leave, for example, the problem of imperfect

induction to other branches of knowledge that are less prepared to cope with it). All it means is that here we have arrived at the (at least, present) limit of logical formulation, and now we have to adopt other methods of investigation, or rather, to complete the methods of mathematical *logic* by others *as well*.[76]

Let us compare a causal proposition to other propositions.

> One of the differences between the conditional proposition and that expressing causality is that the conditional proposition reflects all kinds of necessary dependence of a certain something upon another while the causal proposition is but the causal relation of something to something else.
>
> Conditional propositions, too, can, and do, reflect causal relations. 'If an animal is poisoned by carbone dioxide, then its blood is no longer able to absorb oxygen'—this conditional proposition for example, reflects the causal relation between the poisoning of an animal by carbon dioxide, on the one hand, and the inability of its blood to absorb oxygen, on the other. There is, however, an essential difference between the conditional propositions expressing causal relations and those expressing causality.
>
> The main difference is that the occurrences, or facts, concerned in the propositions expressing causality, always appear as existing in reality—while in conditional propositions they may (and often do) appear as things which are only supposed to be existent.[77]

Let us begin with the second part of this quotation. No doubt the above example refers to occurrences which exist in reality. But does it really imply that causal propositions can refer to such occurrences only? Let us see an example to the contrary: Eve tore off the apple because Lucifer had persuaded her. For my part I doubt the existence of Lucifer. That is, existence has no part in the distinctiveness of the propositions at issue.

Furthermore, the limitations of the view that all conditional propositions express necessary relations, or connections, have already been pointed out before. Accordingly, in my opinion, the following versions should be distinguished:

(1)   if *p*, then *q*

[76] Szalai. p. 399.
[77] Tavanets, p. 125.

(2) p is the cause of q
(3)   it is necessary that if p, then q
(4)   it is necessary that p is the cause of q

III

When examining causality Aristotle was confronted with the problem of interaction which he believed he solved in the following way:

> If, however, they cannot each be the cause of the other (for cause is prior to effect, and the earth's interposition is the cause of the moon's eclipse and not the eclipse of the interposition)—if, then, demonstration through the cause is of the bare fact, one who knows it through the eclipse knows the fact of the earth's interposition but not the reasoned fact. Moreover, that the eclipse is not the cause of the interposition, but the interposition of the eclipse, is obvious because the interposition is an element in the definition of eclipse, which shows that the eclipse is known through the interposition and not vice versa.[78]

That is, if A is prior to B, then B cannot be prior to A. Arguments denying interaction were widely accepted for a long time due to the underdevelopment of sciences.

The study of interaction as a philosophical category can be found first with Kant.

> All substances, in so far as they can be perceived to coexist in space, are in thoroughgoing reciprocity...
>
> Now assuming that in a manifold of substances, as appearances, each of them is completely isolated, that is, that no one acts on any other and receives reciprocal influences in return, I maintain that their *coexistence* would not be an object of a possible perception and that the existence of one could not lead by any path of empirical synthesis to the existence of another. For if we bear in mind that they would be separated by a completely empty space, the perception which advances from one to another in time would indeed, by means of a succeeding perception, determine the existence of the latter, but would not be able to distinguish whether it follows objectively upon the first or whether it is not rather coexistent with it.[79]

[78] Aristotle. 1928. 98b17–27.
[79] Kant, *Kritik*, pp. 302, 304; Kemp Smith, pp. 233, 234.

Hegel took a big step forward in revealing the dialectic of interaction.

> Reciprocal action realises the causal relation in its complete develop-
> ment. It is this relation, therefore, in which reflection usually takes shelter
> when the conviction grows that things can no longer be studied
> satisfactorily from a causal point of view, on account of the infinite progress
> already spoken of. Thus in historical research the question may be raised in
> a first form, whether the character and manners of a nation are the cause of
> its constitution and its laws, or if they are not rather the effect. Then, as the
> second step, the character and manners on one side and the constitution and
> laws on the other are conceived on the principle of reciprocity: and in that
> case the cause in the same connection as it is a cause will at the same time be
> an effect, and *vice versa.*[80]

The importance of proceeding to the concrete notion of a category after
having reached the height of its abstract notion has already been stressed
before. Let me refer here to the process from the abstract universal to the
concrete universal. The same must be kept in mind with regard to the
category of interaction.

> Reciprocity is undoubtedly the proximate truth of the relation of cause
> and effect, and stands, so to say, on the threshold of the notion; but on that
> very ground, supposing that our aim is a thoroughly comprehensive idea,
> we should not rest content with applying this relation. If we get no further
> than studying a given content under the point of view of reciprocity, we are
> taking up an attitude which leaves matters utterly incomprehensible. We
> are left with a mere dry fact; and the call for mediation, which is the chief
> motive in applying the relation of causality, is still unanswered. And if we
> look more narrowly into the dissatisfaction felt in applying the relation of
> reciprocity, we shall see that it consists in the circumstance, that this
> relation, instead of being treated as an equivalent for the notion, ought, first
> of all, to be known and understood in its own nature. And to understand the
> relation of action and reaction we must not let the two sides rest in their
> state of mere given facts, but recognize them, as has been shown in the two
> paragraphs preceding, for factors of a third and higher, which is the notion
> and nothing else. To make, for example, the manners of the Spartans the
> cause of their constitution and their constitution conversely the cause of
> their manners, may no doubt be in a way correct. But, as we have

[80] Hegel, *Encyclopädie* p. 346; Wallace tr., pp. 280–281.

comprehended neither the manners nor the constitution of the nation, the result of such reflections can never be final or satisfactory. The satisfactory point will be reached only when these two, as well as all other, special aspects of Spartan life and Spartan history are seen to be founded in this notion.[81]

For formal logic, interaction, if it is treated at all, is just one of the relations, but from the point of view of dialectic it is a relation of a universal nature.

> ... from this universal reciprocal action we arrive at the real causal relation. In order to understand the separate Phenomena, we have to tear them out of the general inter-connection and consider them in isolation, and *then* the changing motions appear, one as cause and the other as effect.[82]

Just as the universal among quantifiers and the necessary among modalities, interaction is the strongest category among the relations.

> *Reciprocal action* is the first thing that we encounter when we consider matter in motion as a whole from the standpoint of modern natural science. We see a series of forms of motion, mechanical motion, heat, light, electricity, magnetism, chemical union and decomposition, transitions of states of aggregation, organic life, all of which, if *at present* we *still* make an exception of organic life, pass into one another, mutually determine one another, are in one place cause and in another effect, the sum-total of the motion in all its changing forms remaining the same (Spinoza: *substance is causa sui* strikingly expresses the reciprocal action). Mechanical motion becomes transformed into heat, electricity, magnetism, light, etc., and *vice versa*. Thus natural science confirms what Hegel has said....that reciprocal action is the true *causa finalis* of things. We cannot go back further than to knowledge of this reciprocal action, for the very reason that there is nothing behind to know.[83]

This very consideration has made me put the category of interaction into the innermost circle of Figure 38.

---

[81] *Op. cit.*, pp. 346–347; A. V. Miller tr., pp. 281–282.
[82] Engels, *Dialektik der Natur.* pp. 246–247; Clemens Dutt tr. p. 307.
[83] *Op. cit.*, p. 246; Dutt pp. 306–307.

According to the formal approach every interaction (4) is a causal relation (3 *o* 4) but not *vice versa* since there are irreversible causal relations as well. E.g., sunshine is the cause of the development of chlorophyll but it does not hold true the other way round. Mere causality (3) is contrary to interaction but this contrariety is not exclusive.

> Further, we find upon closer investigation that the two poles of an antithesis, positive and negative, e.g., are as inseparable as they are opposed, and that despite all their opposition, they mutually interpenetrate. And we find, in like manner, that cause and effect are conceptions which only hold good in their application to individual cases; but as soon as we consider the individual cases in their general connection with the universe as a whole, they run into each other, and they become confounded when we contemplate that universal action and reaction in which causes and effects are eternally changing places, so that what is effect here and now will be cause there and then, and *vice versa*.[84]

C

I

The middle term has occurred several times in the foregoing. This category played a fundamental role in the Aristotelean syllogistics.

> ... in all our inquiries we are asking either whether there is a 'middle' or what the 'middle' is: for the 'middle' here is precisely the cause, and it is the cause that we seek in all our inquiries. Thus, 'Does the moon suffer eclipse?' means 'Is there or is there not a cause producing eclipse of the moon?', and when we have learnt that there is, our next question is, 'What, then, is this cause?'; for the cause through which a thing *is* —not *is this or that*, i.e. has this or that attribute, but without qualification *is*—and the cause through which it is—not *is* without qualification, but *is this or that* as having some essential attribute or some accident— are both alike the 'middle'.[85]

In this respect the Aristotelean syllogisms can be regarded as special versions of the *causal inference*.

---

[84] Engels, *Anti-Dühring* p. 25; Eng. tr., p. 36.
[85] Aristotle. 1928. 90a5–12.

Upon the above considerations, Aristotle found the search for the middle term of primary importance. In principle he stated the following on this issue:

> Clearly then, if the same term is not stated more than once in the course of an argument, a syllogism cannot be made: for a middle term has not been taken. Since we know what sort of thesis is established in each figure, and in which the universal, in what sort the particular is established, clearly we must not look for all the figures, but for that which is appropriate to the thesis in hand.[86]

Theophrastus made the middle term independent of the real ground by declaring it a go-between of the extreme terms by virtue of its position only. The Stoics were not interested in the middle term at all.

By the end of ancient times, Theophrastus's approach had prevailed and stayed so during the Middle Ages. The invention of the middle term (*inventio medii*) was thought to be necessary for the technical construction of syllogisms rather than for the revelation of the cause. The first comprehensive figure to assist in this operation had been made by Philoponus (6th century) and was developed further by Petrus Tartaretus (15th c.). These figures were called '*pons asinorum*' (bridge of asses) with a special emphasis on their being designed to make things clearer.

The rejection of scholasticism had some effect also on the views concerning the necessity of the middle term.

> *For the perfect understanding of a question we must abstract it from all that is superfluous, rendering it as simple as possible, and, resorting to enumeration, divide it into its minimal parts.*
>
> This is the one respect in which we imitate the dialecticians. Just as, in the treatment of the forms of the syllogism, they assume that the terms or matter of the syllogisms are known, so too we here lay it down as a prerequisite that the question at issue be perfectly understood (i.e. that we are from the start in possession of all the data required for its solution). We do not, however, like them, distinguish two extremes and a middle term. This noted, let us now consider the whole matter afresh. Firstly, there must in every question be something not yet known; otherwise inquiry would be to no purpose. Secondly, the not yet known must be in some way marked

[86] *Op. cit.*, 47b8–13.

out; otherwise we should not in our investigation be determined to it instead of to something else. Thirdly, it can be so marked out only by way of something that is already known.[87]

Descartes' approach, however, did not become generally accepted. The ancient ideas, though in a somewhat modified form, were revived, too. The original standpoint of Aristotle was represented, first of all, by Trendelenburg and Überweg. They defined the middle term as a reflection of the real ground.

The prevailing view, however, remained, as it had been, that of Theophrastus. One of its typical formulations is the following:

> We are in the habit of employing a *middle term* or medium whenever we are prevented from comparing two things together directly, but can compare each of them with a certain third thing. We cannot compare the sizes of two halls by placing one in the other, but we can measure each by a foot rule or other suitable measure, which forms a common measure, and enables us to ascertain with any necessary degree of accuracy their relative dimensions...
>
> The use of a middle term in syllogism is closely parallel to what it is in the above instances, but not exactly the same. Suppose, as an example, that we wish to ascertain whether or not 'Whales are viviparous', and that we had not an opportunity of observing the fact directly; we could yet show it to be so if we knew that 'whales are mammalian animals', and that 'all mammalian animals are viviparous'. It would follow that 'whales are viviparous'; and so far as the inference is concerned it does not matter what is the meaning we attribute to the words 'viviparous' and 'mammalian'. In this case 'mammalian animal' is the middle term.[88]

Concluding this sketchy survey, let me quote Hegel whose position is based on the principles of dialectical logic.

> If from the middle term, that a wall has been painted blue, it is inferred that therefore the wall is blue, this is a correct inference; yet in spite of this syllogism the wall can be green if it has also been painted over yellow, from which latter circumstance taken by itself it would follow that it was yellow

[87] Descartes, *Règles pour la direction de l'esprit*, in *Œuvres*. Vol. II. (Paris 1940) p. 60; Kemp Smith tr., p. 76.
[88] Jevons, 1872. pp. 126–127.

... It is justly held that there is nothing so inadequate as a formal syllogism of this kind, since it is a matter of chance or caprice which middle term is employed. No matter how elegantly a deduction of this kind has run its course through syllogisms, however fully its correctness may be conceded it still leads to nothing of the slightest consequence, for the fact always remains that there are still other middle terms from which the exact opposite can be deduced with equal correctness.[89]

According to Hegel a meaningful middle term is the union of the extreme terms.

We have come across two extreme views in the foregoing: (1) only inferences which have a middle terms are syllogisms, (2) there is no need for the middle term at all. Let us begin with the first one. There are syllogisms where the middle term has no part to play (e.g., conditional, disjunctive, etc.). This is one of the reasons why Aristotle did not deal with such syllogisms. We have also seen that Barbara is the only mode which strictly conforms to the original interpretation of the middle term.

But that is not the end of it. Aristotle's statement that a syllogism may have three terms only is thought to be valid even today. Let us compare the following inference with this thesis:

All animals (M) are living beings (P).
All dogs (S) are mammalian animals (M_1).

All dogs (S) are living beings (P).

In this respect the following thesis can be formulated according to what Aristotle has said about the first figure: if four terms relate to one another in a way that the first (S) is contained in the second (M_1), the second in the third (M), and finally, the third in the fourth (P), then there is a syllogistic relation between the extremes. In consequence it is not compulsory that two of the terms which occur in the premises should be identical. We can arrive at a correct inference even in the case of subordination.

The following inference is also valid:

All P is M
All S is not-M

No S is P

[89] Hegel, *Wissenschaft* p. 128; A. V. Miller tr., p. 671.

Here the following objection can be made: the minor premise can be simply obverted into the proposition 'no S is M', and as a result we end up with a usual Camestres mode. It is true, indeed. But how is, for example, Cesare traced back to Celarent in traditional logic? By a simple conversion of the major premise:

No P is M	No M is P
All S is M	All S is M
No S is P	No S is P

Although a simple conversion has been performed here, Cesare is still regarded as an independent mode. Why could we not think the same of the inference taken as an example despite the fact that it can be simply obverted? Either both of them are valid, or none.

Our starting point should be that there are four terms—two subjects and two predicates— in the premises. Certain logical connections between them allow for valid inferences. Such a connection is, e.g., identity. If identity is present it is enough to consider three terms only. Thus, traditional syllogistic has studied an extreme case.

In accordance with the above remarks, the warning against '*quaternio terminorum*' becomes useless. The mere fact that some categorical syllogism has four terms is not a logical mistake yet. Let us see the following example:

Matter is eternal.
Wool is matter.
Wool is eternal.

What makes this conclusion false is not the fact that it is based on four terms but that the relation, or connection, of the terms fails to lead to a valid inference because two different concepts of matter are involved. Consequently, it is not the number of the terms but the logical connections of the premises that determines the validity of inferences.

Naturally, I do not intend to support the other extreme view by all that has been said above. I admit that there are inferences with middle term (*Mittelbegriffsschlüsse*), and that they are necessary in certain cases. I think, however, that their scope of validity must not be overestimated.

In fact, the introduction of the middle term was due to the recognition of the necessity of intensive inferences. This demand must not be ignored, but to identify it with the middle term is unnecessary.

II

We often make mistakes in everyday thinking. We mix up the various meanings of terms, or fail to take into consideration the consequences of our assertions, or make unfounded statements. These mistakes can be avoided to a great extent by studying the logical laws of thought.

What is usually described as the laws of thought (this expression was first used by Galen) are identity, contradiction, the excluded middle, and the law of sufficient reason. The aim of this brief survey is, first of all, to reveal their interrelations.

The goal to strive for unambiguity in thought was postulated as early as in ancient times. The demand for unambiguity requires of thought to use linguistic expressions with a definite meaning and not to mix up their different meanings.

> ... for not to have one meaning is to have no meaning, and if words have no meaning our reasoning with one another, and indeed with ourselves, has been annihilated; for it is impossible to think of anything if we do not think of one thing...[90]

Ambiguity is a logical mistake which is manifest when a term with various meanings is used in a given reasoning. E.g., the following sentence can be read in one of Heraclitus' fragments: eyes and ears are bad witnesses with people who are callous.

This sentence entails ambiguity. First, it can be interpreted in a way that eyes and ears (that is, the senses) are bad witnesses for people because people are callous. The possibility of such an interpretation is based on the fact that the subordinate clause could be formulated also like this: since they (i.e. people) are callous.

Second, the sentence under discussion can also be understood like this: though eyes and ears (the senses) are suitable for reflecting the truth, those people who are callous misinterpret the evidence of the external senses.

[90] Aristotle. 1928. 1006b7–11.

This interpretation is based on the following possible transformation of the subordinate clause: if they are callous.

This ambiguity allowed for contrary interpretations of Heraclitus's view. One of them made it appear as if Heraclitus had regarded the evidence of the senses as false and unreliable. The other argued that Heraclitus described as false not perception as such but the reliability of the senses of those people who are callous.

The requirement of unambiguity is of primary importance for the validity of thought. By meeting this requirement we are protected against the mistake of taking a seeming identity for a real one. This requirement is usually expressed in everyday language like this: let us stick to the subject. In logic it means that one must always be alert to study the given question and not something else.

In order to establish a theoretical ground for the requirement of unambiguity the schoolmen had formulated the law of identity of formal logic. Its most widely known formula is: every term is identical with itself. Or by symbols: *A est A*.

Aristotle was still of the opinion that things must be identical with themselves in the same aspect only and that in different aspects they can be both identical and different. As early as in scholasticism, however, and even more so when metaphysical thought had become the dominant trend, it was asserted that things must be regarded as identical, and nothing but identical, with themselves even in various aspects. This metaphysical position was based on a thesis of Leibniz according to which identity is the identity of those which are indiscernible from one another (*identitas indiscernibilium*), or in other words: every thing is what it is.

In mathematical logic, among others, the following interpretation can be read:

> So far we have met with the principle of identity in two forms: once as a relationship in propositional logic and then as a relationship in the calculus of classes:
>
> (a) The truth values of two propositions p, q are identical if $(p \rightarrow q) \cdot (q \rightarrow p)$. If we consider a proposition p alone, then it is always the case that $p \rightarrow p$.
>
> (b) In the calculus of classes it is similarly the case that two classes A, B are identical, if $(A \subset B) \cdot (B \subset A)$. For every class A it is the case that $A \subset A$.[91]

[91] Klaus, pp. 278–279.

Relativists proceed from the misconception that two contradictory statements can be true at the same time. As early as in ancient Greece there were such philosophers, namely the Sophists, who were ready to prove of contradictory propositions that both of them were equally true.

Aristotle was the first to formulate as a law the requirement of avoiding logical contradiction:

> ... the same attribute cannot at the same time belong and not belong to the same subject and in the same respect; we must presuppose, to guard against dialectical objections, any further qualifications which might be added.[92]

The formula for the law of contradiction in traditional logic is: A is not-A, and in mathematical logic: $p \cdot \bar{p}$. As far as the law of being free of contradictions is concerned, the negations of these formulae are used.

The law of freedom from contradictions is based on the connection that two propositions contradictory to one another cannot be true at the same time. Nevertheless, it leaves the question open what happens if none of the two statements, but perhaps a third one, is true. Aristotle gave the following answer:

> Of opposites, contradictories admit of no middle term; for this is what contradiction is—an opposition, one or other side of which must attach to anything whatever, i. e. which has no intermediate.[93]

The law of the excluded middle fails to answer which of the two contradictory propositions is true. It only formulates the requirement that one of them has to be accepted as true. The significance of this law is to define the framework within which a given question should be decided: this man is either guilty, or not-guilty. It is not the task of logic to decide the question. That is left to the corresponding special science or practice.

The formula of the excluded middle in traditional logic is: all A is B or not-B, and in mathematical logic: $p \ o \ \bar{p}$.

[92] Aristotle, *Metaphysica* 1005b18–21.
[93] *Op. cit.*, 1057a34–36.

The claim to substantiate our statements can be met as early as with Aristotle.

> We suppose ourselves to possess unqualified scientific knowledge of a thing, as opposed to knowing it in the accidental way in which the sophist knows, when we think that we know the cause on which the fact depends, as the cause of that fact and of no other, and, further, that the fact could not be other than it is ... There may be another manner of knowing as well—that will be discussed later. What I now assert is that at all events we do know by demonstration. By demonstration I mean a syllogism productive of scientific knowledge, a syllogism, that is, the grasp of which is *eo ipso* such knowledge.[94]

He grounded the conclusion of syllogisms in two interrelated ways. On the one hand, with the help of the middle term: *S* is *P* because *M* (e. g., the Greeks are mortal because they are men). This is why he found the invention of the middle term of special importance (see the above discussion of the middle term).

On the other hand, with the help of the premises:

> Assuming then that my thesis as to the nature of scientific knowing is correct, the premisses of demonstrated knowledge must be true, primary, immediate, better known than and prior to the conclusion, which is further related to them as effect to cause. Unless these conditions are satisfied, the basic truths will not be 'appropriate' to the conclusion ... The premisses must be the causes of the conclusion, better known than it, and prior to it; its causes, since we possess scientific knowledge of a thing only when we know its cause; prior, in order to be causes; antecedently known, this antecedent knowledge being not our mere understanding of the meaning, but knowledge of the fact as well.[95]

Adequate, or appropriate, middle terms and premises respectively provide sufficient reason for drawing the conclusion:

> ... follows of necessity from their being so. I mean by the last phrase that they produce the consequence, and by this, that no further term is required from without in order to make the consequence necessary.[96]

[94] Aristotle. 1928. 71b8–18.
[95] *Op. cit.*, 71b19–23, 29–33.
[96] *Op. cit.*, 24b19–22.

The above does not challenge Leibniz's credit for calling attention to the connection under discussion by formulating it as a law of sufficient reason. In traditional logic, opinions were divided as to whether or not the principle of sufficient reason should be regarded as a logical law. Mathematical logic has taken a negative stand on this issue.

Some authors find it necessary to state further laws in addition to those discussed above. Let me quote one of them:

> ... in the operation of discourse or reasoning we need certain additional laws, or axioms, or self-evident truths, which may be thus stated:
>
> (1)   *Two terms agreeing with one another and the same third term agree with each other.*
>
> (2)   *Two terms of which one agrees and the other does not agree with one and the same third term, do not agree with each other.*[97]

The first principle can be called the law of compatibility, by using Sigwart's term (*Prinzip der Übereinstimmung*). E. g., if iron is compatible both with good heat-conduction and good electricity conduction, then the latter two are also compatible with one another.

The majority of logical works merely state the laws of thought. The only problem which has aroused interest with regard to their relations is whether the laws of contradiction and the excluded middle respectively can be regarded as separate principles, or are necessarily the same.

In order to make clear the relations of the above laws let us take the relations of propositions (Figure 32) as a basis. Identity (p is identical with p) can be considered as a specific case of biconditional relation (if p, then, and only then, q). That is, if two propositions are identical, then they are conditional upon one another, but it does not hold good the other way round (e. g., if the opposite sides of a quadrangle are parallel, then, and only then, the diagonals of the quadrangle bisect one another).

[97] Jevons, 1872. p. 121.

Discussing the law of sufficient reason, Klaus made the following remark:

> Although causal relations are in fact a very special kind of 'if-then' relation, the range of the 'if-then' relation extends far beyond the field of causality.[98]

Indeed, if 'q because p' (or p is the cause of q), then 'if p, then q', but it does not hold true vice versa. The law of sufficient reason requires the following: (1) if a conditional proposition is given, it must be checked whether or not the antecedent entails the consequent, (2) if a statement is given, then such an antecedent should be found for it which makes true the conditional proposition constructed with its help.

The relation discussed under the name of law of compatibility is a special case of conjunction. Let me recall that by conjunction I mean 'p is compatible with q'.

That the law of the excluded middle belongs to the field of disjunction is generally accepted. In my interpretation disjunction is 'p is incompatible with q'. Accordingly, the law of contradiction is also a special version of disjunction: if p contradicts not-p, then p is incompatible with not-p, but it does not hold true the other way round. Thus, the problem whether contradiction and the excluded middle are independent laws is of no major importance as regards their relation to disjunction.

In consequence the laws under discussion can be subordinated to the corresponding relations of propositions:

	Relations of propositions	Laws
(1)	disjunction	contradiction—the excluded middle
(2)	conjunction	compatibility
(3)	conditional	sufficient reason
(4)	biconditional	identity

If the above laws are depicted in the usual figure, their relations can be read, but with due precaution. For example, if some proposition is

[98] Klaus, p. 79.

sufficient reason for another proposition it implies their compatibility, but it does not hold true the other way round.

So far these laws have been discussed as the relations of propositions. Since there is an analogy between the relation of propositions and those of terms what has been said here, *mutatis mutandis*, applies also to the relations of terms. Upon these considerations we can state that the laws of thought are but formulations of significant special instances of the relations of terms and propositions.

CHAPTER SIX

17. SOME PROBLEMS OF THE SYLLOGISM

I

It is not indifferent with regard to certain inferences whether the terms are concrete or abstract. Aristotle wrote the following on this issue:

> Men will frequently fall into fallacies through not setting out the terms of the premiss well, e. g. suppose A to be health, B disease, C man. It is true to say that A cannot belong to any B (for health belongs to no disease) and again that B belongs to every C (for every man is capable of disease). It would seem to follow that health cannot belong to any man. The reason for this is that the terms are not set out well in the statement, since if the things which are in the conditions are substituted, no syllogism can be made, e. g. if 'healthy' is substituted for 'health' and 'diseased' for 'disease'. For it is not true to say that being healthy cannot belong to one who is diseased. But unless this is assumed no conclusion results, save in respect of possibility: but such a conclusion is not impossible: for it is possible that health should belong to no man.[1]

Sándor Szalai commented on the quoted text as follows:

> We arrive from apparently true premises at a false conclusion by the syllogism $E^n A E^n$ [n = necessary—Gy. T.], valid in the first figure, in the following case:

> Health can belong to no disease
> (i. e., no disease can be said to be healthy)
> Disease belongs to every man
> (i. e., every man is said—occasionally—to be ill, or diseased)
> _____
> Health can belong to no man
> (i. e., no man can be said to be healthy)

[1] Aristotle. 1928. 48a1–15.

392

The conclusion is false, indeed, but it is due to the fact that the major premise only appears to be true. It will become clear in no time as soon as concrete terms are used instead of the abstract ones since it can*not* be asserted even of an ill man that he cannot be healthy in due time. The correctly formulated premises lead to a true conclusion by the $E^cA^cE^c$ (c = contingent) syllogism, valid in the first figure, namely:

> Contingently no ill man is healthy (in due time).
> Contingently every man is ill.
> ___
> Contingently no man is healthy (in due time).[2]

Let us first analyse the major premise. According to Szalai it is to be understood like this: no disease can be said to be healthy. In Aristotle's example, however, the predicate of the major premise is: health. Can one identify health and healthy? In my opinion no, because, e.g., the proposition 'every man can be healthy' is t while the proposition 'every man can be health' is f. Thus, the major premise correctly—and in traditional transcription—will be: no disease can be health.

The minor premise, according to Aristotle, is: every man can be diseased.* According to the restriction, however, term B is: disease. Therefore the minor premise is: every man can be disease.

The minor premise, according to Szalai, is: every man is said—occasionally—to be ill. There is no need to dwell upon the fact that the predicate has been replaced. What does, however, 'occasionally' mean? As it turns out later on, it is used as a synonym of 'contingent'. But Aristotle used propositions expressing possibility and not contingency. To replace one by the other is also unjustified. What becomes of the syllogism if the propositions are contingent is another question also referred to by Aristotle.

Let us describe this syllogism, keeping in mind the original restrictions:

(1)   No disease can be health (t).
        Every man can be disease (f).
        ___
        No man can be health (t).

---

[2] Aristotelés, *Organon*. Vol. I. (Budapest 1961). pp. 386–387.

* This analysis is based on the Hungarian translation which I find to be closer to the original: there instead of 'capable of disease' is 'can be diseased'—Gy. T.

Let us compare it with Aristotle's reasoning. As we have seen, when constructing the syllogism, despite the restrictions he used 'diseased' as the predicate of the minor premise and the term 'health' as the predicate of the conclusion. In consequence he arrived at an inference where the premises are *t*, but the conclusion is *f*:

(2)   No disease can be health (*t*).
        Every man can be diseased (*t*).
        _____
        No man can be healthy (*f*).

I think the mistake resulted from the fact that the linguistic expressions of the terms are not correct. This view would hold true if 'diseased' were a synonym of 'disease'. They are, in fact, two expressions close in meaning but still not synonymous. Thus, it has no sense to assert that one of them is more valid than the other.

When 'disease' is replaced by 'diseased' and 'health' by 'healthy', what are used now are *not* more valid terms but other terms. They produce the following inference:

(3)   No diseased can be healthy (*f*).
        Every man can be diseased (*t*).
        _____
        No man can be healthy (*f*).

That is, the following happened in the case of formula (2): Aristotle used the terms according to the original restriction in the major premise but failed to do so with regard to the predicate of the minor premise and the conclusion respectively. This inconsistency has led to a formula where the premises are *t* but the conclusion is *f*. If he had stuck to the original restriction (1), or had changed the terms consistently (3), he would not have been confronted with this pseudo-problem.

Let us proceed now to the contingent syllogism. Szalai, in the footsteps of Aristotle, held that the Celarent mode, if contingent propositions are used, leads to a valid inference. An example to the contrary is:

It is contingent that no medical lieutenant-colonel is a pilot (*t*).

It is contingent that all space pilots are medical lieutenant-colonels (*t*).
_____
It is contingent that no space pilot is a pilot (*f*).

However, we might arrive at a syllogism which expresses a fact if, e.g., man is replaced by man who has flu:

> It is a fact that no ill man is healthy (t).
> It is a fact that every man who has flu is ill (t).
> _____
> It is a fact that no man who has flu is healthy (t).

Aristotle examined not only in the first but also in the second and third figures how inference is affected by concrete and abstract terms respectively. I shall not analyse these investigations of his for essentially the same problems are discussed there which have been treated before. Let me quote only the summary:

> It is evident then that in all these cases the fallacy arises from the setting out of the terms: for if the things that are in the conditions are substituted, no fallacy arises. It is clear then that in such premisses what possesses the condition ought always to be substituted for the condition and taken as the term.[3]

Aristotle thought the cause of this problem to be of a linguistic nature and, therefore, to be solved by the appropriate modifications of the expressions. In reality, however, he not only modified the expressions but transposed the terms as well, in fact, without justification, as it has been pointed out before. There are several ways of replacing a term. Therefore it is reasonable to call the mistake seen above by name: hypostatization. It means to make some property into a separate and distinct substance as if it existed independently. An example of hypostatization occurs when disease is replaced by diseased, or ill, man.

The above problem kept on recurring in various forms. Hegel quoted the following example in an existential inference: "... this Rose is red; Red is a colour: this Rose is a coloured object."[4] Are colour and coloured really synonyms? If they were, then the conclusion would be (t) even in this form: hence this rose is a colour. If they are not, then it was not justified to use 'coloured' in the conclusion instead of 'colour' which was used in the minor premise.

---

[3] Aristotle. 1928. 48a24–28.
[4] Hegel, *Encyclopädie* p. 386; Wallace tr. p. 317.

Klaus suggested the following solution:

Let us take the following example:

Red is a colour	R is F
Copper is red	K is R
Copper is a colour	K is F

This obviously brings us to a nonsensical conclusion. Where is the mistake? It is not so that R is in the one case the subject and in the other the predicate. We can spare ourselves the entire series of considerations which traditional logic introduced to avoid this paradoxical result, if we assume that K, F, R *are classes of things* or *predicate*. The 'subjects' are the things which are subsumed under these terms, they are the elements of these classes. If we take this fact into consideration, the paradoxical conclusion of the argument disappears.

The correction of the syllogism yields:

(1)   Every thing that has the property of being red also has the property of being coloured.

(2)   Every thing that has the property of being copper also has the property of being red.

(3)   Every thing that has the property of being copper also has the property of being coloured.

Thus: The class R is contained in the class F

Or: The class K is contained in the class R

Thus   The class K is contained in the class F.[5]

Let us accept Klaus' opinion that the original example consists of universal propositions and transcribe the inference accordingly:

Every red is a colour.

Every copper is red.

Every copper is a colour.

This is an inference of the Barbara type whose conclusion is f. If the conclusion of a valid inference is f, then at least one of its premises is also f. The minor premise is obviously t. Consequently, the major premise is f.

[5] Klaus, p. 218.

By using 'colour' instead of 'coloured' we may seem to have the *same* term in mind, but from the point of view of extension and not intension. In reality, the terms at issue represent *different* types of abstraction. The above problem can, and must be, solved from the point of view of intension since extension has no part to play here.

Relying on the above findings, I think the correct distinction between the concrete and the abstract is of significance for logic.

II

This is what Aristotle wrote about compound syllogisms:

> It is clear too that every demonstration will proceed through three terms and no more, unless the same conclusion is established by different pairs of propositions; e.g. the conclusion *E* may be established through the propositions *A* and *B*, and through the propositions *C* and *D*, or through the propositions *A* and *B*, or *A* and *C*, or *B* and *C*. For nothing prevents there being several middles for the same terms. But in that case there is not one but several syllogisms. Or again when each of the propositions *A* and *B* is obtained by syllogistic inference, e.g. *A* by means of *D* and *E*, and again *B* by means of *F* and *G*. Or one may be obtained by syllogistic, the other by inductive inference. But thus also the syllogisms are many; for the conclusions are many, e.g. *A* and *B* and *C*.
>
> But if this can be called one syllogism, not many, the same conclusion may be reached by more than three terms in this way, but it cannot be reached as C is established by means of *A* and *B*.[6]

Let us take the following example:

> Any body which is warmer than its environment emits more heat than it absorbs.
> Any body which emits more heat than it absorbs loses energy.
>
> Hence any body which is warmer than its environment loses energy.

[6] Aristotle. 1928. 41b36–42a8.

Any body which is warmer than its environment loses energy.
Any body which loses energy increases the energy of other
bodies.

Hence any body which is warmer than its environment
increases the energy of other bodies.

Thus, Aristotle has examined even instances comprising more than
three terms and more than two premises respectively. He even accepted
them as syllogisms but only as additions to the original syllogisms.
Therefore, despite the recognition that a syllogism generally may have any
number of terms, he insisted on the proper syllogism being composed of
not more than three terms.

The very same consideration made him stick to the thesis allowing not
more than three figures:

> If then we must take something common in relation to both, and this is
> possible in three ways (either by predicating A of C, and C of B, or C of
> both, or both of C), and these are the figures of which we have spoken, it is
> clear that every syllogism must be made in one or other of these figures. The
> argument is the same if several middle terms should be necessary to
> establish the relation to B; for the figure will be the same whether there is
> one middle term or many.[7]

I have tried to prove before that the three figures represent an arbitrary
restriction even in the case of simple syllogisms. It is even more so with
compound ones. The fact that the syllogisms concerned can be traced back
to simple ones does mean that they have no figures of their own.

When discussing the fourth figure, I have already referred to the source
according to which Galen distinguished the four figures of compound
syllogism. Let me recall here the consideration this distinction was based
upon. If the three figures of simple syllogism are joined in pairs, we have in
principle nine combinations, but only four of them will remain after a
selection according to the aspects introduced above.

This by itself would be enough to disprove Aristotle's view that
syllogism has but three figures. It is even more obvious if one proceeds
from the thesis that the simple syllogism has four figures. Then, having

---

[7] *Op. cit.*, 41a13–20.

joined them into pairs, sixteen combinations will be arrived at. Aristotle should be criticized not because he neglected their detailed analysis (why it is useless will be explained in the next chapter) but because he *a priori* excluded them from the field of examination.

Traditional logic, ignoring even Aristotle's standpoint, distinguished only two kinds of compound syllogism, namely (1) progressive and (2) regressive syllogisms.

$$(1)\ \frac{\begin{array}{l} M \text{ is } P \\ N \text{ is } M \end{array}}{N \text{ is } P} \qquad (2)\ \frac{\begin{array}{l} S \text{ is } N \\ N \text{ is } M \end{array}}{S \text{ is } M}$$

$$\frac{\begin{array}{l} N \text{ is } P \\ S \text{ is } N \end{array}}{S \text{ is } P} \qquad \frac{\begin{array}{l} S \text{ is } M \\ M \text{ is } P \end{array}}{S \text{ is } P}$$

This distinction was based on the consideration that in case (1) the subject of our propositions is gradually restricted while in case (2) the predicate of the original proposition is generalized.

How does this approach fit into the theory of simple syllogisms? It fits well in case (1) because it is composed of two syllogisms of the Barbara type. Case (2), however, is extremely problematic. As we know, according to the traditional convention the subject of the conclusion should occur in the minor premise. In addition, the terms of the conclusion succeed in a way that the extreme term of the minor premise is followed by the extreme term of the major premise. The regressive syllogism does not satisfy any of these conditions. Then, which original figure does this syllogism belong to?

This problem can be solved apparently by changing the sequence of the premises:

$$\frac{\begin{array}{l} N \text{ is } M \\ S \text{ is } N \end{array}}{S \text{ is } M} \qquad \frac{\begin{array}{l} M \text{ is } P \\ S \text{ is } M \end{array}}{S \text{ is } P}$$

Now it conforms to the convention. In this case, however, the formula consists of two inferences of the Barbara type just as in the progressive syllogism, and thereby the distinction becomes senseless.

Let us proceed further. It is obviously not only syllogisms of the two first figures which lead to valid compound inferences. Let us see an example:

$$P\,a\,M \qquad\qquad N\,e\,P$$
$$N\,e\,M \qquad\qquad N\,a\,S$$
$$\overline{} \qquad\qquad \overline{}$$
$$N\,e\,P \qquad\qquad S\,o\,P$$

The former syllogism belongs to figure (II) and the latter to figure (III). Several more versions can be seen in Drobisch.[8] That is, two out of the many possible combinations have been isolated here under a special aspect in a way which is inconsistent even within the theory of syllogistic.

III

There are still more deviations from the original syllogism:

> ... if both premisses have not been stated, we must ourselves assume the one which is missing. For sometimes men put forward the universal premiss, but do not posit the premiss which is contained in it, either in writing or in discussion: or men put forward the premisses of the principal syllogism, but omit those through which they are inferred, and invite the concession of others to no purpose. We must inquire then whether anything unnecessary has been assumed, or anything necessary has been omitted, and we must posit the one and take away the other, until we have reached the two premisses: for unless we have these, we cannot reduce arguments put forward in the way described. In some arguments it is easy to see what is wanting, but some escape us, and appear to be syllogisms, because something necessary results from what has been laid down ...[9]

Two deviations are mentioned in the above text. (1) The inference may include premises which are not necessary for drawing the conclusion and, therefore, should be avoided. As far as I know this issue has not been dealt with so far in logic. (2) One of the propositions is not explicitly stated in the inference. Such inferences were later called 'enthymemes' (simple abridged syllogism). For example: chemical elements may turn into one another, hence radium can be turned into lead. The proposition that radium and lead are chemical elements is omitted here.

[8] Drobisch, pp. 122–127.
[9] Aristotle. 1928. 47a14–24.

As to the filling in of the missing proposition traditional logic established the following principles: (1) If one of the premises and the conclusion are given, then the missing premise is a proposition which, together with the given premise, necessarily entails the conclusion. (2) If two premises are given the conclusion is a proposition which necessarily follows from the premises.

Thus, the aim of the above operation is to find a proposition which, together with the given propositions, would result in a valid inference.

> In everyday discourse most logical inference is enthymematic. We are constantly sparing ourselves the reiteration of known facts, trusting the listener to supply them where needed for the logical completion of an argument. But when we want to analyze and appraise a logical inference which someone has propounded, we have to take such suppressed premisses into account. At this point two problems demand solution simultaneously: the problem of filling in the details of a logical deduction leading from premises to desired conclusion, and the problem of eking out the premisses so that such a deduction can be constructed. Solution of either problem presupposes solution of the other; we cannot set up the deduction without adequate premisses, and we cannot know what added premisses will be needed until we know how the deduction is to run.[10]

The filling in of enthymemes has a special importance when one of the propositions is omitted from the inference with fraudulent intention.

An enthymeme can be not only simple but also compound.

> ... when the premisses are even, the terms must be odd; when the terms are even, the premisses must be odd: for along with one term one premiss is added, if a term is added from any quarter. Consequently since the premisses were (as we saw) even, and the terms odd, we must make them alternately even and odd at each addition.[11]

Aristotle's starting point was that a proper syllogism has three terms and two premises. Any number of terms but always the same number of premises can be added to this syllogism. Let us see the simplest case when only one term and one premise are added:

---

[10] Quine, p. 186.
[11] Aristotle. 1928. 42b12–17.

> Any body which is warmer than its environment emits more
> heat than it absorbs.
> Any body which emits more heat than it absorbs loses energy.
> Any body which loses energy increases the energy of other
> bodies.
> _____
> Hence any body which is warmer than its environment
> increases the energy of other bodies.

According to Aristotle's conditions we find here four terms and three
premises. However, on the ground that each premise has a subject and a
predicate and that it is compulsory for every two of them to be identical,
six terms and three premises can be detected here.

Inferences of this type were called '*soriticus syllogismus*' first by Marius
Victorinus (in the middle of the 4th c.). As an abbreviation the term
'sorites' (*chain of syllogisms*) became widely accepted later. The sorites is
an abridged syllogism where only the last conclusion occurs while the in-
between conclusions with their corresponding premises are omitted.

We have met progressive and regressive modes in the treatment of
compound syllogisms. Accordingly, chains of syllogisms are also
distinguished into (1) progressive and (2) regressive.

(1) M is P	(2) S is N
N is M	N is M
S is N	M is P
S is P	S is P

In this respect let me refer to what has been said about the
corresponding modes of compound syllogisms. To make this reasoning
more colourful I shall quote Themistocles' humourous inference:

> My son dominates his mother,
> his mother dominates me,
> I dominate Athens,
> Athens dominates Greece,
> Greece dominates Europe,
> Europe dominates the whole world,
> thus my son dominates the whole world.

Apart from such chains of syllogisms, *epicheirema* is also discussed in logical works. Epicheirema is an incomplete syllogism where one or both of the premises are enthymemes. For example:

> The school equipment should be taken care of because it belongs to public property.
> The desks belong to the school equipment.
> ___
> Hence the desks must be taken care of.

> M is P because K
> S is M
> ___
> S is P

In this inference the major premise is an enthymeme:

> What belongs to public property must be taken care of.
> School equipment belongs to public property.
> ___
> School equipment must be taken care of.

IV

Having analysed the modes of syllogism, Aristotle set out to solve the following problem:

> We must now state how we may ourselves always have a supply of syllogisms in reference to the problem proposed and by what road we may reach the principles relative to the problem: for perhaps we ought not only to investigate the construction of syllogism, but also to have the power of making them.[12]

Further on Aristotle discusses the aspects of establishing a syllogism at length. I think it is unnecessary to follow him on this way for the essence of the procedure concerned has already been summed up by Sándor Szalai:

> ... Aristotle who has defined syllogistical inference as a decisive form of demonstration tries to find the premises in a way that is rather strange to our way of thinking. Namely, the premise to be demonstrated is divided into two terms, i.e., a subject term and a predicate term, and then he finds a third

___

[12] *Op. cit.*, 43a20–24.

27

term which can be related to the two former ones as the subject of a true
proposition. That is, his reasoning went as follows: (a) it is to be
demonstrated that some mammals are aquatic animals. (b) What is the term
which can be equally related to the term 'mammal' and the term 'aquatic
animal' as a subject or a predicate? Perhaps the term 'seal'? Of the seal it
can, indeed, be stated that it is a mammal and also an aquatic animal. (c)
But if the subject and the predicate term of the conclusion can equally be
asserted of a middle term, then it is the third figure, and, indeed, these two
universal affirmative premises lead to a particular affirmative conclusion.
Consequently, the inference he looked for is:

> All seals are aquatic animals.
> All seals are mammals.
> _____
> Some mammals are aquatic animals.

It is not easy to decide whether Aristotle has actually used this reasoning,
which seems to us rather artificial, in his scientific research, or whether he
suggested it only in principle to establish the syllogistic proof of
propositions recognized as true. We generally find it entirely different to
recognize the truth of a proposition and to find its proof, on the one hand,
and to establish the proof in a syllogistical form, on the other. In fact, we
tend to avoid syllogistical constructions even when it would be possible to
demonstrate proof in the form of a syllogism.[13]

It is a fact, however, that scholastic controversialists in the Middle Ages
set forth their argumentations in the form of syllogisms, and those who
could subtly construct syllogisms were greatly respected. This is why, for
example, Duns Scotus was called 'doctor subtilis'.

It is common knowledge that in the modern age many have opposed this
method. Nevertheless, syllogistic has played a dominant part in logic, at
least until the end of the 19th century. One of the reasons for it is, no
doubt, comfortable conservatism, that is, sticking to traditions. But the
more important reason is that syllogistic could not be replaced by
anything more satisfactory. Therefore, logicians could not go beyond the
stage of 'critical realism'.

It may seem surprising that arguments which are met with in books or
conversations are seldom or never thrown into the form of regular

---

[13] Aristotelés, Organon. (Budapest 1961) pp. 380–381.

syllogisms. Even if a complete syllogism be sometimes met with, it is generally employed in mere affectation of logical precision. In former centuries it was, indeed, the practice for all students at the Universities to take part in public disputations, during which elaborate syllogistic arguments were put forward by one side and confuted by precise syllogisms on the other side. This practice has not been very long discontinued at the Universities of Oxford, and is said to be still maintained in some continental Universities; but except in such school disputations it must be allowed that perfectly formal syllogisms are seldom employed.[14]

Let us take the following into consideration: the construction of a syllogism aims at the unequivocal demonstration of the truth of a proposition. For simplicity's sake let us examine the traditional syllogism in which four kinds of proposition (*A, E, I, O*) occur. Their table of values is already known. How can one evaluate, for example, the proposition 'some mammals are aquatic animals' with the help of it? Since the terms of this proposition of type *I* are in a relation of crossing, its value is *t*.

Our job is, of course, not as easy in all cases. It is a well-known fact, e.g., that the proposition 'all swans are white' had been considered *t*, that is, its terms had been thought to be in a relation of subordination, but later it turned out to be a relation of crossing. Consequently, the knowledge of the relation between the terms of a proposition provides sufficient reason to decide its value.

What can, however, guarantee that this relation has been defined correctly? Nothing more nor less than the method applied so far, based on the following principle: if the premises are true and the inference is valid, then the conclusion is inevitably *t*. Logic can guarentee neither the truth of the premises nor the relation of terms being correctly stated. Its function is only to make clear what the result will be under certain conditions.

What is then the use of innovation? I can give the following answers:

(1) According to the prevailing opinion, one or more propositions which have already been proved are needed to demonstrate the truth of a proposition. That means that external means, namely other propositions, are used in order to guarantee sufficient reason. The relations of terms, on the contrary, belong to the *inner basis* of the propositions concerned.

[14] Jevons, 1872. p. 152.

27*

Thus, the proposition at issue, so to say, determines itself in the final analysis.

As we have seen, Aristotle built his system on the relations between terms. In this respect I share his view. He was, however, of the opinion that to reveal the logical relation of the terms a third one, the middle, is needed. I think this external term can be avoided.

One could object to the suggested method on the ground that it relies on experience and is not of a logical nature. Indeed, it is eventually based on experience, but not directly. The relation is indirect since terms and their relations belong to the domain of formal content. That is why Aristotle relied on them in his syllogistic.

(2) The method of term relations can be applied primarily in simple cases. What is the use of looking for the premises in order to demonstrate that the proposition 'every apple is fruit' is true? To use compound syllogism in this case would be like cracking nuts with a steam hammer, not to speak of the difficulties of finding premises within the framework of syllogistics to define the truth of a proposition like 'all birds are birds'.

18. THE INTERRELATIONS OF MEDIATE INFERENCES

A

I

The attempts to grasp the general have made the study of induction necessary.

> ... for two things may be fairly ascribed to Socrates — inductive arguments and universal definition...[15]

Socrates regarded induction as the way to proceed from the singular to the general.

These categories were still conditional upon one another in his theory: "... but Socrates did not make the universals or the definitions exist apart; *they*, however, gave them separate existence, and this was the kind

---

[15] Aristotle, *Metaphysica*. 1078b27–28.

of thing they called Ideas".[16] That is, it was Plato and his followers who introduced the interpretation of the general as something which exists independently and thereby separated it from the singular.

Aristotle attached great importance to induction.

> It is also clear that the loss of any one of the senses entails the loss of a corresponding portion of knowledge, and that, since we learn either by induction or by demonstration, this knowledge cannot be acquired. Thus demonstration developed from universals, induction from particulars; but since it is possible to familiarize the pupil with even the so-called mathematical abstractions only through induction—i.e. only because each subject genus possesses, in virtue of a determinate mathematical character, certain properties which can be treated as separate even though they do not exist in isolation—it is consequently impossible to come to grasp universals except through induction. But induction is impossible for those who have not sense-perception. For it is sense-perception alone which is adequate for grasping the particulars: they cannot be objects of scientific knowledge, because neither can universals give us knowledge of them without induction, nor can we get it through induction without sense-perception.[17]

He held that cognition proceeds from sense-perception. This opinion was often interpreted later as evidence for his being a sensualist. In fact, he regarded sense-perception only as the starting point, or the lowest level, of cognition for it gives no answer why things are the way they are. E.g., our senses tell us that fire is hot but not why.[18]

Consequently, cognition has also a higher level, namely thought which can grasp the universal.

> If there is nothing apart from individuals, there will be no object of thought, but all things will be objects of sense, and there will not be knowledge of anything, unless we say that sensation is knowledge.[19]

Of ancient logicians it was, first of all, Epicurus and his school who dealt with induction. Philodemus discussed this method in his work *On Signs*

---

[16] *Op. cit..* 1078b30–32.
[17] Aristotle. 1928. 81a38–81b9.
[18] Cf.: Aristotle, *Metaphysica.* 981b16–20.
[19] *Op. cit.*, 999b1–4.

and Designations. Lucretius Carus argued that cognition is based on perception for if the senses fail arguments will fail as well.

The interest of medieval logic was centered on syllogistic. Of induction, if treated at all, mainly perfect induction was studied.

The boom of production and trade in modern times forced the sciences to develop at a higher rate, too. That is how observation, or rather experimentation, and as a concomitant, the inductive method came to the front.

> But not only is a greater abundance of experiments to be sought for and procured, and that too of a different kind from those hitherto tried; an entirely different method, order, and process for carrying on and advancing experience must also be introduced. For experience, when it wanders in its own track, is, as I have already remarked, mere groping in the dark, and confounds men rather than instructs them. But when it shall proceed in accordance with a fixed law, in regular order, and without interruption, then may better things be hoped of knowledge.[20]

Bacon made an attempt at a detailed elaboration of the inductive method in the second volume of the *Novum Organum*. Apart from certain useful aspects to be discussed later, this attempt as a whole failed. J. S. Mill, by having developed the system of inductive logic, made a more significant progress in this field. The following statement is typical of his approach:

> In every induction we proceed from truths which we knew, to truths which we did not know; from facts certified by observation, to facts which we have not observed, and even to facts not capable of being now observed; future facts, for example; but which we do not hesitate to believe on the sole evidence of the induction itself.[21]

While the adherents of induction overestimated the procedure concerned, other empiricists underestimated it. Russell defined the function of logical analysis as the revelation of the simplest—as he put it, atomic—facts and their registration in elementary propositions. If two or more elementary propositions were united by simple logical relations (e.g., conjunction), what he called molecular propositions were arrived at.

[20] Bacon, *Selection* p. 362 (*The New Organon*, 1 (C)).
[21] Mill, p. 163.

Nevertheless, there are universal propositions as well which should be distinguished from the former ones. Generalization must proceed from the fact that there are infinitely many things for if they were but of a finite number, then all universal propositions could be composed from elementary ones and would thus become common molecular propositions. Whoever regards, for example, the proposition 'all men are mortal' as a proposition of the subject and predicate type presupposes that there exists an 'all men', too, apart from the individual men. The class of men, however, does not exist in reality, it is but a logical fiction. In order to avoid mistaken philosophical speculations the propositions concerned should be formulated like this: it holds true of all individuals that if they are men, then they are mortal.

What is achieved by this new formulation? In Cornforth's opinion the following: the only way to state the truth of any generalization is to examine all of its instances one by one in order to find out whether each of them is true. If, for example, we want to assert that all men are mortal, we have to state that John has died, James has died, Steve has died, and so on in the case of all men. Since, however, the instances of generalization are more often than not of an infinite number and a generalization very often continues to refer to the future as well, however many be the instances we have already demonstrated, always new demonstration will be needed. That is, it is not only practically, but often even logically, impossible to decide the truth of a generalization. Consequently, truth in the absolute and strict sense cannot be applied to generalizations, or universal propositions, in the same way as to elementary theses.[22]

According to Marxism: (a) the universal exists as primary, though not independently, in material reality (thus, not only individual men but also the class of men exist objectively), (b) the universal cannot be separated from the particular.

Marxist logicians, however, disagree among themselves as far as the authenticity, methodology, and evaluation of induction is concerned. That we shall see later.

Let us proceed now to the study of the various kinds of induction.

[22] Cf.: Cornforth.

II

Aristotle wrote the following about the perfect inductive inference:

> Now induction, or rather the syllogism which springs out of induction, consists in establishing syllogistically a relation between one extreme and the middle by means of the other extreme, e.g. if B is the middle term between A and C, it consists in proving through C that A belongs to B. For this is the manner in which we make inductions. For example let A stand for long-lived, B for bileless, and C for the particular long-lived animals, e.g. man, horse, mule. A then belongs to the whole of C: for whatever is bileless is long-lived. But B also ('not possessing bile') belongs to all C. If then C is convertible with B, and the middle term is not wider in extension, it is necessary that A should belong to B. For it has already been proved that if two things belong to the same thing, and the extreme is convertible with one of them, then the other predicate will belong to the predicate that is converted. But we must apprehend C as made up of all the particulars. For induction proceeds through an enumeration of all the cases.[23]

It is only in the case of Darapti that, according to the rules of syllogism, a valid inference results from the given premises:

Man, horse, mule, etc. are long-lived.
Man, horse, mule, etc. are bileless.
___
Some bileless living beings are long-lived.

With the signs of the Aristotelean and the traditional logic, respectively:

All C is A	$M_1, M_2, M_3 \ldots M_n$ is P
All C is B	$M_1, M_2, M_3 \ldots M_n$ is S
Some B is A	Some S is P

Yet, an inference can be said to be inductive only if the conclusion is a universal proposition. Aristotle, therefore, had recourse to the following argument:

> ... when A and B belong to the whole of C, and C, is convertible with B, it is necessary that A should belong to all B: for since A belongs to all C, and C to B by conversion, A will belong to all B.[24]

[23] Aristotle. 1928. 68b15–29.
[24] *Op. cit.*, 68a22–25.

This passage is worth attention from two points of view. On the one hand, Aristotle, in principle, took the following position as to the conversion of universal affirmative propositions: "... the terms of the affirmative [universal—Gy. T.] must be convertible, not however universally, but in part ..."[25] In the above reasoning, however, he went beyond this limitation by admitting in a certain case the terms of a proposition of type A can be transposed, that is, can be converted not by limitation only but also simply.

On the other hand, Aristotle made use of *additional knowledge* this time. He did not restrict the proposition 'all C is B' to C being a part of B, but considered also the possibility that B at the same time is also a part of C, i.e. that the two terms have the same extension.

All C is A	$M_1, M_2, M_3 \ldots M_n$ is P
All C is B	$M_1, M_2, M_3 \ldots M_n$ is S
C is identical with B	$M_1, M_2, M_3 \ldots M_n$ altogether is identical with S

This is the very case when simple conversion is justified as an indispensable condition for the inference under discussion to lead to a universal conclusion:

All C is A	$M_1, M_2, M_3 \ldots M_n$ is P
All B is C	S is $M_1, M_2, M_3 \ldots M_n$
All B is A	All S is P

We have arrived at an inference of the Barbara type. Now we can state with good reason that there is an induction of a syllogistical nature.

The above example already carried the germs of a problem which has become more acute in modern logic, namely: whether or not it is correct to rank perfect induction among inductive inferences. The opinion that perfect induction is one kind of syllogism has become prevailing.

> *Perfect induction* consists in observing all the possible particular instances and deriving from them a universal truth as a result. This

induction is similar to inference as they coincide in certain aspects but they also differ from one another.

What induction and inference have in common is that they are both procedures of demonstration leading to a certain truth; but they differ in so far as inference proceeds from the universal to the particular while induction, the other way round, starts from the particular to arrive at the universal.[26]

Mathematical logicians are of a similar opinion but relegate perfect induction to the domain of deductive inferences and not to that of syllogisms.

Their conception could be formulated like this:

> All necessary inferences are deductions.
> A perfect induction is a necessary inference.
> _____
> A perfect induction is deduction.

Perfect induction can be ranked among syllogisms for the fact that its conclusion necessarily follows from the premises. Nevertheless, inferences are divided into deductive and inductive according to as our cognition proceeds and not as the premises and the conclusion are related. The reason why perfect induction is not deductive is that in this case thought proceeds not from the universal to the particular but from the particular to the universal.

Mill assessed this problem in a way different from the usual:

> If we were to say, All the planets shine by the sun's light, from observation of each separate planet, or All the Apostles were Jews, because this is true of Peter, Paul, John, and every other apostle,—these, and such as these, would, in the phraseology in question, be called perfect, and the only perfect, Inductions. This, however, is a totally different kind of induction from ours; it is not an inference from facts known to facts unknown, but a mere short-hand registration of facts known. The two simulated arguments which we have quoted, are not generalizations; the propositions purporting to be conclusions from them, are not really general propositions. A general proposition is one in which the predicate is affirmed or denied of an unlimited number of individuals; namely, all, whether few or many ... [27]

[26] Huszár, p. 273.
[27] Mill, pp. 288–289.

It is an arbitrary restriction to think that the subject of a universal, or general, proposition must be indefinite. The universal character of a proposition is independent of whether the number of objects reflected in the subjects is finite or infinite. Mill needed this arbitrary restriction in order to sort out perfect induction from among inductive inferences. As the historical survey has already revealed, he opposed induction to syllogism. Since perfect induction has a syllogistical character he had to give some kind of explanation why it cannot be regarded as induction. By the way, he contradicted even himself by stating elsewhere the following:

> When the predicate is affirmed or denied of all and each of the things denoted by the subject, the proposition is universal; when of some undefined portion of them only, it is particular.[28]

The assertion that nothing but a progress from known facts to unknown can be regarded as inference also implies an arbitrary restriction. This interpretation is true only of heuristic arguments. If Mill's restricted definition were accepted the presence of inferences in the procedures of demonstration should also be denied.

But does perfect induction really fail to provide us with new knowledge? Let us see Fogarasi's answer:

> Perfect induction presupposes that we know and take into consideration *all* individual cases which belong to a universal ('planet'). Therefore the universal statement expressed in a perfect induction yields nothing new for knowledge.[29]

With regard to syllogism he wrote the following:

> We obtain the universal (the general term, the universal proposition) by generalizing numerous or countless individual experiences. But *after* we are in possession of universal statements, propositions, we can in the further course of knowledge draw conclusions about the individual. That is the function of the syllogism in the process of knowledge.[30]

---

[28] *Op. cit.*, p. 84.
[29] Fogarasi, p. 261.
[30] *Op. cit.*, p. 219.

The view that a perfect inductive inference does not give new knowledge results from the failure to understand the difference between a mechanical summary of propositions asserted of particular objects, on the one hand, and universal propositions, on the other. The conclusion of a perfect induction implies also that the particular objects referred to in the premises belong to the same class and exhaust this very class. This thesis, of course, requires more than the assertions of these particular objects, that is, certain additional knowledge is needed as well.

Let us take the following familiar example.[31]

> The Earth is of spherical form. Mercury is of spherical form. Mars is of spherical form. Venus is of spherical form, etc.
>
> ———————————————
>
> Consequently, all planets are of spherical form.

The above induction in its present form still cannot be regarded as an inference because it lacks the expression of the knowledge that only these planets belong to the planets of our solar system. Without this knowledge we cannot be sure whether or not our solar system has a planet which is not of spherical form. If, however, this knowledge is contained in the premises, the result will be a syllogistical inference:

> The Earth is of spherical form. Mercury is of spherical form. Mars is of spherical form. Venus is of spherical form, etc.
>
> Only the above listed planets make up the planets of our solar system.
>
> ———————————————
>
> Consequently, all planets of our solar system are of spherical form.
>
> M_1 is P, M_2 is P, M_3 is P ... M_n is P
> Only M_1, M_2, M_3 ... M_n is S
> ———————————————
> All S is P

Perfect induction plays an important role in compound demonstrations where the class of the given objects of which some thesis is to be proved is

———————

[31] Cf.: *op. cit.*, p. 261.

divided into independent groups. Then, the truth of the thesis at issue is demonstrated in each group one by one. In this way a perfect induction provides a universal conclusion which extends the truth of the demonstrated thesis to all objects of the given class.

This procedure is widely used in mathematics. When, for example, the following geometrical thesis is to be proved: 'all circumferential angles in a circle are one half of the central angle resting on the arc drawn between their sides', first the three possible alternatives for drawing an angle within a circle are selected. In the first case the centre of the circle is between the sides of the angle drawn into the circle, in the second the centre of the circle lies on one side of the angle, and in the third it is outside both sides of the angle. These three instances exhaust all the possibilities of how an angle drawn within a circle can be situated. Having established these alternatives, the truth of the thesis has to be demonstrated in the first, the second, and the third case one by one. Then all the three instances are combined as a whole, and the conclusion offers a general thesis which holds true of all angles drawn within a circle.

A perfect induction, however, can be used to a limited extent only for it requires the examination of all the objects one by one. In the majority of cases one has to do with facts of infinite number, or if their number is finite, usually not each of them can be studied directly. In such cases the universal conclusion is drawn with the help of an imperfect induction.

III

As to imperfect induction, the following can be read in Aristotle:

> ... perception must be of a particular, whereas scientific knowledge involves the recognition of the commensurate universal. So if we were on the moon, and saw the earth shutting out the sun's light, we should not know the cause of the eclipse: we should perceive the present fact of the eclipse, but not the reasoned fact at all, since the act of perception is not of the commensurate universal. I do not, of course, deny that by watching the frequent recurrence of this event we might, after tracking the commensurate universal, possess a demonstration, for the commensurate universal is elicited from the several groups of singulars.[32]

[32] Aristotle. 1928. 87b38–88a4.

It was perfect induction which was favoured in the Middle Ages due to its syllogistic nature while all induction based on an indefinite number of instances were rather despised. Contrary to Grazia who represented this view, Galileo argued as follows:

> I would only like to point out what a bad logician he (Grazia) turns out to be; for he does not understand that *introduction would be impossible or superfluous if it had to handle every particular case*. Impossible, if the particular (individual) cases are countless; but when they can be counted, the consideration of every single case would make the induction superfluous, or better, what could be learned through induction would be nothing. Thus if there were only three men in the world, the determination: Andrew runs, James runs, and John runs, therefore all men run—would be a superfluous inference and simply the repetition of one and the same thing, just as if we were to say: Since Andrew runs, James runs, and John runs, thus Andrew, James, and John run. And since the number of particular (individual) cases is usually infinite, an argument using induction has sufficient force if the proof appeals to those particulars, which are to the greatest extent suitable.[33]

Galileo was right in criticizing the overestimation of syllogistic induction but he made a mistake by rejecting it. The example in the above quotation throws light on the source of this mistake. Let us examine the following versions:

(1)   Andreas runs, Jakob runs, Johann runs.
         _____
         Hence all men run.

The given premise does not provide sufficient reasons for drawing this conclusion.

(2)   Andreas runs, Jakob runs, Johann runs.
         _____
         Andreas, Jakob, and Johann run.

The subject of the premise and the conclusion respectively has the same extension, therefore we cannot speak of induction here.

---

[33] Quoted in Fogarasi, pp. 262–263.

(3)   Andreas runs, Jakob runs, Johann runs.
            Only Andreas, Jakob, and Johann are men.
            _____

            Andreas, Jakob, and Johann run.

The middle term must not occur in the conclusion of the syllogism.

(4)   Andreas runs, Jakob runs, Johann runs.
            Only Andreas, Jakob, and Johann are men.
            _____

            All men run.

In this case we have a perfect induction, and to identify it with any of the previous versions would be a misinterpretation. In order to demonstrate the importance of imperfect induction it is unnecessary to deny the merits of the perfect induction.

In everyday life one often generalizes merely on the ground that some occurrence has been observed in several cases and has had the same property in all these cases. This way of generalization is called *induction by simple enumeration* (*inductio per enumerationem simplicem*), or *popular induction* (*inductio popularis*).

However, the study of several particular instances in itself is not a sufficient reason yet to make a universal assertion.

> ... inductio mala est, quae per enumerationem simplicem principia concludit scientiarum, non adhibitis exclusionibus et solutionibus, sive separationibus naturae debitis.[34]

Mill made a similar statement:

> Popular notions are usually founded on induction by simple enumeration; in science it carries us but a little way. We are forced to begin with it; we must often rely on it provisionally, in the absence of means of more searching investigation. But, for the acute study of nature, we require a surer and a more potent instrument.[35]

The premises of an induction by simple enumeration can be formulated in this way:

_____

[34] Francis Bacon *Works*. p. 179 (*Novum Organum*, 1 (LXIX)).
[35] Mill, p. 313.

$$M_1, M_2, M_3 \ldots M_n \text{ is } P$$
$$M_1, M_2, M_3 \ldots M_n \text{ is } S$$

Since the number of instance is infinite there is no way to identify the extension of M and S as in the case of perfect induction. Therefore, only the proposition 'some S is P' follows from the premises (it is again a Darapti mode). The conclusion 'all S is P' can be accepted as a more or less probable supposition which requires further evidence. If no more is expected from this operation it can be used not in everyday thought only, but also in science. It is not always possible at the initial stages of an investigation into occurrences to find sufficient reason for generalization. Nevertheless, generalization, even if in a form of supposition, is indispensable for further research. Induction by simple enumeration has a great part to play in the creation of hypotheses.

We must bear in mind, however, that the majority of such inductions leads to false theses. For centuries people have come across such phenomena as that swans are white and metals sink in water. It has been observed also that wherever a void occurs the air or fluid around it fills this void up immediately. It has been concluded from all these observations that all swans are white, all metals sink in water, nature abhors emptiness, etc. These, and many other, popular inductions have been disproved by facts contradicting them in the course of further research. Consequently, the results of induction by simple enumeration must not be regarded as authentically true theses.

Bacon formulated the following requirement for scientific induction:

> In establishing axioms another form of induction must be devised than has hitherto been employed, and it must be used for proving and discovering not first principles (as they are called) only, but also the lesser axioms, and the middle, and indeed all. For the induction which proceeds by simple enumeration is childish: its conclusions are precarious and exposed to peril from a contradictory instance, and it generally decides on too small a number of facts, and on those only which are at hand. But the induction which is to be available for the discovery and demonstration of sciences and arts must analyse nature by proper rejections and exclusions,

and then, after a sufficient number of negatives, come to a conclusion on the affirmative instances; which has not yet been done or even attempted, save only by Plato, who does indeed employ this form of induction to a certain extent for the purpose of discussing definitions and ideas. But in order to furnish this induction or demonstration well and duly for its work, very many things are to be provided which no mortal has yet thought of; insomuch that greater labour will have to be spent in it than has hitherto been spent on the syllogism. And this induction must be used not only to discover axioms, but also in the formation of notions. And it is in this induction that our chief hope lies.[36]

Bacon made an attempt at the development of those methods which contribute to the generalization of empirical data. He was the first to point out a method which tries to register the presence of the phenomenon to be investigated. He called it the table of essence and presence (*Tabula Essentiae et Presentiae*).—The text quoted above reveals that Bacon attached great importance to the exclusion of negative instances. After a detailed description he summed up this procedure in the table of declinations or close absence (*Tabula Declinationis, sive Absentiae in proximo*), and, finally, stated the third mode of induction called the table of grades and comparison (*Tabula Graduum, sive Comparativa*) which is based on concomitant alterations.

Let us take a simple example of scientific induction. Having observed that copper, iron, and mercury are good conductors of electricity, we are confronted with the question what is the reason for this phenomenon. We can assert that this property of the listed materials is produced by their free electrons. Knowing that the presence of free electrons is a feature that metals have in common, we can infer in an inductive way the fact that all metals are good conductors of electricity. In an expanded form it looks like this:

> Copper, iron, mercury, etc. are good conductors of electricity because they have free electrons.
>
> Copper, iron, mercury, etc. are metal.
> All metals have free electrons.
> ___
> All metals are good conductors of electricity.

[36] Bacon, *Selection* p. 364 (*The New Organon*, 1 (CV)).

28

$$M_1, M_2, M_3 \dots M_n \text{ is } P, \text{ because } K$$
$$M_1, M_2, M_3 \dots M_n \text{ is } S$$
$$\underline{\text{All } S \text{ is } K}$$
$$\text{All } S \text{ is } P$$

In order to increase the probability of the conclusion being true, books on logic call attention to the following requirements which should be kept in mind: first of all, one must find as many facts supporting the conclusion as possible. The conscientious selection of facts plays a big role in increasing the probability of the conclusion. Such typical objects of a given class should be observed in which the properties manifest themselves in a characteristic way. At the same time it must be born in mind that these objects should be examined under the most varied conditions.

What the popular and the scientific inductions have in common is that they are based on the study of a more or less great number of instances. But is enumeration really indispensable to ground a general thesis? Mill gave the following answer:

> Not all the instances which have been observed since the beginning of the world, in support of the general proposition that all crows are black, would be deemed a sufficient presumption of the truth of the proposition, to outweigh the testimony of one unexceptionable witness who should affirm that in some region of the earth not fully explored, he had caught and examined a crow, and had found it to be grey.
>
> Why is a single instance, in some cases, sufficient for a complete induction, while in others, myriads of concurring instances, without a single exception known or presumed, go such a very little way towards establishing an universal proposition? Whoever can answer this question knows more of the philosophy of logic than the wisest of the ancients, and has solved the problem of induction.[37]

The requirement of taking into consideration the number of instances examined introduces a factor of uncertainty into induction. Since the number of facts is infinite however many instances have been studied the occurrence of a contradictory instance must always be reckoned with. Is there any criterion to decide whether or not the number of instances

[37] Mill, p. 314.

examined is sufficient to justify the conclusion? As far as I know nobody has found anything like it, and thus the opinion that scientific induction is also a probable inference can be shared.

On this ground some argue as follows:

> ... all inductive inferences have problematic elements. The switch over from particular and often accidental facts to general and necessary statements is always uncertain to a certain extent, or rather, nothing but probable. Its verification and logical ground raises an important question which has created many difficulties for logic from the very beginning.[38]

This statement, however, is a mistaken generalization because (1) it obviously does not hold true either of a perfect induction, or (2) of all imperfect inductions. The latter will be treated in the following.

IV

Let us examine the following terms: 'man'—'living being'. Their relation is characterized, on the one hand, by the former being a specific term and the latter a generic one. On the other, the given terms are in a relation of the subalternate and the subalternant. Let us construct a proposition from these terms, e.g., 'this man is a living being'. Both the subject and the predicate of this proposition can be generalized, even both of them at the same time. Accordingly, the following versions are obtained:

(1)   This man is a living being.
(2)   All men are living beings.
(3)   This man is mortal.
(4)   All men are mortal.

Aristotle commented on this issue in his work '*On the Categories*' as follows:

> When one thing is predicated of another, all that which is predicable of the predicate will be predicable also of the subject. Thus 'man' is predicated of the individual man; but 'animal' is predicated of 'man'; it will, therefore, be predicable of the individual man also: for the individual man is both 'man' and 'animal'.[39]

[38] Marković, p. 51.
[39] Aristotle. 1928. 1b10–16.

28*

The above relation was formulated in medieval logic in two different ways. (1) The property of the property of an object is at the same time the property of the object as well, and what contradicts the property of an object contradicts the object itself (*nota notae est rei ipsius, repugnans notae, repugnat rei ipsi*). (2) A predicate which belongs to a predicate belongs to the subject as well (*regula de quocunque*).

The question to what extent these theses can be regarded as axioms of the syllogism aroused a controversy in modern logic. Fogarasi wrote the following:

> According to some logicians the *dictum de omni* is the axiom of the syllogism, and the principle of the character of the object is its consequence. According to Kant the situation is just the reverse: the *dictum de omni* follows from the latter principle. In my opinion we are dealing with the same axiom, which is applied in the one case to the extension of the term and in the other to its content. For this reason it seems as if the one could be inferred from the other. In fact the axiom expresses actually existing relations in reality—namely the relation of a whole to its parts. If a part (*A*) is part of a whole (*B*), which itself in turn forms part of a greater part (*C*), then this part (*A*) is itself also a part of *C*. The middle term (*B*) is in the one case a whole and in the other a part.[40]

Fogarasi is right in so far as the above comparison is forced. Still it does not solve the problem. In the history of logic a trend developed claiming that all syllogisms are deductive inferences. It has played a dominant role even in our time. But let us have a closer look at Aristotle's example:

> An individual man is man.
> Man is a living being.
> _____
> An individual man is a living being.

The predicate of the conclusion is the generalization of the predicate of the major premise. Thus, from the point of view of the predicate, it is an inductive inference. However, the above inference could be said to contain deduction as well. I do not deny this possibility since I admit the unity of induction and deduction. In a given aspect, however, one of them prevails, and it is induction in the present case.

[40] Fogarasi, p. 224.

Why was induction reduced to the generalization of the subject in traditional logic? It is common knowledge that Aristotle attributed a privileged role to Barbara among the modes of syllogisms. This mode can be described also like this: if S is subordinated to M, and M is subordinated to P, then S is subordinated to P. On this ground Aristotle defined the valid modes in a way that the subject of the conclusion was identical with the extreme term of the minor premise which, therefore, became called later the minor term. Let us take an inference of the Darapti type as an example:

All M is P (major term)
All M is S (minor term)
———————————————
Some S is P

Aristotle has stated nowhere that the major term of the premise cannot be used as the subject of the conclusion. He himself, however, ignored this version in the analysis of syllogisms.

His successors followed his example as far as simple syllogisms were concerned but disagreed with him in the treatment of compound syllogisms. That is, they distinguished two kinds of the latter. What they called progressive syllogism was defined as a gradual limitation of the subject of the original proposition, e.g.:

The industrialization of our country makes the supply of tractors and other machines for agriculture possible.

The supply of agricultural machines is an indispensable condition for increasing the crop capacity of agriculture.

———————————————————————————————————————

The industrialization of our country makes the increase of the crop capacity of agriculture possible.

The ever increasing demand of the population can, and must, be met primarily by increasing the crop capacity of agriculture.

———————————————————————————————————————

Hence the industrialization of our country is indispensable in order to better fulfil the demands of the population.

Its formula is:

$$M \text{ is } P$$
$$N \text{ is } M$$
$$\overline{N \text{ is } P}$$
$$S \text{ is } N$$
$$\overline{S \text{ is } P}$$

In what is called regressive syllogism, on the contrary, the predicate of the original propositions will be generalized:

Any body which is warmer than its environment emits more heat than it absorbs.

Any body which emits more heat than it absorbs loses energy.

Hence any body which is warmer than its environment loses energy.

Any body which loses energy increases the energy of other bodies.

Hence any body which is warmer than its environment increases the energy of other bodies.

$$S \text{ is } N$$
$$N \text{ is } M$$
$$\overline{S \text{ is } M}$$
$$M \text{ is } P$$
$$\overline{S \text{ is } P}$$

The above example illustrates that the progressive syllogism uses a deductive inference and the regressive an inductive one. The traditional designations are misleading.

As we have seen, traditional logic acted inconsistently in the analysis of simple and compound syllogisms respectively. In order to overcome this inconsistency let us proceed from the comparison of the following instances:

(I)	(II)
All men (M) are mortal (P).	All Greeks (S) are men (M).
All Greeks (S) are men (M).	All men (M) are mortal (P).
All Greeks (S) are mortal (P).	All Greeks (S) are mortal (P).

In case (I), it is the limitation of the subject of the original proposition which stands in the foreground, while in case (II) it is the generalization of the predicate of the original proposition. As can be observed at the same time, in the former the predicate and in the latter the subject of the original proposition and that of the conclusion are identical, and consequently, have not changed in the course of inference. Case (I) is a deduction, and case (II) an induction.

According to traditional logic case (II) can be traced back to case (I) by a transposition of the premises (*mutare*).

> In whatever sequence the three propositions of an inference follow, it does not affect the validity of the inference for there must be only a logical and not a sequential connection in an inference. Consequently, one can combine first S with M and then M with P. And, indeed, this form of inference is employed by several Arab philosophers who take the middle as the first proposition and the major premise as a second one. It is a good sequence of the propositions because it makes the nature of the middle thesis clearer both in thought and in the form of inference. For example:
>
> soul is spirit (middle premise)
> spirit is immortal (major premise)
> _____
> hence soul is immortal (conclusion)
>
> Nevertheless, the best of all is to follow Aristotle. He put the major premise, i.e., the combination of M and P, in the first place, the middle, i.e., the combination of S and M, in the second, and closed the whole by a conclusion, i.e., by a combination of S and P. This arrangement of the propositions is the best of all because in this way the reasoning becomes more clear and understandable.[41]

But there is a hitch somewhere in this argumentation. The generally accepted operations of tracing back are the following: conversion, *reductio ad absurdum*, and selection. The justification of transposition has been contested even in traditional logic, not to mention that the validity of this operation has not been proved yet. It is certain, however, that transposition makes the difference of the above two cases disappear, i.e., that in case (1) the subject is limited (man—Greek) while in case (II) the

[41] Huszár, p. 188.

predicate is generalized (living being—mortal). To do away with this difference is in the interest of the adherents of deductive logic.

As the historical survey has revealed, syllogisms as a whole have been identified with deduction. In consequence what did not fit in the Procrustean bed of this interpretation was either ignored, or distorted. Having broken out from this vicious circle, we may conclude that inductive inferences can be just as well syllogistic as deductive.

B

I

This is what Aristotle wrote of the relation of induction and deduction:

> There is on the one hand Induction, on the other Reasoning. Now what reasoning is has been said before: induction is a passage from individuals to universals, e.g. the argument that supposing the skilled pilot is the most effective, and likewise the skilled charioteer, then in general the skilled man is the best at this particular task. Induction is the more convincing and clear: it is more readily learnt by the use of the senses, and is applicable generally to the mass of men, though Reasoning is more forcible and effective against contradictious people.[42]

Later in ancient times a polarization took place. The dominant school of thought concentrated on deduction and was represented primarily by the Peripatetics and the Stoics. As we have seen, the adherents of induction were, first of all, Epicurus and his followers.

The skeptics were critical of the above trends. Sextus Empiricus made the following objection to syllogism: the truth of the major premise can be demonstrated by induction only, but this induction implies the truth of the conclusion. The proposition, for example, that all men are living beings, can be regarded as valid only if we (already) know that Socrates is a living being. In this way, however, syllogisms turn into a vicious circle, and the conclusions are in advance contained in the major premises.

In the Middle Ages, universal theses were not based on experience for experience was generally despised. This approach, no doubt, was of some

---

[42] Aristotle. 1928. 105a11–17.

help in avoiding the above mentioned vicious circle. But at what price? The main goal of scholasticism was to develop a uniform and comprehensive system of theological and philosophical theses. The schoolmen tried to achieve this goal by the following means: the system was built on the statements of authorities which became the axioms of the system. By elaborating the works of the authorities they selected and defined categories. The axioms, the categories, and the definition served as a starting point for further research.

Deduction was supposed to be the main method in the further development of this system. It was applied in various ways, e.g., one of these procedures, explication (*lectio*), implied the study and the commentary of a text by an author. What they were striving for in the process was to demonstrate the theses occurring in the text in a pedantic way.

Disputes (*disputatio*) became common practice in the period of scholasticism. Deduction prevailed since the disputants set forth their arguments in syllogistical forms. Those who were good at constructing syllogisms were held in high esteem. Authoritative proofs had a considerable part in these disputes. This method involved frequent reference to various quotations from the works of authorities from which the theses to be demonstrated could be deduced.

To avoid giving a one-sided survey let me note that induction was studied in medieval logic, among others, by Occam and Buridan.

Bacon fiercely opposed the scholastic view:

> The syllogism consists of propositions, propositions consist of words, words are symbols of notions. Therefore if the notions themselves (which is the root of the matter) are confused and overhastily abstracted from the facts, there can be no firmness in the superstructure. Our only hope therefore lies in a true induction.[43]

It would be a mistake, however, to regard Bacon as a one-sided propagator of induction.

---

[43] Bacon *Works* Vol. I. p. 153 (*Novum Organum*, 1 (XIV)); F. H. Anderson tr, p. 41.

> Those who have handled sciences have been either men of experiment or men of dogmas. The men of experiment are like the ant: they only collect and use; the reasoners resemble spiders, who make cobwebs out of their own substance. But the bee takes a middle course: it gathers its material from the flowers of the garden and of the field, but transforms and digests it by a power of its own. Not unlike this is the true business of philosophy, for it neither relies solely or chiefly on the powers of the mind, nor does it take the matter which it gathers from natural history and mechanical experiments and lay it up in the memory whole, as it finds it, but lays it up in the understanding altered and digested.[44]

Now let us follow the reasoning of a follower of Bacon, namely Mill. Before proceeding to the discussion of induction, he first analysed the syllogism. His basic objection coincided with that of the skeptics: all syllogisms, as far as they serve as a proof for a conclusion, involve a *petitio principii*.

> From this difficulty there appears to be but one issue. The proposition that the Duke of Wellington is mortal, is evidently an inference; it is got at as a conclusion from something else; but do we, in reality, conclude it from the proposition, All men are mortal? I answer, no.[45]

But, then, what from?

> ... if we have a collection of particulars sufficient for grounding an induction, we need not frame a general proposition; we may reason at once from those particulars to other particulars.[46]

The majority of logicians mean by inference a logical operation by which the conclusion necessarily follows from the premises. Mill was right in pointing out that there are also probable inferences which do not require universal propositions. What happens, however, when a necessary inference is the goal?

> When, therefore, we argue from a number of known cases to another case supposed to be analogous, it is always possible, and generally advantageous, to divert our argument into the circuitous channel of an induction

---

[44] Bacon, *Selection* p. 360 (*The New Organon,* 1 (XCV)).

[45] Mill, p. 186.

[46] *Op. cit.,* p. 196.

from those known cases to a general proposition, and a subsequent application of that general proposition to the unknown case.[47]

What result have we obtained? If we know that Socrates, Cicero, Napoleon, etc., are mortal we may conclude that the Duke of Wellington is mortal. This inference, however, will be valid only if the proposition 'all men are mortal' is used as an intermediate. The latter proposition presupposes the conclusion. In this way the universal proposition proves to be an intermediate in, and not the starting point of, the inference. Yet, the fact that the universal proposition is indispensable remains unchallenged and so the problem of the vicious circle is still there.

Fogarasi wrote the following in this respect:

> Stuart Mill achieved a great effect with the following argument: In the syllogism: All men are mortal etc., it is already comprehended, already presupposed that Caius is mortal. If we were not convinced that every single man is mortal, we could not assert the major premiss. The major premiss already anticipates the conclusion...
>
> Mill is correct in his interpretation of the given example. but he is mistaken in maintaining that the syllogism as such is a senseless repetition and that it is not adequate to the real needs of thought. It is not the syllogism itself but rather the example which is bad and useless. Unfortunately, the examples used in recent textbooks on logic are not much better.[48]

I have two objections to raise: (1) Mill has not asserted that syllogism is a senseless repetition.

> ...I must yet enter a protest, as strong as that of Archbishop Whately himself, against the doctrine that the syllogistic art is useless for the purposes of reasoning. The reasoning lies in the act to generalization, not in interpreting the record of that art; but the syllogistic form is an indispensable collateral security for the correctness of the generalization itself.[49]

(2) Fogarasi failed to offer an example better than the one he had criticized, although that would have been in vain, of course, since the solution to this problem is independent of what example is chosen.

[47] *Op. cit.*, p. 197.
[48] Fogarasi, p. 219.
[49] Mill, p. 196.

Neither Mill, nor his followers could demonstrate in what way a universal proposition follows necessarily from particular facts (supposing the number of instances is infinite). Therefore Engels had good reasons to write the following:

> According to the inductionists, induction is an infallible method. It is so little so that its apparently surest results are every day overthrown by new discoveries......If induction were really so infallible, whence come the rapid successive revolutions in classification of the organic world? They are the most characteristic product of induction, and yet they annihilate one another.[50]

These shortcomings contributed considerably to the increasing influence of deductive logic. The schoolmen who continued their activity, though to a lesser degree, even in modern times (e.g., the neo-Thomists) should be ranked among the adherents of deductivism. Others have rejected scholasticism and preserved the Aristotelean syllogistic (e.g., Überweg).

The new trend was represented, first of all, by those who meant by deduction a system of axioms of the geometrical type and also the inferences which could be established on this ground.

> We were saying that of all the disciplines yet known, arithmetic and geometry alone are free from any taint of falsity or uncertainty; let us now consider more carefully the reason why this is so. First we must note that there are two ways by which we arrive at the knowledge of things, viz. either by experience or by deduction, i.e. the pure illation of one thing from another. We must further note that while our experiences of things are often fallacious, deduction, though it may, through failure to take advantage of it, be omitted, can never be wrongly performed by an understanding that is in the least degree rational.[51]

That was the opinion of Descartes who countered medieval dogmatism and took up the position of systematic skepticism. At the same time he also rejected the approach of the skeptics for whom doubt is a goal in itself. Descartes strove for certainty and thought he had found the ground for it

---

[50] Engels, *Dialektik der Natur*. (Berlin 1952). pp. 242–243: Clemens Dutt tr., pp. 302–303.
[51] Descartes, *Œuvres* Vol. II. p. 11; Kemp Smith tr., pp. 8–9.

in what he called our innate ideas and in proof. His attempt to replace belief by knowledge, and the irrational by the rational, is, no doubt, positive. At the same time, however, his view is of a subjective idealistic nature for it presupposes the existence of ideas in our mind which are independent of reality. It is also metaphysical because the universal is in this way isolated from the particular, and thought from experience.

Kant was motivated by a similar consideration:

> ... experience never confers on its judgments true or strict, but only assumed and comparative *universality*, through induction. We can properly only say, therefore, that, so far as we have hitherto observed, there is no exception to this or that rule. If, then, a judgment is thought with strict universality, that is in such manner that no exception is allowed as possible, it is not derived from experience, but is valid absolutely *a priori*.[52]

The selection of self-evident axioms, independent of experience, protects one from a vicious circle, indeed, but it cannot prevent the selection of propositions which in reality are false as the starting point for a deduction. Consequently, Hegel could repeat his old objection with good reason:

> ... In the favorite perfect syllogism:
>
> > All men are mortal
> > Now Caius is a man
> > Therefore Caius is mortal,
>
> the major premiss is correct only because and in so far as the *conclusion is correct*: if Caius should chance to be not mortal, the major premiss would not be correct. The Proposition which was supposed to be the conclusion must already be immediately correct on its own account, because otherwise the major premiss could not embrace all individuals; before the major premiss can pass as correct, there is the *prior* question whether the conclusion itself may not be an *instance* against it.[53]

In mathematical logic it is deduction which has again come into prominence. There are, however, certain attempts at studying non-deductive operations with the means of mathematical logic.

[52] Kant, *Kritik* p. 49; Kemp Smith tr.,p. 44.
[53] Hegel, *Wissenschaft* Zweiter Teil. p. 151.

Logic is essentially a deductive science and, pursued consistently, it excludes reductive methods and inferences from its subject matter. Nonetheless due to the importance which reductive methods and inferences have for human knowledge, we think it justified not to abstain from discussing at least a few of the most important methods of this kind.[54]

The most concise summary of the Marxist view on these matters has been given by Engels:

Induction and deduction belong together as necessarily as synthesis and analysis. Instead of one-sidedly lauding one to the skies at the expense of the other, we should seek to apply each of them in its place, and that can only be done by bearing in mind that they belong together, that they supplement each other.[55]

I shall try to adopt this approach in what follows.

II

I think what has been said so far is a sufficient reason to disprove the view that induction is merely an inference concerning probability. It requires further explanation, however, why deduction must not be identified with necessary inferences. Let us proceed from the following statement of Aristotle:

If there is the species 'water-animal', there will be the genus 'animal', but granting the being of the genus 'animal', it does not follow necessarily that there will be the species 'water-animal'.[56]

That is, the existence of the species necessarily entails the existence of the genus, but does not necessarily follow from its existence. This rule applies, of course, to the relation of the particular and the universal just as much as to that of the species and the genus. Accordingly, induction seems to be necessary here while deduction is not-necessary. Let us see these relations also in the form of mediate inferences:

---

[54] Händel-Kneist, p. 9.
[55] Engels, *Dialektik der Natur.* p. 242.
[56] Aristotle. 1928. 15a6–7.

All water-animals are animals (*t*).	All water-animals are animals (*t*).
There are water-animals (*t*).	There are animals (*t*).
There are animals (*t*).	There are water-animals (*t/f*).

Aristotle's thesis should be amended as follows: the non-existence of the genus necessarily implies the non-existence of the species, but this inference ceases to be necessary the other way round:

| There are no animals (*t*). | There are no water-animals (*t*). |
| There are no water-animals (*t*). | There are no animals (*t/f*). |

But the above connections are not only valid for existential propositions:

> ... there is no necessity that all the attributes that belong to the genus should belong also to the species; for 'animal' is flying and quadruped, but not so 'man'. All the attributes, on the other hand, that belong to the species must of necessity belong also to the genus; for it 'man' is good, then animal also is good.[57]

(1)   There is an animal which is flying (*t*).

   Man is flying (*f*).

(2)   This man is good (*t*).

   There is an animal which is good (*t*).

One could say that these instances cannot be described as a deductive (1) and an inductive (2) inference respectively for the relation of the universal and the particular manifests itself between the terms only and not between the propositions. As an answer let us examine a usual syllogism:

All men are mortal (all *M* is *P*).
All Greeks are men (all *M* is *P*).

All Greeks are mortal (all *S* is *P*)

---

[57] *Op. cit.*, 111a25–29.

Although this inference consists of universal propositions only, it is still regarded as deductive. This statement is true only if it is based on the relations of the terms.

Now let us find the origin of the view which identifies deduction with necessary inference. Aristotle wrote the following with regard to the combination *AE* in the first figure:

> But if the first term belongs to all the middle, but the middle to none of the last term, there will be no syllogism in respect of the extremes; for nothing necessarily follows from the terms being so related; for it is possible that the first should belong either to all or to none of the last, so that, neither a particular nor a universal conclusion is necessary. But if there is no necessary consequence, there cannot be a syllogism by means of these premisses.[58]

Aristotle selected those modes for which the conclusion necessarily follows from the premises, and the rest was of minor importance to him. Let us see an example for one of the ignored modes:

> All metals are heat-conductors.
> All irons are heat-conductors.
> _____
> All irons are metals.

Many people ignorant of logic would describe this inference as correct. This opinion is strongly influenced by the fact that the conclusion is true. But would they think it to be true only for this?

According to an expert in logic this inference is mistaken because the premises do not provide sufficient reason for drawing the conclusion. But what kind of conclusion is it for that they fail to provide sufficient reason? It is a necessary conclusion. It is, however, common knowledge that syllogistic and non-syllogistic inferences are usually distinguished. Induction, analogy, etc., are ranked among the latter, but deductive inferences are not even mentioned there. Why is it justified to regard induction by simple enumeration as not-syllogistic while the inference in the above example is not?

[58] *Op. cit.*, 26a2–7.

In the history of logic the following opinion has prevailed: some deductive inference is valid in so far as it is necessary inference, and if it is not necessary, then it is false. This view does not hold true in two respects. On the one hand, because it leads to the following unjustified distinction:

TABLE 68

	Syllogistic	Not-syllogistic
Deduction	correct	mistaken
Induction	correct	can be correct and mistaken respectively

This double standard is the product of one-sided 'deductivism'.—On the other hand, if a deductive inference is not-syllogistic it still can be correct.

> I had become increasingly aware of the very limited scope of deductive inference as practised in logic and pure mathematics. I realised that all the inferences used both in common sense and in science are of a different sort from those in deductive logic, and are such that, when the premises are true and the reasoning correct, the conclusion is only probable.[59]

The acceptance of non-syllogistic deductive inferences makes it possible to use more than 19 out of the 256 modes of categorical syllogism, on the one hand, and apart from extending the field of deductive inferences, it contributes to the liquidation of the mistaken identification of deduction and syllogistic inference, on the other.

III

Axioms play an important role in grounding a deduction. As far as they are concerned, I find the study of two issues necessary: (1) How can one recognize an axiom? (2) Can axioms be demonstrated?

Aristotle gave an ambiguous answer to the first question. He thought, on the one hand, that axioms must be known in advance:

[59] Russell, *My Philosophical Development.* p. 141.

... he who knows best about each genus must be able to state the most certain principles of this subject ... For a principle which every one must have who understands anything that is, is not a hypothesis; and that which every one must know who knows anything, he must already have when he comes to a special study.[60]

On the other hand, he declared the following:

All instructions given or received by way of argument proceed from pre-existent knowledge. This becomes evident on a survey of all the species of such instruction. The mathematical sciences and all other speculative disciplines are acquired in this way, and so are the two forms of dialectical reasoning, syllogistic and inductive; for each of these latter makes use of old knowledge to impart new, the syllogism assuming an audience that accepts its premises, induction exhibiting the universal as implicit in the clearly known particular.[61]

Accordingly, universal theses, axioms included, are obtained by inference and not by some previous knowledge.

Things do not stand better with regard to the second issue. As the above quotation reveals, a universal proposition can be demonstrated by proceeding from the knowledge of particulars. Elsewhere, however, the following can be read:

I call the basic truths of every genus those elements in it the existence of which cannot be proved. As regards both these primary truths and the attributes dependent on them the meaning of the name is assumed. The fact of their existence as regards the primary truths must be assumed; but it has to be proved of the remainders, the attributes.[62]

He supported the above thesis by two arguments.

(1) ... for not to know of what things one should demand demonstration, and of what one should not, argues want of education. For it is impossible that there should be demonstration of absolutely everything (there would be an infinite regress, so that there would still be no demonstration) ...[63]

---

[60] Aristotle, *Metaphysica*. 1005b7–10, 15–18.
[61] Aristotle. 1928. 71a1–8.
[62] *Op. cit.*, 76a31–34.
[63] Aristotle, *Metaphysica*. 1006a6–10.

(2) ... the upholders of circular demonstration are in the position of saying that if *A* is, *A* must be—a simple way of proving anything ...

Propositions the terms of which are not convertible cannot be circularly demonstrated at all, and since convertible terms occur rarely in actual demonstrations, it is clearly frivolous and impossible to say that demonstration is reciprocal and that therefore everything can be demonstrated.[64]

But undemonstrated fundamental theses are problematic. Aristotle was aware of this, too:

... for if a man is going to refuse to admit (the premiss) and claim that you shall argue to it as well, he will give the signal for a harder undertaking than was originally proposed: if, on the other hand, he grants it, he will give the original thesis credence on the strength of what is less credible than itself. If, then, it is essential not to enhance the difficulty of the problem, he had better grant it; if, on the other hand, it is essential to reason through premisses that are better assured, he had better refuse. In other words, in serious inquiry he ought not to grant it, unless he be more sure about it than about the conclusion...[65]

Aristotle did not restrict himself only to theoretical statements but elaborated the axiomatic system of syllogistic which was the first of its own kind according to the sources at our disposal. He accepted as axioms four modes (or, as a result of his later considerations, two) in the first figure. Then he traced all the valid modes back to these axioms by using conversion, *reductio ad impossibile*, and selection.

It is well-known how much these basic principles have been misused in medieval philosophy. Therefore Bacon could rightly write the following:

The axioms now in use, having been suggested by a scanty and manipular experience and a few particulars of most general occurrence, are made for the most part just large enough to fit and take these in; and therefore it is no wonder if they do not lead to new particulars. And if some opposite instance, not observed or not known before, chance to come in the way, the axiom is rescued and preserved by some frivolous distinction; whereas the truer course would be to correct the axiom itself.[66]

[64] Aristotle. 1928. 73a4–6, 17–20.
[65] *Op. cit.*, 159a7–13.
[66] Francis Bacon *Works*. Vol. I. p. 161 (*Novum Organum*, I(XXV)); F. H. Anderson tr., p. 44.

In order to avoid the mistakes resulting from superficiality he suggested starting with the particulars and then learning to know, and also to prove, the axioms, by employing the inductive method.

Descartes, on the contrary, held the following view:

> ... that those propositions which are immediately gathered from primary data are, according to our differing manner of arriving at them, known sometimes by intuition and sometimes by deduction—the primary data themselves by intuition alone, the remote conclusions not otherwise than by deduction.
>
> These two paths are the most certain of the paths to knowledge, and in respect of powers native to us no others should be admitted. All other paths should be regarded as dangerous and liable to error.[67]

It raises the question: if, e.g., the intuitions of two people differ from one another, what is the criterion to decide their validity? Furthermore, why is experience more suspicious than intuition?

Engels criticized the method of constructing axioms as follows:

> The general results of the investigation of the world are obtained at the end of this investigation, hence are not *principles*, points of departure, but *results*, conclusions. To construct the latter in one's head, take them as the basis from which to start, and then reconstruct the world from them in one's head is *ideology*, an ideology which tainted every species of materialism hitherto existing; because while in *nature* the relation of thinking to being was certainly to some extent clear to materialism, in history it was not, nor did materialism realize the dependence of all thought upon the historical material conditions obtaining at the particular time.
>
> As Dühring proceeds from 'principles' instead of facts, he is an ideologist, and can screen his being one only by formulating his propositions in such general and vacuous terms that they appear *axiomatic, flat*. Moreover, nothing can be concluded from them; one can only read something *into* them.[68]

This and other arguments related to it, however, failed to persuade those who doubted the authenticity of induction proceeding from

[67] Descartes, *Œuvres*, II. p. 15; Kemp Smith tr., p. 14.
[68] Engels, *Anti-Dühring*. (Berlin 1948) p. 419 (From Engels' *Vorarbeiten zum 'Anti-Dühring'*); Clemens Dutt tr., p. 468.

experience. The prevailing opinion still propagates the cognition of axioms in an intuitive way (by direct insight, on the ground of their evidence, etc.). Nevertheless, the discursive cognition of axioms also had adherents. As an example let me quote the following reasoning:

> In the transition from crude fact to science, we need forms of inference additional to those of deductive logic. Traditionally, it was supposed that induction would serve this purpose, but this was an error, since it can be shown that the conclusions of inductive inferences from true premises are more often false than true. The principles of inference required for the transition from sense to science are to be attained by analysis. The analysis involved is that of the kinds of inference which nobody, in fact, questions: as, for example, that if, at one moment, you see your cat on the hearth-rug and, at another, you see it in a doorway, it has passed over intermediate positions although you did not see it doing so. If the work of analysing scientific inference has been properly performed, it will appear that concrete instances of such inference are (a) such as no one honestly doubts, and (b) such as are essential if, on the basis of sensible facts, we are to believe things which go beyond this basis.[69]

Admitting the importance of analysis, I still dispute that it could replace induction.

The prevailing opinion concerning the other question is that axioms cannot be proved, or demonstrated.

> Deduction requires such theses which are clear in themselves, therefore do not need any proof and, in fact, cannot be proved.[70]

On the basis of the supposition that the thesis used as an argument should be more universal than the one to be demonstrated, the above statement holds good. Otherwise either the more universal thesis should be accepted as a genuine axiom, or a *regressus ad infinitum* would occur. But why should we not use singular theses as arguments?

And now let us consider the ideas of those who think that axioms can be proved:

---

[69] Russell, *My Philosophical Development*. p. 153.
[70] Huszár, p. 270.

Axioms are not the kind of theses which cannot be proved. Sciences seek to prove all their theses. Therefore some theses and fundamental terms are selected, and the others are built on them. They are the axioms which cannot be demonstrated in a logical way though, but have been proved many times by the thousands of years of the social and historical practice of mankind.[71]

Practical proof as an objective criterion of cognition is of fundamental importance. That is how one can go beyond the vicious circle on which the approach of the skeptics is founded. Skepticism cannot be disproved by logical arguments only.

I definitely deny, however, that axioms could be demonstrated only in an extralogical way. On this question I share the following view:

In light of dialectical materialism, what is the meaning of the Aristotelian principle that axioms *cannot be proved*? And if they cannot be proved, what reasons can we have for accepting them? The answer is the following: The axioms have arisen in the course of historical and logical generalization of the results of experience. In this sense their correctness has received proof and support. But since they have become axioms, the necessity of constantly checking them out disappears. For the practice of logical thinking they are thus basic postulates.[72]

We have come to the following conclusions: axioms play an important role in grounding a deduction. Induction is an indispensable means for the cognition and demonstration of axioms. Whoever underestimates or tries to deny induction its place builds deduction on a shaky foundation.

IV

So far inferences based on a subaltern relation have been discussed. Analogical inferences are usually referred here as a third kind. In this respect Aristotle maintained the following:

We have an 'example' when the major term is proved to belong to the middle by means of a term which resembles the third. It ought to be known both that the middle belongs to the third term, and that the first belongs to

[71] *Logika.* (Budapest 1956). p. 344.
[72] Fogarasi, p. 337.

that which resembles the third. For example let *A* be evil, *B* making war against neighbours, *C* Athenians against Thebans, *D* Thebans against Phocians. If then we wish to prove that to fight with the Thebans is an evil, we must assume that to fight against neighbours is an evil. Evidence of this is obtained from similar cases, e.g. that the war against the Phocians was an evil to the Thebans. Since then to fight against neighbours is an evil, and to fight against the Thebans is to fight against neighbours, it is clear that to fight against the Thebans is an evil. Now it is clear that *B* belongs to *C* and to *D* (for both are cases of making war upon one's neighbours) and that *A* belongs to *D* (for the war against the Phocians did not turn out well for the Thebans): but that *A* belongs to *B* will be proved through *D*. Similarly if the belief in the relation of the middle term to the extreme should be produced by several similar cases. Clearly then to argue by example is neither like reasoning from part to whole, nor like reasoning from whole to part, but rather reasoning from part to part, when both particulars are subordinate to the same term, and one of them is known. It differs from induction, because induction starting from all the particular cases proves (as we saw) that the major term belongs to the middle, and does not apply the syllogistic conclusion to the major term, whereas argument by example does make this application and does not draw its proof from all the particular cases.[73]

The term analogy is used in a double sense. On the one hand, a relation of similarity, on the other, a method based on this very relation.

Aristotle defined the concept of similarity in two ways:

> Those things are called 'like' which have the same attributes in every respect, and those which have more attributes the same than different, and those whose quality is one; and that which shares with another thing the greater number or the more important of the attributes (each of them one of two contraries) in respect of which things are capable of altering, is like that other thing.[74]

And elsewhere:

> Things are like if, not being absolutely the same, nor without difference in respect of their concrete substance, they are the same in form; e.g. the larger square is like the smaller, and unequal straight lines are like; they are like,

[73] Aristotle. 1928. 68b38–69a19.
[74] Aristotle, *Metaphysica*. 1018a15–19.

but not absolutely the same. Other things are like, if, having the same form, and being things in which difference of degree is possible, they have no difference of degree. Other things, if they have a quality that is in form one and the same—e.g. whiteness—in a greater or less degree, are called like because their form is one. Other things are called like if the qualities they have in common are more numerous than those in which they differ—either the qualities in general or the prominent qualities; e.g. tin is like silver, *qua* white, and gold is like fire, *qua* yellow and red.[75]

In so far as the above definitions refer to the same concept they ar' incompatible with one another. We should, however, make the following distinction: let us call 'similarity' the relation which allows for total identity and 'only-similarity' the relation which excludes this alternative and admits of identity in a certain respect only. With this distinction the two definitions are not only justifiable but also necessary.

During the history of logic analogy has been treated even less than induction. If dealt with at all, it was warned against rather than made use of:

> What we have chiefly to guard against is the wasting of our time in guessing unmethodically, at random. For although the answer can often be obtained without method, and sometimes, if fortune favors, more quickly than by method, yet in so proceeding we are bound to weaken the mind's powers of insight, accustoming ourselves to what is puerile and trifling, and acquiring the habit of attending always only to the surface-appearance of things, unable to penetrate more deeply. We must not, however, fall into the counter-error of those who occupy themselves only with things lofty and momentous; they reap as the reward of their manifold labours nothing but confusion of mind, not the profound knowledge to which they are aspiring. This is why we ought to train ourselves first in those easier matters, but methodically ... and so, as easily as though we were at play, to penetrate ever more deeply into the truth of things.[76]

Indeed, one must beware of superficial analogies. It would be a mistake, however, to ignore the fact that the similarities between objects and occurrences make it possible for us, though knowing only some of their

---

[75] *Op. cit.*, 1054b4–13.
[76] Descartes, *Œuvres* II. (Paris 1940) p. 40.

particular properties, to draw conclusions about the properties still unknown on the ground of comparison. Analogies are often used in science. For example, when the characteristics of electric current are to be described, electricity is compared to the flow of fluids in a river. It has been demonstrated that the difference of potentials in the wires is similar to the difference of water levels in the various parts of a river, and the intensity of current is similar to the quantity of water flowing through the cross-section of the river during one time unit.

The dominance of subjective idealism has affected also the judgement of analogy.

> All likeness and unlikeness of which we have any cognizance, resolve themselves into likeness and unlikeness between states of our own, or some other, mind. When we say that one body is like another, (since we know nothing of bodies but the sensations which they excite) we mean really that there is a resemblance between the sensations excited by the two bodies, or between some portions at least of those sensations. If we say that two attributes are like one another, (since we know nothing of attributes except the sensations or states of feeling on which they are grounded), we mean really that those sensations, or states of feeling, resemble each other. We may also say that two relations are alike. The fact of resemblance between relations is sometimes called *analogy*, forming one of the numerous meanings of that word.[77]

In complete contradiction to this approach, materialism defines similarity as an objectively existing relation.

By analogical inference, we conclude from the characteristics two objects have in common and also from other known properties of one of them that the other object has still other properties in common with the former. The structure of this inference can be formulated, among others, in this way:

$$P \text{ is } M_1, M_2, M_3$$
$$\underline{S \text{ is } M_1, M_2}$$
$$S \text{ is } M_3$$

[77] Mill, pp. 70–71.

Opinions vary on the question whether or not the analogical inference offers an authentic result. One group of logicians regards it as an inference with probability only. The other group, however, whose opinion I share, also accepts other such analogical inferences whose premises necessarily entail the conclusion.

One can come to such a conclusion, e.g., by a perfect analogical inference.

> Two systems of mathematical elements, say R and R', are related in a way that certain relations between the elements of system R are subject to the same laws as the relations between the corresponding elements of system R'.
>
> This kind of analogy can be well illustrated by the example discussed in point (1). Let us choose as R the sides of a triangle, and as R' the surfaces of a tetrahedron.
>
> There is a mutually equivalent correspondence between the elements of systems R and R' which leaves certain relations unaltered. In other words, if there is such a relation between the elements of one of the two systems, there must be the same between the corresponding elements of the other system. Such a relation of two systems is a very well defined kind of analogy which is called isomorphism (or holomorphic isomorphism).[78]

On the other hand, an authentic conclusion results from what is called strict analogical inference. Its formula is:

$$P \text{ is } M_1, \text{ consequently } M_2$$
$$\underline{S \text{ is } M_1}$$
$$S \text{ is } M_2$$

For example:

> Plants are living beings, consequently they multiply.
> Animals are living beings.
> _____
> Animals multiply.

Analogical inferences prove to be extremely valuable and very useful in cases where they do not provide an authentic result. They often inspire the

---

[78] Pólya, *A gondolkodás iskolája* (School of Thinking). Budapest 1957. p. 64.

most unexpected conjectures and suppositions which push scientific research in a certain direction and lead to important discoveries. The theory of the wave-nature of matter, of great importance in modern physics, was developed on the ground of an analogy between the radiating atom and musical instruments. This analogy and the one that compared waves of matter and electromagnetic vibration have served as a basis for the development of an entirely new branch of physics, namely wave mechanics. At another time it was the analogy between electricity and heat which helped to reveal many properties and laws of electric phenomena.

V

Aristotle already made an attempt at systematization of the inferences discussed above:

> Deductive inference: from the whole to the part.
> Inductive inference: from the part to the whole.
> Analogical inference: from the part to the part.

Here I shall not yet dwell upon the question as to whether or not the ground for the classification—the relation of whole and part—is correct. However, I want to note here already that the fourth version is missing: inference from the whole to the whole.

It is common knowledge that the study of inductive and analogical inferences has been pushed into the background for two thousand years. Therefore, the systematization of mediate inferences did not create a problem.

In modern logic, however, these inferences have gradually gained ground and therefore the necessity to clear up their interrelations has also made itself felt.

> Although, therefore, all processes of thought in which the ultimate premises are particulars, whether we conclude from particulars to a general formula, or from particulars to other particulars according to that formula, are equally Induction; we shall yet, conformably to usage, consider the name Induction as more peculiarly belonging to the process of establishing the general proposition, and the remaining operation, which is substantially that of interpreting the general proposition, we shall call by its usual name,

Deduction. And we shall consider every process by which anything is inferred respecting an unobserved case, as consisting of an Induction followed by a Deduction; because, although the process needs not necessarily be carried on in this form, it is always susceptible of the form, and must be thrown into it when assurance of scientific accuracy is needed or desired.[79]

Accordingly, there are three kinds of mediate inferences: induction, deduction, and the combination of the two which has not been named. The classification is based on the relations of quantifiers (universal, particular, singular) instead of the relationship of the whole and the part.

The inference which Mill mentioned as a third alternative has later been interpreted in the following way:

... we may reason without rendering our conclusion either more or less general than the premises, as in the following: —

Snowdon is the highest mountain in England or Wales.
Snowdon is not so high as Ben Nevis.
Therefore the highest mountain in England or Wales is not so high as Ben Nevis.

Again:

Lithium is the lightest metal known.
Lithium is the metal indicated by one bright red line in the spectrum.
Therefore the lightest metal known is the metal indicated by a spectrum of one bright red line.

In these examples all the propositions are singular propositions, and merely assert the identity of singular term, so that there is no alteration of generality. Each conclusion applies to just such an object as each of the premises applies to. To this king of reasoning the apt name of *traduction* has been given.[80]

Two kinds of traduction are usually distinguished: the analogical and the relative inferences. The above considerations lead us to the following classification:

[79] Mill, p. 203.
[80] Jevons, 1872. pp. 211–212.

(1)   Deductive inference: from the general to the singular.
(2)   Inductive inference: from the singular to the general.
(3)   Traductive inference: from the singular to the singular.

Many have remarked on the shortcomings of the above interpretation. Namely, we can conclude from the general not only to the singular but also to something less general, and the general can be arrived at not only from the singular but also from the particular, etc. It is easy to overcome these shortcomings:

(1)   From wider extension to narrower extension.
(2)   From narrower extension to wider extension.
(3)   From the same extension to the same extension.

The problem seems to be solved. Let us see, however, the following inference:

> Every iron is metal.
> No stone is metal.
> _____
> No stone is iron.

This syllogism was regarded as a deductive inference in traditional logic. The snag was, however, that it inevitably contradicted the accepted criterion of deduction. Can we speak of traductive inference in this case? Let us consider the following in order to answer this question.

Stone is foreign to iron. This relation is independent of the extension of the terms concerned. Whatever quantity of either the one, or the other is examined, their relation remains unchanged. Let us extend this solution to the inference concerned in a way that extensively indefinite terms and propositions are used:

> Iron is metal.
> Stone is not metal.
> _____
> Stone is not iron.

The term 'stone' can be quantified in any way under the condition that the minor premise stays true, and that the conclusion will necessarily follow from the premises. Thus, this inference is based on a relation of incompatibility involved in it rather than on extensive relations.

Let us see now the following formula:

> M is subordinated to P.
> S is subordinated to M.
> ___
> S is subordinated to P.

Can we describe this formula as a deductive inference only if there is some kind of quantification in it? To disprove this supposition let us take the following example:

> A proposition asserting facts is subordinated to a proposition of necessity.
> A proposition of possibility is subordinated to a proposition asserting facts.
> ___
> A proposition of possibility is subordinated to a proposition of necessity.

It is an error to force deduction into all inferences which proceed from a wider extension to a narrower one. In my view this inference is based on a relation of subordination. Continuing this way of thinking, we shall find the ground of inductive inference in a relation of superordination. Since both subordination and superordination imply entailment, this relation proves to be the common basis of both inductive and deductive inferences.

As to the relation of compatibility, let me refer to the operations already discussed. One of them is analogical inference which is based on a relation of similarity. The other operation was defined like this: if P coincides with M, and S coincides with M, then S is compatible with P. The inferences based on the relation of identity have also been met before.

So we have come to the following conclusion: the basis of deductive, inductive, and traductive inferences is not the relation between the whole and the part or between the quantifiers, but the relation of terms. Therefore, these operations, too, can be regarded as relative inferences.

In the foregoing, I have demonstrated the connection of the discussed inferences with the fundamental relations of terms, but with one exception. The traductive inference which cannot be traced back to any of these relations has been left out. As the term implies, it is neither induction, nor deduction. Traduction as a collective term had a positive role in the traditional classification by calling attention to the fact that, apart from inductive and deductive inferences, there are also other operations to be reckoned with. Its significance in a differentiated classification, however, comes to no more than, for example, the term 'not-analogical inferences'.

19. THE VALIDITY OF INFERENCE

I

Let us proceed from the following reasoning:

> A syllogism is discourse in which, certain things being stated, something other than what is stated follows of necessity from their being so. I mean by the last phrase that they produce the consequence, and by this, that no further term is required from without in order to make the consequence necessary.[81]

In Chapters 4–6 of the first part of the *Prior Analytic*, the Stagirite examined the relations of *t* propositions only. To explicitly declare this was not necessary since the propositions *AEIO* are *t* in his interpretation.[82] Now he raised the question whether there is a necessary relation between *t* propositions only, or also when one, or each, of the propositions is *f*. In order to answer this question the relation of the two original propositions and the third proposition had to be explored.

Aristotle, on the ground of the values of the propositions involved in the inference, distinguished the following alternatives:

> It is possible for the premisses of the syllogism to be true, or to be false, or to be the one true, the other false. The conclusion is either true or false necessarily.[83]

---

[81] Aristotle. 1928. 24b18–22.
[82] *Op. cit.*, 24b26–30.
[83] *Op. cit.*, 53b4–6.

Accordingly, the following combinations are obtained:

TABLE 69

	(1)	(2)	(3)	(4)	(5)	(6)	(7)	(8)
Major premise	t	t	t	t	f	f	f	f
Minor premise	t	t	f	f	t	t	f	f
Conclusion	t	f	t	f	t	f	t	f

If the premises are taken as one proposition, the following classification will be the result (Aa = antecedent, Bc = conclusion):

TABLE 70

	(1)	(2)	(3)	(4)
Aa	t	t	f	f
Bc	t	f	t	f

Aristotle commented on the versions summed up in Table 70 as follows: in case (1) the conclusion follows from the antecedent *of necessity*. In case (2) it is *impossible* to draw a conclusion because t cannot entail f. In case (3) it is *possible* to conclude. Although he failed to explicitly discuss case (4), the context unequivocally implies that he regarded it also an inference *of possibility*.[84]

His above remarks were closely connected with his conviction that there is a causality between the antecedent and the consequent in an inference.

> For the letters are the cause of syllables, and the material is the cause of manufactured things, and fire and earth and all such things are the causes of the whole, and the hypotheses are causes of the conclusion, in the sense that they are that out of which these respectively are made ...[85]

Let us describe causal relations, sticking to the sequence in Table 70:

[84] *Op. cit.*, 53b4–57b17.
[85] Aristotle, *Metaphysica*. 1013b17–28.

(1)   If a cause exists, then its effect must necessarily exist.

(2)   If a cause exists, then it is impossible for its effect not to exist.

(3)   If a cause does not exist, then the effect can still exist.

(4)   If a cause does not exist, then the effect does (or may) not exist.

Let us see now in more detail Aristotle's commentaries on the combination in Table 70:

> From true premises it is not possible to draw a false conclusion, but a true conclusion may be drawn from false premises, true however only in respect to the fact, not to the reason. The reason cannot be established from false premises: why this is so will be explained in sequel.
>
> First then that it is not possible to draw a false conclusion from true premises, is made clear by this consideration. If it is necessary that B should be when A is, it is necessary that A should not be when B is not. If then A is true, B must be true: otherwise it will turn out that the same thing both is and is not at the same time. But this is impossible. Let it not, because A is laid down as a single term, be supposed that it is possible, when a single fact is given, that something should necessarily result. For that is not possible. For what results necessarily is the conclusion, and the means by which this comes about are at the least three terms, and two relations of subject and predicate or premises. If then it is true that A belongs to all that to which B belongs, and that B belongs to all that to which C belongs, it is necessary that A should belong to all that to which C belongs, and this cannot be false: for then the same thing will belong at the same time. So A is posited as one thing being two premises taken together.[86]

First let us examine the thesis that if the antecedent is *t* the conclusion must inevitably also be *t*. (It was put into the following words in medieval logic: *posita conditione ponitur conditionatum* = 'the postulation of the condition postulates also the consequence'. Or in another formulation: *a principio ad principiatum* = 'from the antecedent to the consequent'.)

I have two remarks with regard to the above thesis. (1) Aristotle's starting point was *ontological*: if there is *A*, then there is necessarily also *B*. This is the reason why the truth of *B* follows from the truth of *A*. (2) The

[86] Aristotle. 1928. 53b6–24.

30

thesis holds true also if there is a *necessary* relation between *A* and *B*. Therefore, the exact formulation is: if the premises are *t* and in a necessary relation with the conclusion, then the conclusion is inevitably *t*.

As we have seen, it was by analysing such combinations of propositions that Aristotle selected the 14 valid modes where only *t* propositions occurred. Thus, the truth of the propositions has implicitly been included in the Aristotelean criteria of valid inference. Having considered the above said, how should one interpret the following thesis: if the premises are *t* and the inference is valid, then the conclusion is necessarily *t*. This assertion can also be expressed like this: if the premises are *t* and the inference is such that the truth of the conclusion necessarily follows from *t* propositions, then the conclusion is necessarily *t*.

It follows from the thesis discussed above that if the premises are *t* the conclusion cannot be *f*. In this respect let us proceed from the statement that the antecedent must consist of two propositions at least. It means, among other things, that immediate inferences whose antecedent consists of one proposition only cannot be considered as inferences.

In the *Hermeneutics*, however, the following can be read:

> Since the contrary of the proposition 'every animal is just' is 'no animal is just', it is plain that these two propositions will never both be true at the same time or with reference to the same subject. Sometimes, however, the contradictories of these contraries will both be true, as in the instance before us: the propositions 'not every animal is just' and 'some animals are just' are both true.
>
> Further, the proposition 'no man is just' follows from the proposition 'every man is not-just' and the proposition 'not every man is not-just', which is the opposite of 'every man is not-just', follows from the proposition 'some men are just'; for if this be true, there must be some just men.[87]

This passage obviously refers to these inferences whose antecedent consists of one proposition only, and which are necessary inferences since if it is *t* that 'every man is not-just', then the proposition 'no man is just' must also be *t*.

[87] *Op. cit.*, 20a16–23.

What can be the reason that Aristotle excluded these operations from the necessary inferences in the above quotation from the *Prior Analytic*? Let us examine the following example:

> *I*: Some men are just (*t*).
> _____
> *E*: No man is just (*f*).

This is, no doubt, a necessary inference but its antecedent is *t* while its conclusion is *f*. This relation, however, could not be reconciled with the principle that 'from true premises it is not possible to draw a false conclusion'. In order to maintain this principle Aristotle had to deny in the *Prior Analytic* what he admitted in the *Hermeneutics*, namely, that immediate inferences are justifiable as necessary inferences.

If the truth of *B* is supposed to follow from the truth of *A* of necessity, then it is, of course, impossible that the falsity of *B* should necessarily follow from the truth of *A*. But: (1) This relation does not exclude the possibility that the truth of *A* necessarily entails the falsity of *C* (see the above example). (2) Considering also that the truth of *B* does not follow necessarily from the truth of *A*, it is possible that the falsity of *B* necessarily follows from the truth of *A*. Thus, the thesis that *f* cannot follow from *t* does not hold true in general, but *only if the truth of A implies the truth of B*.

Why could Aristotle declare that the truth of the conclusion necessarily follows from the truth of the premises? Because he had defined the valid modes by excluding those cases where the antecedent was *t* but the conclusion *f*. That is, if all the instances where the original pair of propositions is *t* and the third proposition is *f* are disregarded, *then a f* conclusion, indeed, cannot follow from *t* premises.

II

In the foregoing I have analysed the *t*—*t* and *t*—*f* combinations of the values of the premises and the conclusion respectively. Let us consider now the version *f*—*t*. Aristotle, on the ground of the first figure, wrote, among others, the following:

> ... from what is false a true conclusion may be drawn, whether both the premises are false or only one, provided that this is not either of the

premisses indifferently, if it is taken as wholly false: but if the premiss is not taken as wholly false, it does not matter which of the two is false . . . Let *A* belong to the whole of *C*, but to none of the *B*s, neither let *B* belong to *C*. This is possible, e. g. animal belongs to no stone, nor stone to any man. If then *A* is taken to belong to all *B* and *B* to all *C*, *A* will belong to all *C*; consequently though both the premisses are false the conclusion is true: for every man is an animal.[88]

Further on[89] Aristotle analysed in detail numerous instances of the first figure which can be classified as follows:

	Major premise	Minor premise	Conclusion
(1)	wholly *f*	wholly *f*	can be *t*
(2)	partly *f*	partly *f*	can be *t*
(3)	wholly *f*	*t*	cannot be *t*
(4)	partly *f*	*t*	can be *t*
(5)	*t*	wholly *f*	can be *t*
(6)	*t*	partly *f*	can be *t*

The classification reveals that two cases escaped Aristotle's attention:

(7)	wholly *f*	partly *f*	can be *t*
(8)	partly *f*	wholly *f*	can be *t*

Aristotle called a proposition wholly *f* when its contrary is *t*, and partly *f* when its contradictory is *t*. Having summed up the above six versions, Aristotle came to the conclusion that there are three alternatives in the first figure to arrive at a *t* conclusion from *f* premises:

(1)   both propositions are *f* (partly or wholly)
(2)   the minor premise is wholly *f* while the major one is *t*,
(3)   one of the propositions is partly *f* but the other is *t*.

According to Aristotle the inference from *f* premises to *t* conclusion is not of equal rank with the one from *t* premises to *t* conclusion. While he

[88] *Op. cit.*, 53b26–35.
[89] *Op. cit.*, 54a1–55b2.

described the former as a syllogism asserting a fact only, he wrote the following about the latter:

> ... the premisses of demonstrated knowledge must be true, primary, immediate, better known than and prior to the conclusion, which is further related to them as affect to cause. Unless these conditions are satisfied, the basic truth will not be 'appropriate' to the conclusion. Syllogism there may indeed be without these conditions, but such syllogism, not being productive of scientific knowledge, will not be demonstration.[90]

. He supported his thesis that *one cannot necessarily conclude* to a *t* proposition from *f* ones with the following argument:

> The reason is that when two things are so related to one another, that if the one is, the other necessarily is, then if the latter is not, the former will not be either, but if the latter is, it is not necessary that the former should be. But it is impossible that the same thing should be necessitated by the being and by the not-being of the same thing. I mean, for example, that it is impossible that *B* should necessarily be great since *A* is white and that *B* should necessarily be great since *A* is not white. For whenever since this, *A*, is white it is necessary that that, *B*, should be great, and since *B* is great that *C* should not be white, then it is necessary if *A* is white that *C* should not be white. And whenever it is necessary, since one of two things is, that the other should be, it is necessary, if the latter is not, that the former (viz. *A*) should not be. If then *B* is not great *A* cannot be white. But if, when *A* is not white, it is necessary that *B* should be great, it necessarily results that if *B* is not great, *B* itself is great. (But this is impossible.) For if *B* is not great, *A* will necessarily not be white. If then when this is not white B must be great, it results that if *B* is not great, it is great... [91]

Let us make clear this demonstration by formulae. It starts with this *modus tollens*:

(1)   If *A* is white, then *B* is great.
        B is not great.
        _____
        Hence *A* is not white.

---

[90] *Op. cit.*, 71b20–24.
[91] *Op. cit.*, 57b1–16.

Let us suppose that B is great if A is not white. In this case the result will be the following *modus tollens*:

(2)   If A is not white, then B is great.
           B is not great.
           _____
           Hence A is white.

The following conditional syllogism can be established from the minor premise and the conclusion of this inference and from the major premise of the former inference:

(3)   If B is not great, then A is white.
           If A is white, then B is great.
           _____
           Hence if B is not great, then B is great.

And that, being a contradiction, is a nonsense. It is impossible that something should be the case and should not be the case at the same time. The propositions of the above example should be replaced by these:

           A is white   — both premises are t
           A is not white   — not both of the premises are t
           B is great   — the conclusion is t
           B is not great   — the conclusion is f

Then inference (3) would run like this:

           If the conclusion is f, then both premises are t.
           If both premises are t, then the conclusion is t.
           _____
           If the conclusion is f, then the conclusion is t.

According to Aristotle the conclusion being nonsense is evidence for the fact that if t premises necessarily lead to a t conclusion, then f premises cannot lead to it, of necessity.

Łukasiewicz thought this evidence not to be valid on the ground of the rule of Clavius. This rule asserts that if the proposition concerned validly follows from the negation of another proposition, then it is t. In modern formulation: $(\bar{p} \rightarrow p) \rightarrow p$. The proposition 'if B is not great, then B is great' does not break the law of contradiction. This law is violated only if conjunction is used instead of implication: B is not great and B is great.[92] I leave it to the reader to accept this paradox of interpretation or not. Anyway, let me refer to what has already been said in this respect before.

And now let us see—*mutatis mutandis*—what has been explained about the combination t—f. If we proceed from the truth of B necessarily following from the truth of A, then it is impossible that the truth of B could necessarily follow from the falsity (or rather, not-truth) of A. But: (1) this relation does not exclude the possibility that the falsity of A necessarily leads to the truth of C (no man is just — some men are just). (2) If the relation that the truth of B does not necessarily follow from the truth of A is also taken into consideration (which possibility has been admitted by the Stagirite), then there is a possibility that the truth of B should necessarily follow from the falsity of A. Thus, the thesis that t cannot of necessity follow from f does not hold good in general, but only when the truth of A implies the truth of B.

Aristotle stated the valid modes of syllogism by excluding all instances where the premises were f but the conclusion t. Accordingly, if all those cases are disregarded where the premises are f and the conclusion is t, then a t conclusion cannot necessarily follow from a f antecedent, indeed.

Now a few words about the combination f—f. Although Aristotle did not deal with it specifically, he was confronted with this problem while discussing the combination f—t:

> Let A belong to no B, and B to all C. If then the premiss BC which I take is true, and the premiss AB is wholly false, viz. that A belongs to all B, it is impossible that the conclusion should be true: for A belonged to none of the Cs, since A belonged to nothing to which B belonged, and B belonged to all C.[93]

Let me illustrate this reasoning with the following example:

> Every living being (B) is stone (A).
> Every man (C) is a living being (B).
> _____
> Every man (C) is stone (A).

[92] Cf.: Łukasiewicz, p. 50.
[93] Aristotle. 1928. 54a7–11.

That is, if one premise is f, then there is a case where the conclusion is necessarily f.

The *reductio ad impossibile* leads to the same result, too. Let us choose this syllogism:

$$\frac{\begin{array}{l}\text{All } M \text{ is } P \ (t)\\ \text{All } S \text{ is } M \ (t)\end{array}}{\text{All } S \text{ is } P \ (t)}$$

The negation of the proposition 'All S is P' is 'Some S is not P'. Let us add to it the major premise of this syllogism, namely, 'All M is P'. These premises make up a valid syllogism of the Baroco type:

$$\frac{\begin{array}{l}\text{All } M \text{ is } P \ (t)\\ \text{Some } S \text{ is } P \ (f)\end{array}}{\text{Some } S \text{ is not } M \ (f)}$$

This conclusion contradicts the minor premise of the original syllogism, namely, 'All S is M'. If this reasoning is correct why could not we draw a necessary conclusion from f to f?

To sum it up: Aristotle has revealed certain connections from an important, but rather restricted, point of view. They hold true within the given restrictions. Aristotle, however, was of the opinion that these connections are generally characteristic of inferences. Thereby he distorted his theory of inferences to a great extent. If one goes beyond this narrow circle it becomes clear that necessary inferences can be arrived at for cases of any value combination. That is, the necessity of the inference is independent of the values of the propositions involved in it.

III

It was the Megaric Stoic school which raised the question as to what results would be obtained if causality were disregarded in an inference and any proposition could be substituted for the premises and the conclusion. Philo (about 300 B.C.) was among the first who attempted to give an answer to this question.

This is how Sextus Empiricus reported on it:

Philo said that the connected (proposition) is true when it is not the case that it begins with the true and ends with the false. So according to him there are three ways in which a true connected (proposition) is obtained, only one in which a false. For (1) if it begins with true and ends with true, it is true, e. g. 'if it is day, it is light'; (2) when it begins with false and ends with false, it is true, e. g. 'if the earth flies, the earth has wings'; (3) similarly too that which begins with false and ends with true, e. g. 'if the earth flies, the earth exists'. It is false only when beginning with true, it ends with false, e. g. 'if it is day, it is night'; since when it is day, the (proposition) 'it is day' is true — which was the antecedent; and the (proposition) 'it is night' is false, which was the consequent.[94]

In medieval logic the Megaric Stoic approach gained adherents mainly among the nominalists. Pseudo-Scotus, for example, wrote as follows: inference is a hypothetical statement which is impossible if the premises are t and the conclusion is f.

Traditional logic restricted itself mainly to the elaboration of the interrelations between t premises and a t conclusion. Überweg's following remark is very typical of this approach:

> The proof of the material truth of a conclusion correctly inferred from true premises lies in the logical correctness of the inference itself; for since the logical norms of deduction, just as logical norms in general, are founded on the idea of truth..., an inference which led to something untrue would turn out to be contrary to the logical norms and consequently to be incorrect—thus not fulfilling our initial assumption.[95]

Mathematical logic rediscovered and developed further the approach of the Stoics:

> Whether a truth-functional schema S_1 implies another S_2, can be decided always by taking S_1 as antecedent and S_2 as consequent of a conditional, and testing the conditional for validity. For, according to our definition, S_1 implies S_2 if and only if no interpretation makes S_1 true and S_2 false, hence if and only if no interpretation falsifies the material conditional whose antecedent is S_1 and whose consequent is S_2. In a word, *implication is validity of the conditional.*[96]

[94] Quoted in: Bochenski, *Formale Logik*. p. 134; Ivo Thomas tr., p. 117.
[95] Überweg, p. 363.
[96] Quine, pp. 33–34.

What is the relation between the positions of traditional and mathematical logic?

> In traditional logic it is often demanded that in every inference the presuppositions must be true propositions. For modern logic such a condition does not apply to all kinds of inferences, for our examples have already shown that a logically compelling conclusion can also be drawn from false propositions. From the falsehood of p follows with logical consistency the truth of $\bar{p}$; from the falsehood of O follows with the same lawfullness the truth of A. Accordingly, this opinion must be corrected to the effect that in such inferences the presuppositions need not be true propositions, but their truth values must be known.[97]

This criticism of the traditional logic holds true but applies to mathematical logic as well to a certain extent. What I have in mind here is the operation whereby some f proposition is replaced by another, contradictory, one which is t. To illustrate it with the example quoted above: it is f that O—it is t that not O. That is, the goal agrees with the traditional one: to avoid f propositions in the premises.

It is a question of principle for certain mathematical logicians that one cannot conclude from false premises. Frege, for example, wrote the following in a letter to Jourdain: a mere thought which is not accepted as true cannot be a premise, for only what is recognized as true can be used as a premise. Although mere hypotheses cannot function as premises, one can still investigate what consequences would follow from the supposition if we accepted the truth of A. In this case, however, the result should involve this condition, namely: in so far as A is true. And then, this very condition proves A not to be a premise for a genuine premise does not occur in the final result of the inference.[98]

One question is still to be answered, namely: why should we bother with f propositions when all sciences strive to arrive at necessarily t conclusions from t premises? Is it not merely a demand for formal completeness?

Let us bear in mind, first of all, that in many cases we cannot decide whether a proposition is t or f. Propositions concerning the living beings of

---

[97] Händel-Kneist, Kurzer Abriss der Logik. [Berlin 1960.] p. 111.
[98] Cf. Jourdain, p. 240.

numerous planets of the solar system or the causes of cancer are good examples. Should we, then, give up all inferences which concern them, or should we rather state what the results are, if they are true and if they are false?

Many propositions which had been regarded as *t* proved to be *f* in the course of scientific development. It is enough to refer to the Ptolemaic system here. If inference is restricted only to such propositions which are, or thought to be, *t*, we only contribute to dogmatism.

IV

Now let us disregard the values of the propositions involved in an inference and examine the modes of relations between the premises and the conclusion. These interrelations will be discussed as the *modalities of inference*.

Aristotle was mainly interested in inferences for which the conclusion *necessarily* (syllogistically) follows from the premises. During his research, however, he came across *not-necessary* inferences as well, as we have seen above.

The study of the syllogism played a dominant role in ancient and medieval logic. A certain change occurred only in the modern age.

> The syllogism consists of propositions, propositions consist of words, words are symbols of notions. Therefore if the notions themselves (which is the root of the matter) are confused and overhastily abstracted from the facts, there can be no firmness in the superstructure. Our only hope therefore lies in a true induction.[99]

Bacon, however, failed to find a procedure demonstrating the necessity of the inductive inference. Nevertheless, he achieved so much that induction could not be ignored, or neglected, any more. Under the given circumstances there evolved a trend of thought which admitted induction, but did not regard it as of equal rank with deduction.

---

[99] Francis Bacon. *Works*. Vol. I. p. 158 (*Novum Organum*, 1(XIV)); F. H. Anderson tr., p. 41.

> Every inference of reason must give necessity. *Induction and analogy* are therefore not inferences of reason, but rather merely logical *presumptions* or even empirical inferences...[100]

This opinion has been shared by Marxist logicians, too, though mainly in a moderate form.

> Why do we maintain that the danger of error is greater and that the certainty about the correctness of an inference is less in an inductive inference than in a syllogism? The syllogism receives the conditions of its truth so to speak from outside: the premises must be true. If the premises are true, then in conformity with the rules of the syllogism the truth of the conclusion follows with necessity, *unconditionally*, excluding every objection. Although the cost of attaining this certainty of the conclusion is that it provides no new knowledge which goes beyond the extension of the premises, nonetheless, the structure of the syllogism undoubtedly reduces the chance of fallacy. On the other hand, in induction we infer from the known to the unknown, from a part of the cases to all cases. The chances for knowledge of new relationships are greater, but the chances of error are also increased. Inferring the conclusion from the premises always involves a certain amount of risk. Incomplete induction—which for simplicity's sake we shall just call induction from now on—never achieves unconditional certainty.[101]

Basically two problems arise regarding the above quotations: (1) whether or not the view accepting only necessary and probable inferences holds good, (2) whether or not the validity of inference depends on the modality of this operation. What the authors have stated concerning the relation between the truth of premises and the correctness of inference can also be challenged, of course, but this issue has already been discussed before.

It would be an error to restrict the modality of inferences to necessary and probable only. Let us suppose that the following propositions are given: (1) The AB side of the ABC triangle is equal to the A_1B_1 side of the $A_1B_1C_1$ triangle. (2) The CAB and ABC angles of the first triangle are

---

[100] Kant, *Schriften* p. 565.
[101] Fogarasi, pp. 264–265.

equal to the corresponding $C_1A_1B_1$ and $A_1B_1C_1$ angles. It *actually follows* from the above propositions that the triangles concerned are congruent.

In order to arrive at a necessary inference we have to make use also of the general thesis on the coincidence of triangles:

> If two angles of a triangle and the side belonging to them are equal to two angles of another triangle and to the side belonging to them, the triangles at issue are congruent.
>
> The two angles of the ABC triangle and the side belonging to them are equal to the corresponding two angles of the $A_1B_1C_1$ triangle and the side belonging to them.

Therefore the ABC and the $A_1B_1C_1$ triangles are congruent.

Traditional logic made a distinction between immediate inferences, according to whether or not the value of the conclusion could be decided. An inference was described as correct in the former case and fallacious in the latter. Thus, e.g., to conclude from the truth of proposition (I) to proposition (A) was regarded as a logical fallacy. From a modal point of view it is a *contingent inference*: proposition (A) may and may not be t. Why would the fact that the value of proposition (A) cannot be decided on the ground of proposition (I) make it a fallacy to conclude from the truth of proposition (I) to proposition (A)?

There can be no doubt that such inferences are of very little use for cognition. Naturally, it is the establishment of necessary inferences that one must primarily strive for. However, this consideration is no excuse for arbitrarily restricting the modal variety of inferences. Furthermore, the applicability of inferences should not be chosen as a ground to establish the criterion of correct inference. In my view this criterion is independent of the modality of inferences.

Finally, let us study the requirement that a correct inference should be of a universal character. Aristotle wrote the following on this subject:

> Seeing, therefore, that demonstrations are commensurately universal and universals imperceptible, we clearly cannot obtain scientific knowledge by the act of perception: nay, it is obvious that even if it were possible to perceive that a triangle has its angles equal to two right angles, we should

still be looking for a demonstration—we should not (as some say) possess
knowledge of it; for perception must be of a particular, whereas scientific
knowledge involves the recognition of the commensurate universal.[102]

While probable inferences gained some ground along the necessary
ones, the requirement of universality remained unchallenged. Let us,
however, think over the following: in the traditional view, conditional
categorical syllogisms admit concluding from the affirmation of the
antecedent to that of the consequent and from the negation of the
consequent to that of the antecedent. Accordingly, the following inference
is fallacious (*fallacia ex consequenti*):

> If the wind is blowing, the weather-cock is revolving.
> The weather-cock is revolving.
> _____
> Hence the wind is blowing.

There can be no doubt that it is not justified to draw this conclusion in
all cases since some other circumstance can just as well be the cause for the
weather-cock revolving. Yet normally this conclusion is justified. And as
in the present example so it is in the majority of cases. Why should it not be
accepted as a correct *inference of partial validity*? Let us examine the
following version as an argument for this way of putting the question:

> If the weather-cock is revolving the wind is blowing.
> The weather-cock is revolving.
> _____
> Hence the wind is blowing.

Is it not strange to declare this inference correct and to completely reject
the former one? Why could the former inference not be correct as well with
due emphasis on the corresponding restriction? Just because these two
inferences are not equally valid? In my view the correctness of inferences is
erroneously identified with their universal validity.

In scientific research one often concludes from effect to cause. The
occurrence of the rainbow, for example, has been observed for centuries.
For a long time, specific supernatural forces were sought behind it. At the
same time, contrary to religious beliefs, many scientists tried to give a

[102] Aristotle. 1928. 87b33–39.

natural explanation for this phenomenon. It was stated that a phenomenon similar to the rainbow can also be detected on drops of dew, on hexagonal crystals, on the drops of waterfalls, etc. Comparison of all these instances revealed that the only thing they have in common is light going through a refractive body. Since a rainbow occurs only if rain is followed by sunshine it was concluded that the cause of this phenomenon is the light going through the raindrops. Although such inferences did not lead to a *t* conclusion in every case, they were still accepted as correct in traditional logic. But this being so, why should one reject *ex consequenti?*

Let us see now what Hegel said:

> If any one, when awaking on a winter morning, hears the creaking of the carriages on the street, and is thus led to conclude that it has frozen hard in the night, he has gone through a syllogistic operation: — an operation which is every day repeated under the greatest variety of conditions. The interest, therefore, ought at least not to be less in becoming expressly conscious of this daily action of our thinking selves, than confessedly belongs to the study of the functions of organic life, such as the processes of digestion, assimilation, respiration, or even the processes and structures of the nature around us.[103]

I think the above example is *ex consequenti,* on the one hand, and an inference concerning facts, on the other.

V

The questions raised above lead to the problem of the relation between *inference* and proof, or *demonstration.* This was commented on by Aristotle in the *Prior Analytic* as follows:

> After these distinctions we now state by what means, when, and how every syllogism is produced; subsequently we must speak of demonstration, because syllogism is the more general: the demonstration is a sort of syllogism, but not every syllogism is a demonstration.[104]

Why syllogism is of wider scope than demonstration was explained in the *Topics*:

[103] Hegel, *System* p. 387.
[104] Aristotle. 1928. 25b26–30.

> Now reasoning is an argument in which, certain things being laid down,
> something other than these necessarily comes about through them. (a) It is a
> 'demonstration', when the premises from which the reasoning starts are
> true and primary, or are such that our knowledge of them has originally
> come through premises which are primary and true...[105]

These statements make it obvious that those who regard the truth of
premises as an indispensable condition for a correct inference restrict
inference to demonstration. In a textbook recently published the following
can be read:

> *The premises of an inference must be true...*
> Logic which would not care about the truth of the premises could become
> an instrument of demonstrating fallacies in an apparently logical form.[106]

What does the author want to warn of? In scholastic logic there was,
indeed, a tendency to make wild speculations accepted as demonstrated
facts with the help of syllogism. As a counterreaction, many logicians
rejected the syllogism in modern times. Others as, e.g., the author cited
above, defended the syllogism but only under the above mentioned
condition. The common mistake of all three views is the supposition that
the syllogism and demonstration coincide.

Let us go further. Is it really a condition of correct inference to be
necessary, as Aristotle and others believed? I have tried to demonstrate
that it is not the case. This requirement is, of course, justified in the case of
syllogistic inference. And finally, universal validity is not a condition of
correct inference, either.

What is left after all this? In the history of logic there evolved a principle
which, *free* of all the restrictions attached to it, is well-suited as definition. *An inference is correct if the antecedent serves as a sufficient reason for
drawing the conclusion.* Sufficient reason is the summary of all the
conditions which justify the declaration of a proposition.

Traditional logic interpreted the connection between sufficient reason
and correct inference in the following way: in the everyday practice of
thinking, people often conclude from the existence of particular

---

[105] *Op. cit.*, 100a25–30.
[106] Marković, p. 67.

occurrences to that of other occurrences without being convinced as to whether or not it is justifiable. For example, if we see that the thermometer shows 20 degrees we conclude that the temperature of the air surrounding us is the same. But does the temperature of the thermometer always coincide with that of the air? If not, then the above inference is not complete since the conditions under which the conclusion would necessarily follow are not revealed.

In order to make it complete, the following antecedent should be formulated: if two bodies are in contact for a sufficient period, their temperatures will converge. Consequently, the former premise must be amended. In order to conclude as to the temperature of the air, not only the temperature of the thermometer has to be known but also whether it has been in contact with the surrounding air for long enough. This inference would look like this in an expanded form:

> If a body has been in contact with another for sufficient time, then its temperature is identical with the temperature of this body.
> The thermometer which premanently shows 20 degrees has been in contact with the surrounding air for sufficient time.
> ___
> Hence the temperature of the thermometer is identical with the temperature of the air.

For simplicity's sake it is implied here that the thermometer is in the shade. In general the premises of this inference ought to express also that the thermometer is not directly reached by the sun's rays.

In an incomplete inference—due to the lack of sufficient reason—the conclusion is at most probably, but not necessarily, t. All incomplete inferences can be transformed into syllogisms if certain additional knowledge is ranked among the premises. Transformation into a syllogism is of great importance because it entails the revelation of all the knowledge necessary for the given conclusion and in this way protects us against fallacies resulting from unchecked premises. If an inference has a syllogistic form, this proves that its premises involve the knowledge sufficient for drawing the given conclusion.

The transformation into a syllogism is important, on the other hand, because it makes the process of thought exact and convincing from a logical point of view. That is, in this way it becomes possible to state in a clear form the logical necessity that one proposition follows from the other. In so far as the necessary character of the relation between the conclusion and the premises is not clearly expressed in incomplete inferences, the logical power is lost. In such cases one has no logical criterion at all to decide whether or not the given conclusion follows from the corresponding propositions.

This was the traditional approach. Let us ask now what is meant by the lack of sufficient reason. According to the above survey, it is defined primarily from the point of view of logical necessity. If the conclusion does not necessarily follow from the premises, then the sufficient reason is missing. Let us have a closer look at the above example. If the thermometer shows 20 degrees it probably follows that the temperature of the air is 20 degrees. Is it indeed justifiable to speak of a lack of sufficient reason in this case? The given inference can be considered as duly founded.

Such, and similar, inferences are not only correct but also play a great part in everyday thought and are of great importance in preparing for scientific discoveries. Whoever allows only inferences of demonstrated knowledge isolates logic from practical thinking. In order to overcome this state of affairs the field of correct inferences should be defined more broadly than has been the case so far.

CHAPTER SEVEN

20. CATEGORIES OF ONTOLOGY

In the previous parts of this work I have analysed mainly such categories of philosophy as have already been treated in logic to some extent. Now I will try to demonstrate how to include in logical analysis categories which lie beyond this circle.

I

The first problem of philosophy, both historically and in principle, is to account for the fundamentals and substance of the universe. To start with, ancient Greek philosophers identified the substance of the universe as some particular external form of matter. Water, air, fire, etc., represented the substance of reality to them.

Any answer to the question of the substance of the universe offers at the same time an explanation for the origin of the world. How the world has come into being is organically linked with the question, what is the fundamental nature of reality? Naive materialism did not assume the beginning of the world to be an absolute beginning but postulated a procession from eternity of some concrete material phenomenon.

Naive materialism, especially the pre-Socratic type, was right in looking for a common basis to explain the working of the universe, and also in finding this common substance in matter. The idea that the substance of the universe is to be found in some particular form of matter is, however, unjustifiable.

Substance has a central place in Aristotle's philosophy. In my view two of his numerous statements on substance are worth noting here:

> The subject of our inquiry is substance; for the principles and the causes we are seeking are those of substances. For if the universe is of the nature of a whole, substance is its first part; and if it coheres merely by virtue of serial

succession, on this view also substance is first, and is succeeded by quality, and then by quantity. At the same time these latter are not even being in the full sense, but are qualities and movements of it,—or else even not-white and the not-straight would be being; at least we say even these *are,* e.g. 'there is a not-white'. Further, none of the categories other than substance can exist apart. And the early philosophers also in practice testify to the primacy of substance; for it was of substance that they sought the principles and elements and causes. The thinkers of the present day tend to rank universals as substances (for genera are universals, and these they tend to describe as principles and substances, owing to the abstract nature of their inquiry); but the thinkers of old ranked particular things as substances, e.g. fire and earth, not what is common to both, body.[1]

Further on he wrote as follows:

Some things can exist apart and some cannot, and it is the former that are substances. And therefore all things have the same causes, because, without substances, modifications and movements do not exist. Further, these causes will probably be soul and body, or reason and desire and body.[2]

That is: (1) substance is the first category. (2) even if this term has not been defined unambiguously the fact that—according to Aristotle—it contains both the material and heavenly existent, in general is obvious.

Later in the history of philosophy the view prevailed that substance means exclusively, or primarily, the heavenly existent. In the Middle Ages special emphasis was given to the idea of God as absolute substance. This view was held for some time even by those who opposed scholasticism in every other way. Descartes, for example, defined substance as something which exists without depending on any other thing for its existence; and by such a substance which does not depend on any other thing in any way only one thing can be meant, namely God.[3]

Spinoza identified substance with nature. In so far as Spinoza meant by nature, first of all, matter, he was a materialist. But when he used the expression 'God, or nature' (*deus, sive natura*) he had taken up a pantheistic position.

---

[1] Aristotle, *Metaphysica.* 1069a 17–29.
[2] *Op. cit.,* 1071a 1–4.
[3] Descartes, *Principia philosophiae.* (Amsterdam 1672) Part I. § 48.

Spinoza's concept of substance has often been criticized for being abstract. In this respect it is Hegel's standpoint I find worth noting:

> Absolute necessity is absolute relation because it is not *being* as such, but *being* that is *because* it is, being as absolute self-mediation. This being is *substance*; as the final unity of essence and being it is the being in *all* being; it is neither the unreflected immediate, nor an abstract being standing behind Existence and Appearance, but it is immediate actuality itself . . . [4]

The term 'substance' has been pushed into the background in Marxist philosophy, mainly due to its ambiguity. It would be unjustified, however, to underestimate its historical significance.

> The category of substance has played an important role in the history of philosophy. Since substance is very closely connected with the interrelation of our terms 'matter' and 'spirit', 'nature' and 'god'. The way substance is interpreted reveals the level of development of knowledge and science in a particular age. Both the social and economic system, and the political and ideological superstructure, typical of the given period, exert a remarkable influence on the interpretation of substance. The latter has always been a point at issue in the great philosophical debates which reflected the various social struggles, i.e., class struggle. [5]

Substance, if it still occurs in Marxist terminology, means matter as such, or matter as an abstraction.

> . . . matter, as it has been said before, differs not from the mind only but also from its own formations, states, and qualities. In this respect matter functions as substance in relation to its own manifestations, i.e., its own concrete states and qualities. Matter as substance is the basis of all that exists. [6]

To find our way among the categories mentioned in this historical survey, let us have a closer look at the figure below:

[4] Hegel, *Wissenschaft*. Erster Teil. pp. 697–698; A. V. Miller tr., p. 555.
[5] Fogarasi, p. 381.
[6] Septulin, p. 72.

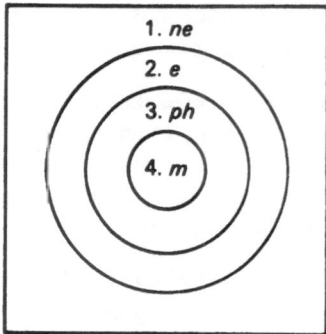

Fig. 39.

The designations are to be understood as follows: m = matter,
ph = phenomenon, e = existent, ne = non-existent. Ring 2 and ring 3
together represent what exists but is not matter, and by this I mean what
excludes both matter and everything that does not exist.

II

It was typical of the views of the first philosophers concerning the relation
of matter (Figure 39, ring 4) and phenomena (2 v 4) that they identified
matter with some concrete phenomenon (water, fire, etc.). In other words,
intuitive materialists tried to find the common feature of various
occurrences at the level of phenomenon.

Aristotle made an attempt to go beyond occurrences:

> By matter I mean that which in itself is neither a particular thing nor of a
> certain quantity nor assigned to any other of the categories by which being
> is determined. For there is something of which each of these is predicated,
> whose being is different from each of the predicates (for the predicates other
> than substance are predicated of substance, while substance is predicated of
> matter). Therefore the ultimate substratum is of itself neither a particular
> thing nor of a particular quantity nor otherwise positively characterized;
> nor yet is it the negations of these, for negations also will belong to it only by
> accident.[7]

[7] Aristotle, *Metaphysica*. 1029a 20–26.

At the same time, however, he considered matter only as a potential which becomes reality through a form of a spiritual kind. In this light and with some oversimplification, the following comparison can be made. Naive materialists rendered matter independent of spirit but failed to separate it from phenomena by abstraction while Aristotle sought to abstract matter from phenomena but subordinated it to spirit. Later on, the position of intuitive materialism was represented, among others, by Epicurus, Lucretius, Bacon, and Gassendi, while Aristotle's thesis was propagated by Avicenna and Thomas Aquinas.

Modern science made more and more urgent the demand implied as early as in ancient philosophy, namely the demand for a more general and more abstract concept of matter. To meet this demand became a primary goal, most of all, in Spinoza's world of thought. He was the one to take the first decisive step toward an abstract matter concept. In Spinoza's philosophy materialism practically got rid of its tendency to consider one form, or one part, of matter as the fundamental nature of all existing things. It is in the philosophy of Spinoza that the materialistic approach has first reached a stage of development where matter can be postulated in general, without any kind of formal or partial restriction. We should not, however, forget one drawback of Spinoza's approach: he defined the essence, or substance, of the universe not as matter only but also as nature which in his pantheistic interpretation included matter.

According to Spinoza matter is an abstraction to which one can trace back all the manifold occurrences of the universe as well as the multiplicity of its properties—in a way similar to the method of Euclidean geometry, by which all the theorems can be deduced from its axioms. It was the first philosophy to allow materialism—theoretically, of course—to take a new shape with every scientific discovery, as Engels put it.

At that time, however, there were no scientific discoveries which could have been instrumental in interpreting the evolution of the superior forms of matter, or in scientifically demonstrating the genesis of consciousness. Consequently, there was a big gap between the concept of abstract matter and the phenomena in Spinoza's philosophy, and it could be bridged only by artificial constructions at best.

Thus, the concept of abstract matter concept developed by Spinoza had some obscure points which could easily be used by idealism as issues to criticize, and thereby to attack materialism. As we have seen, Spinoza could not follow along with the idea of abstract matter splitting into a multiplicity of occurrences. He lacked the necessary preconditions to do so. As opposed to Spinoza's philosophy, idealism emphasizes this very multiplicity. This special stress on diversity accounts for the introduction of idealistic principles of spiritual nature in, for example, the philosophy of Leibniz. He argued that the universe is composed of monads, of individual elementary units, different in character and spiritual in nature. In this way the unsolved aspect of Spinoza's thought was made the main concern of idealism. On the other hand, however, matter played but a minor role in his philosophy.

While Leibniz merely underestimated the role of matter, Berkeley plainly denied its existence:

> Now in that which you call the obscure indefinite sense of the word *matter*, it is plain, by your own confession, there was included no idea at all, no sense except an unknown sense, which is the same thing as none. You are not therefore to expect I should prove a repugnancy between ideas where there are no ideas; or the impossibility of matter taken in an *unknown* sense, that is no sense at all.[8]

Materialism could successfully stand up against idealistic arguments only if it relied on the new development of natural sciences. Therefore it followed the first steps of specialized natural sciences with attention and generalized their results. It was Diderot who made the first move of vital importance for the further development and modernization of materialism.

Diderot was interested in the problems of biology, a science just being born. For it was only with the aid of biology that the unity of the realms of the organic and the inorganic could be proved. As long as biology had not appeared as a science and life was interpreted on a theological rather than a scientific basis, there was an unabridgeable gap between the organic and the inorganic. In Diderot's philosophy, however, the faculty of perception is seen as one of the properties of highly developed matter, and so inanimate things and living beings are not in flat opposition any more.

---

[8] Berkeley, *Three Dialogues between Hylas and Philonous* pp. 225–226.

Holbach tried to define matter in relation to perception:

> ... generally speaking, matter is everything that affects our senses in one
> way or another; and the qualities that we attribute to different kinds of
> matter are based on the different impressions or changes which they
> produce in us.[9]

There were still certain fields left which materialism failed to gain a firm
hold on. First among these came the problem of the genesis, role, and
range of consciousness. Questions of social existence, too, were treated
with inconsistency, and even materialist philosophers otherwise consistent
arrived at idealistic conclusions in this field. At that time it was impossible,
indeed, to cover these issues with a uniform concept of matter.

As a result of the activity of Marx and Engels, the concept of matter has
undergone a complete change. Dialectical materialism was developed and
disproved mystical and idealistic ideas, on the one hand, and overcame the
limitations of naive materialism, on the other. Marxism managed to arrive
at a concrete concept of matter through the abstract form it gave to it.

Materialism, as we have seen before, had two closely connected
problems. How to trace back the multiplicity of phenomena to the general
idea of matter was one of them. The other resulted from the necessity to
present this very matter as the source of the great variety of occurrences.

Lenin drew a distinction between philosophical and physical concepts
of matter. From the point of view of philosophy, matter is reality,
independent of our consciousness. This definition denies that consciousness had any part in the substance and genesis of the universe. It does not
associate the notion of matter with any particular form of it, either. This
approach contradicts all views which reduce matter to what exists in
perception, and, moreover, has a tangible and definite extension. What is
essential from the point of view of philosophy is the objectivity of matter
as a ground for the study of occurrences.

It leads to logical mistakes, or rather, deceptions, if the concepts of
concrete and abstract matter get mixed up. The statement 'matter is
imperishable', e.g., holds true of abstract matter only. On the other hand,
'matter has extensions' proves to be a judgement valid only for concrete
matter.

[9] Holbach, p. 38.

III

On the relation of existence and non-existence, numerous basically different opinions were formed in the 6th and early in the 5th centuries B.C. Heraclitus held the view that existence and non-existence are both identical and nonidentical: it is the same and still not the same river we step into, we exist and do not exist. On the other hand, according to the followers of Phythagoras, existence and non-existence exclude one another.

Aristotle's approach can be characterized by the following:

> ... there cannot be an intermediate between contradictories, but of one subject we must either affirm or deny any one predicate ... he who says of anything that it is, or that it is not, will say either what is true or what is false, but neither what is nor what is not is said to be or not to be.[10]

Elsewhere, however, he writes as follows:

> By 'as the man comes from the boy' we mean 'as that which has come to be from that which is coming to be, or as that which is finished from that which is being achieved' (for as becoming is between being and not being, so that which is becoming is always between that which is and that which is not...)[11]

Epicurus was of the opinion that nothing can be born out of something which does not exist *(ex nihilo nihil fit)* and that what exists cannot cease to do so. The recognition that the world cannot be created, or destroyed, but exists forever is a positive achievement of this approach. But he is mistaken in the total denial of transition between the existent and nonexistent.

Descartes approached what does not exist in a negative way only by identifying it with not-good and not-true.

Opposing the underestimation of the nonexistent, Hume argued as follows:

> Whatever *is* may *not be*. No negation of a fact can involve a contradiction. The non-existence of any being, without exception, is as clear

---

[10] Aristotle, *Metaphysica*. 1011b24–26, 28–29.
[11] *Op. cit.*, 994a25–28.

and distinct an idea as its existence. The proposition, which affirms it not to be, however false, is no less conceivable and intelligible, than that which affirms it to be.[12]

Hegel wrote the following:

> No great expenditure of wit is needed to make fun of the maxim that Being and Nothing are the same, or rather to adduce absurdities which, it is erroneously asserted, are the consequences and illustrations of that maxim.
>
> If Being and Nought are identical, say these objectors, it follows that it makes no difference whether my home, my property, (the air I breathe, this city, the sun, the law, mind, God) *are* or *are not*. Now in some of these cases, the objectors foist in *private aims*, the *utility* a thing has for *me*, and then ask, whether it be all the same to *me* if the thing exist and if it do not... A *substantial* distinction is in these cases secretly substituted for the empty distinction of Being and Nought ... When a concrete existence is disguised under the name of Being and non-Being, empty-headedness makes its usual mistake of speaking about, and having in mind an image of, something else than what is in question: and in this place the question is about Being and Nothing.[13]

Whoever ignores the difference between concrete and abstract existence simply cannot escape the mistake of identifying existence (circle 2) with bare existence (ring 2).

This is how Engels made a distinction between abstract and concrete existence:

> When we speak of *being* and *purely* of being, unity can only consist in that all the objects to which we are referring—*are*, exist. They are comprised in the unity of this being, and in no other unity, and the general dictum that they all *are* not only cannot give them any additional qualities, whether common or not, but provisionally excludes all such qualities from consideration. For as soon as we depart even a millimeter from the simple basic fact that being is common to all these things, the *differences* between things begin to emerge ... and whether these differences consist in the circumstance that some are white and others black, that some are animate and others inanimate, that some may be of this world and others of the

---

[12] Hume, p. 164.
[13] Hegel, *Encyclopädie* (Leipzig 1905) pp. 110–111; Wallace tr., pp. 164–165.

world beyond, cannot be decided by us from the fact that mere existence is in equal manner ascribed to them all.

The unity of the world does not consist in its being, although its being is a pre-condition of its unity, as it must certainly first *be* before it can be *one*. Being, indeed, is always an open question beyond the point where our sphere of observation ends. The real unity of the world consists in its materiality, and this is proved not by a few juggled phrases, but by a long and wearisome development of philosophy and natural science.[14]

Some Marxist commentators overemphasized the idea expressed in the last sentence of the above quotation and lost sight of the problem of abstract existence. They ignored Lenin's remark, too:

Being in general?—means such indeterminateness that Being = not-Being. All-sided, universal flexibility of concepts, a flexibility reaching to the identity of opposites,—that is the essence of the matter. This flexibility, applied subjectively = eclecticism and sophistry. Flexibility, applied *objectively,* i. e., reflecting the all-sidedness of the material process and its unity, is dialectics, is the correct reflection of the eternal development of the world.[15]

Let us have a look at the circular line between *e* and *ne* in Figure 39. If interpreted formally, it is a dividing line which illustrates that the categories concerned exclude one another. In a dialectical interpretation, however, this line both divides and unites. To draw a parallel, it is like a river which does not merely separate the opposite banks but connects them as well. In this sense the line mentioned above serves to demonstrate that existence and nonexistence exclude and at the same time precondition and imply one another. Dialectic does not reject the formal approach but goes beyond it.

Thus, one of the many possible ways to illustrate the above interpretation is:

---

[14] Engels, *Anti-Dühring* p. 51; Eng. tr. pp. 65–66.
[15] Lenin, *Philosophical Notebooks* p. 110.

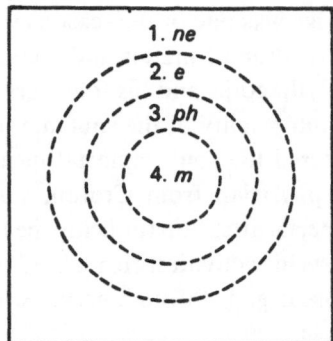

Fig. 40.

In the realm of living beings it is the relation of life and death which corresponds to that of existence and nonexistence. In metaphysical theories categories are thought to be mutually exclusive only:

> We say, for instance, that man is mortal, and seem to think that the ground of his death is in external circumstances only; so that if this way of looking were correct, man would have two special properties, vitality and— also—mortality. But the true view of the matter is that life, as life, involves the germ of death, and that the finite, being radically self-contradictory, involves its own self-suppression. 'The living die, simply because as living they bear in themselves the germ of death'.[16]

IV

Prehistoric man distinguished body and soul under the influence of dreams. Later the opposition of matter (Figure 39, circle 4) and consciousness (2 *o* 3) was based on this distinction.

Intuitive materialism correctly recognized that matter is prior to consciousness. However it could not reveal the complex interrelation between matter and consciousness. This inevitable failure of intuitive materialism which resulted from the relative underdevelopment of

---

[16] Hegel, *Encyclopädie* pp. 190–191; Wallace tr., p. 148. [the final sentence is taken from the Wallace tr. p. 174.—Ed.]

contemporary knowledge was one of the reasons for the development of idealistic views of the relation of matter and consciousness.

First among them, Pythagoras and his followers should be mentioned who propounded the immortality of the soul and its independence from the body. They considered the soul immortal in its state of permanent motion. Alcmaion, a physician from Croton, was one of the Pythagoreans. He deserves special attention for being the first to declare that the brain is the source of psychic activities. He realized that the brain is linked with the individual sense organs through nerves which are empty inside. He called them channels.

In opposition to idealistic notions, Democritus was of the opinion that all existing things are of material nature, the soul included. The soul, which is just as mortal as the body, is composed of tiny spherical atoms which get into the body by breathing. When the inhaled soul-atoms are less in number than those exhaled one is asleep. When all the soul-atoms have left, death sets in (cf. 'to expire').

In the 4th century B.C., objective idealism became dominant. According to Plato the ideas which exist independently of the material world make up the substance of all things. Material occurrences are but imperfect copies of the ideas and therefore secondary. In his effort to explain the interrelation of these two spheres Plato resorted to what he called reminiscence, on the one hand, and to participation, on the other.

In the course of argument, however, he came up against serious difficulties which he tried to overcome by creating myths. He claimed that one can learn to know the higher world in the form of myths only, i.e. through images, allegories, and symbols. Myths endowed his ideas with mystery and dignity. Thus, Plato fell back into the old mythological way of thinking.

Aristotle approached this fundamental issue of philosophy by assuming that the source of our knowledge exists independently of our consciousness. Nevertheless, he interpreted this reality beyond our consciousness in an idealistic way. He believed the substance of reality to be a form of spiritual nature, and that nothing but this form could transform matter existing only potentially into reality. In cognition, as he saw it, mind, similarly to a wax tablet, takes on the form but not the matter of different objects.

Now summing up what we have said about the soul, let us assert once more that in a sense the soul is all existing things. What exists is either sensible or intelligible; and in a sense knowledge is the knowable and sensation the sensible. We must consider in what sense this is so. Both knowledge and sensation are divided to correspond to their objects, the potential to the potential, and the actual to the actual. The sensitive and cognitive faculties of the soul are potentially these objects, viz., the sensible and the knowable. These faculties, then, must be identical either with the objects themselves or with their forms. Now they are not identical with the objects; for the stone does not exist in the soul, but only the form of the stone.[17]

No remarkable progress was made in this subject for the two thousand years after Aristotle's death. That is why we are content to refer only to the major trends in this brief survey. From the 3rd century B.C. to the 7th century of our era, objective idealism was represented by the Platonists and the Peripatetics while the philosophers adhering to the middle and late Stoa took the position of subjective idealism. In the centuries of the rise of feudal society the predominating philosophy was that of the Fathers of the Church, i. e. the patristics, who tried to adjust Platonic mysticism to Christianity. Three periods can be distinguished in the history of scholasticism. The first period (11th—12th cc.) was dominated by extreme (Platonic) realism, the second (13th c.) and the third (14th—15th cc.) by moderate (Aristotelean) realism and nominalism respectively.

In modern times the materialist conception, so long in the background, entered the battlefield of ideas again.

Man, being the servant and interpreter of Nature, can do & understand so much and so much only as he has observed in fact or in thought of the course of nature, beyond this he neither knows anything nor can do anything.[18]

It took a long time, however, to overcome its naive character. Gassendi, for example, considered soul as some pure, transparent, and subtle substance.

---

[17] Aristotle, *On the Soul.* pp. 179–181.
[18] Bacon, *Selection* p. 331 (The *New Organon,* 1(I)).

Idealists continued their efforts to make consciousness independent of matter:

> Next I had described the rational soul and shown that it can in no wise be derived from the power of matter as can the other things of which I have spoken, and must be due to a special act of creation.[19]

While Descartes postulated individual consciousness as the basic substance, Leibniz maintained the same of monads which exist independently of consciousness.

The first success during the rise of the exact sciences was achieved in the field of celestial and terrestrial mechanics. Mechanical materialism was born out of the generalization of these achievements. This school of thought brought some progress as compared to naive materialism in so far as it was based on scientific discoveries:

> I, physicist and chemist who finds bodies in nature and not in his head, see them existing as diverse entities, clothed with properties and actions; they move in the universe as they do in the laboratory, where a spark being found next to three molecules of saltpeter, carbon and sulfur, an explosion will necessarily follow.[20]

At the same time, mechanical materialism was also a regression as compared to naive materialism. For while breaking down nature into its components, it lost sight of the uniformity of the whole. Another mistake of this approach to be noted was the extension of principles of mechanics beyond the domain in which they hold true.

Fighting materialism, Berkeley defined his position as follows:

> Wood, stones, fire, water, flesh, iron, and the like things, which I name and discourse of, are things that I know. And I should not have known them, but that I perceived them by my senses; and things perceived by the senses are immediately perceived; and things immediately perceived are ideas; and ideas cannot exist without the mind; their existence therefore consists in being perceived...[21]

---

[19] Descartes, *Œuvres Choisies*. (Paris) pp. 51–52. Kemp Smith tr. from *Discourse on Method*, Part V in *Descartes' Philosophical Writings*, pp. 148–149.
[20] Diderot, *Œuvres Philosophiques*. p. 395.
[21] Berkeley, *Three Dialogues between Hylas and Philonous* p. 230.

In Hegel's view the basic substance is what he called world spirit (*Weltgeist*). Nature is the offspring of this very spirit. The Hegelian world spirit is but the idea transformed into an absolute, supernatural, self-existent substance. So far, Hegel's approach coincides with Plato's objective idealism. There are differences however; e. g. the Platonic idea exists beyond the universe while Hegel's absolute spirit lives in it.

We have briefly surveyed some characteristic views of the relation of matter to consciousness. It was not by accident, or independently of each other, but rather as a result of regular development that these views developed. Engels summed up this process of development as follows:

> ... the philosophy of antiquity was primitive, natural materialism. As such, it was incapable of clearing up the relation between mind and matter. But the need to get clarity on the question led to the doctrine of a soul separable from the body, then to the assertion of the immortality of this soul, and finally to monotheism. The old materialism was therefore negated by idealism. But in the course of the further development of philosophy, idealism, too, became untenable and was negated by modern materialism. This modern materialism, the negation of the negation, is not the mere re-establishment of the old, but adds to the permanent foundations of this old materialism the whole thought-content of two thousand years of development of philosophy and natural science, as well as of the history of these two thousand years.[22]

As far as priority is concerned, matter and consciousness are opposed to one another in dialectical materialism. Matter is prior to consciousness in two—closely interrelated—aspects: in historical development just as in principle. Matter is prior historically because the objects of our knowledge do exist before they are reflected in our consciousness. As to principles, it is the objects and occurrences of objective reality which determine the content of our consciousness, and not the other way round, our consciousness projecting its own patterns of thinking into existence.

It would be mistaken, however, to overemphasize the opposition of matter to consciousness, for these categories are conditional upon one another. Matter makes it possible for consciousness to develop; the

[22] Engels, *Anti-Dühring*. pp. 169–170. Eng. tr. pp. 191–192.

objective potentially implies the subjective. On the other hand, consciousness is not exclusively subjective, for as far as its genesis, or rather its
achievement, is concerned it is objective.

21. CATEGORIES OF STATE

According to the Milesian philosophers the world is in a state of
permanent motion, change, and development. Judged by the still extant
sources, these categories appeared to them as some undifferentiated unity.
Thales thought that everything originates from water, i. e. 'the moist', and
then returns there. Anaximander taught that the first principle is the
'*apeiron*', an infinite matter out of which arise the elementary contraries,
warm and cold, moist and dry, which then evolve to the multiplicity of
phenomena. Anaximenes traced all changes back to the air getting thicker
and thinner respectively.

Heraclitus was of the opinion that the transformations of fire account
for the great variety of material phenomena: fire is born out of the death
of the earth, air lives through the death of fire, while its death gives life to
water, the death of which keeps the earth alive. He described the
movement from earth to fire by an upward curve and the opposite process
by a downward one.

Heraclitus's dialectic was transformed into relativism by the Sophists.
Protagoras imagined matter as a pure flow.

Aristotle set the following argument against the above approach:

> . . . it would be fair to criticize those who hold this view for asserting about
> the whole material universe what they saw only in a minority even of
> sensible things. For only that region of the sensible world which
> immediately surrounds us is always in process of destruction and
> generation; but this is—so to speak—not even a fraction of the whole, so
> that it would have been juster to acquit this part of the world because of the
> other part, than to condemn the other because of this. — And again,
> obviously we shall make to them also the same reply that we made long ago;
> we went to show them that there is something whose nature is changeless.
> Indeed, those who say that things at the same time are and are not, should in

consequence say that all things are at rest rather than that they are in movement ...[23]

This can be accepted as a criticism of the relativist conception. What Aristotle rejected was, however, not the excesses only but also the central thesis of intuitive dialectic on the universality of motion and change: he introduced the idea of the unmoved mover as a fundamental cause. The consequences of this view of his will be discussed later.

His attempt at the differentiation of changes should be noted as well. He described four different types of change: changes in substance (generation and destruction), in quality and quantity (increase and diminution), and in place (motion in space).[24] The schoolmen adopted this classification without any modification.

Bacon was among the first to challenge this view:

Nor is there any value in those vulgar distinctions of motion which are observed in the received system of natural philosophy, as generation, corruption, augmentation, diminution, alteration, and local motion. What they mean no doubt is this: if a body in other respects not changed be moved from its place, *this is local motion*; if without change of place or essence, it be changed in quality, this is *alteration*; if by reason of the change the mass and quantity of the body do not remain the same, this is *augmentation* or *diminution*; if they be changed to such a degree that they change their very essence and substance and turn to something else, this is *generation* and *corruption*. But all this is merely popular, and does not all go deep into nature; for these are only measures and limits, not kinds of motion.

... if, leaving all this, anyone shall observe (for instance) that there is in bodies a desire of mutual contact, so as not to suffer the unity of nature to be quite separated or broken and a vacuum thus made; or if anyone say that there is in bodies a desire of resuming their natural dimensions or tension, so that if compressed within or extended beyond them, they immediately strive to recover themselves, and fall back to their old volume and extent; or if anyone say that there is in bodies a desire of congregating toward masses of kindred nature—of dense bodies for instance, toward the globe of the earth, of thin and rare bodies toward the compass of the sky; all these and the like

[23] Aristotle, *Metaphysica*. 1010a26–38.
[24] *Op. cit.*, 1069b9–15.

32*

are truly physical kinds of motion—but those others are entirely logical and scholastic, as is abundantly manifest from this comparison.[25]

Bacon, and especially his followers, focused their attention on mechanical motion at the cost of other possible states.

> All change, all development in nature, was denied. Natural science, so revolutionary at the outset, suddenly found itself confronted by an out-and-out conservative nature, in which even today everything was as it had been from the beginning and in which—to the end of the world or for all eternity—everything would remain as it had been since the beginning.
>
> High as the natural science of the first half of the eighteenth century stood above Greek antiquity in knowledge and even in the sifting of its material, it stood just as deeply below Greek antiquity in the theoretical mastery of this material, in the general outlook on nature. For the Greek philosophers the world was essentially something that had emerged from chaos, something that had developed, that had come into being. For the natural scientists of the period that we are dealing with it was something ossified, something immutable, and for most of them something that had been created at one stroke.[26]

At a further stage of progress the importance of change and evolution was gradually recognized. Such discoveries as the transformation of energy, the differentiation of cells, and the development of species, were especially instrumental in this process. In the field of philosophical abstraction, it was the achievements of classical German idealism which served as a basis for developing the theory of evolution within the framework of dialectical materialism.

> Thus we have once again returned to the mode of outlook of the great founders of Greek philosophy, the view that the whole of nature, from the smallest element to the greatest, from grains of sand to suns, from Protista to man, has its existence in eternal coming into being and passing away, in ceaseless flux, in unresting motion and change. Only with the essential difference that what in the case of the Greeks was a brilliant intuition, is in

[25] Francis Bacon *Works* Vol. I. pp. 177–178 (*Novum Organum*, 1(LXVI)); F.H. Anderson tr., pp.64–65.
[26] Engels, *Dialektik der Natur* p. 12. Clemens Dutt tr., p. 35.

our case the result of strictly scientific research in accordance with experience, and hence also it emerges in a much more definite and clear form.[27]

The interrelations of the categories discussed are illustrated in Figure 41.

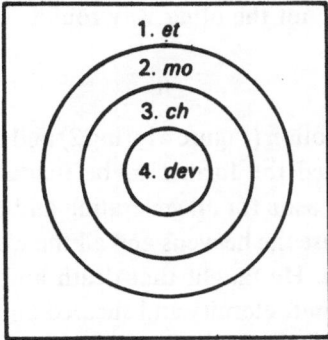

1. *et*

2. *mo*

3. *ch*

4. *dev*

Fig. 41.

The designations are to be understood as follows: *dev* = development, *ch* = change, *mo* = motion, *st* = test [standstill]. Ring 2 stands for motion without change while ring 3 represents change without development. Ring 2 and circle 4 together indicate a motion which is more than change. Rings 2 and 3 together illustrate repetition which excludes both development and standstill. The summary of all possible variations is:

0	— not-*u*		2 *o* 3	—	*mo*, but not *dev*
1	— *st*		2 *o* 4	—	*mo*, but not only *ch*
2	— *mo*, but not *ch*		3 *o* 4	—	*ch*
3	— *ch*, but not *dev*		1 *o* 2 *o* 3	—	not *dev*
4	— *dev*		1 *o* 2 *o* 4	—	not only-*ch*
1 *o* 2	— not *ch*		1 *o* 3 *o* 4	—	not only-*mo*
1 *o* 3	— *st o* only *ch*		2 *o* 3 *o* 4	—	*mo*
1 *o* 4	— *st o dev*		1 *o* 2 *o* 3 *o* 4	—	*u*

[27] *Ibid.* p. 43 (Englısn).

One can assert various propositions using the above variations: the clock-hand moves; the weather changes; society develops, etc. The relation of two propositions of this type can be detected from Figure 41. E.g. the following two propositions, 'S develops' and 'S moves', are in a relation of subordination. The fact that the former holds true implies the same of the latter but not the other way round.

II

As to the relation of motion (Figure 41, ring 2) and rest (circle 1), Milesian philosophers considered the former to be fundamental. Anaximander argued that the only reason for all generation and destruction is given by infinity from which arise the heavens and all the worlds in general, which are infinite in number. He taught that death and, long before it, birth originate from this infinite eternity and succeed one another in an endless cycle. The adherents of an intuitive dialectic inevitably deserve credit for recognition of all-involving motion. It was a mistake, however, to confine motion to its individual concrete forms only, such as expansion and contraction, heating and cooling, condensation and rarefaction.

The relativists, not contented with this emphasis on the vital importance of motion as compared to rest proposed its exclusive role. While Heraclitus thought that one cannot step twice into the same river, Cratylus went further and argued that one can do so not even once. The Eleatics, however, held the view that the universe, by its nature, is immobile.

Aristotle rejected both approaches:

> ... those who say all things are at rest are not right, nor are those who say all things are in movement. For if all things are at rest, the same statements will always be true and the same always false,—but this obviously changes; for he who makes a statement, himself at one time was not and again will not be. And if all things are in motion, no thing will be true; everything therefore will be false. But it has been shown that this is impossible...
>
> But again it is not the case that all things are at rest or in motion *sometimes*, and nothing *for ever*; for there is something which always moves the things that are in motion; and the first mover is itself unmoved.[28]

---

[28] Aristotle, *Metaphysica*. 1012b23–32.

The Stagirite, as seen also in other parts of his work, criticized mainly those who thought everything is in motion. Although he did not share the opinion that everything is unmoved, either, he believed the final instance to be unmoved. This final instance, the prime mover, is of a spiritual nature.

On the relation of motion, immobility, and rest he took the following position:

> The immobile is either that which is wholly incapable of being moved, or that which is moved with difficulty in a long time or begins slowly, or that which is of a nature to be moved and can be moved but is not moved when and where and as it would naturally be moved. This alone among immobiles I describe as being at rest; for rest is contrary to movement, so that it must be a privation in that which is *receptive of movement*.[29]

The way scholastic theology interpreted motion was affected to a great extent by feudal conditions. This social system, where everything was *a priori* settled, was, by its own nature, a closed, immobile hierarchy. It was only natural for such a society to adopt, e.g., Ptolemy's theory about the immobile earth in the centre of the universe. The religious and philosophical dogma of the time tried to prevent all kind of change.

This is not to say, however, that the category of motion was totally ignored. Thomas Aquinas, e.g., drew a distinction between perfect and imperfect motion. He ranked circular motion as the former and the rectilinear one as the latter. Nevertheless it can be generally stated that the contemporary ideas of motion, rather insignificant by themselves, were overshadowed by the principle of constancy and immutability.

With the rise of the Renaissance, this field, too, underwent some major changes. Copernicus proved the appearance underlying Ptolemy's theory false. In reality it is not the sun and the planets which revolve round the earth, but the earth which is in motion, revolving round the sun, similarly to other planets. Moreover, he proved wrong all those who made an unjustified distinction between celestial and earthly motion.

It was Galileo who took the next important step. He successfully confuted the idea that bodies affected by the same forces move at the same

[29] *Op. cit.*, 1068b20–26.

speed. Having proved the uniform acceleration of bodies, he enunciated the law of inertia, namely that a body keeps its uniform rectilinear motion until it is affected by some external force. It was Galileo again who formulated the law of free fall. His life-work laid the foundations for mechanics.

Metaphysical thought separates motion from rest, opposing them to one another. Motion is seen as an interruption of rest. A body is either in motion or at rest.

> The fact that motion can be expressed by its opposite, i.e rest, is no problem for the dialectical approach. For the latter this whole opposition is relative since there is no absolute standstill, or perfect equilibrium. Every particular motion strives for a state of equilibrium which is, again, upset by the total motion. In this way rest and equilibrium, wherever they occur, result from a limited motion, and it goes without saying that motion can be measured and expressed by, and reproduced in one form or another, by its result.[30]

Motion and rest, though opposites, are at the same time conditional upon one another. There is no motion without rest, and *vice versa*. Still, motion and rest are not coordinated notions. Their relation is determined by the fact that motion is absolute while rest is relative. In other words, motion is fundamental, and rest is subordinated to it. Motion involves rest as a relative circumstance. This is not to say, however, that rest should be understood as something secondary.

> Equilibrium is inseparable from motion. In the motion of the heavenly bodies there is *motion in equilibrium* and *equilibrium in motion*, (relative). But all 'specifically relative motion, i.e. here all separate motion of individual bodies on one of the heavenly bodies in motion, is an effort to establish relative rest, equilibrium. The possibility of bodies being at relative rest, the possibility of temporary states of equilibrium, is the essential condition for the differentiation of matter and hence for life.[31]

---

[30] Aristotelés, *Metafizika*. (Budapest 1957). p. 63.
[31] Engels, *Dialektik der Natur*. p. 262; Clemens Dutt tr., p. 326.

III

The representatives of intuitive dialectic attached great importance to the category of change (Figure 41, 3 *o* 4). Its detailed analysis, however, was made by Aristotle only. He distinguished three kinds of change: Something can change accidentally (e.g. when a learned man is walking), or in some of its parts (e.g. when an eye is healed the body becomes healthy), and finally, there is also what moves by its original nature, so to say, by itself.[32]

Now let us see how he assessed the relation of change and motion.

> There are six sorts of movement: generation, destruction, increase, diminution, alteration, and change of place.
>
> It is evident in all but one case that all these sorts of movement are distinct each from each. Generation is distinct from destruction, increase and change of place from diminution, and so on. But in the case of alteration it may be argued that the process necessarily implies one or other of the other five sorts of motion. This is not true, for we may say that all affections, or nearly all, produce in us an alteration which is distinct from all other sorts of motion, for that which is affected need not suffer either increase or diminution or any of the other sorts of motion.[33]

His predecessors usually confused the terms 'motion' and 'change' and used them as synonyms. In the above quotation Aristotle proved that (1) becoming something else is different from all the other kinds of motion, and (2) becoming something else is not identical with motion, although it is a kind of motion, i.e., becoming something else always implies motion, but not the other way round.

The above paragraph is a quotation from the *Organon*. In the *Metaphysics* we can read the following on the same subject:

> Since every movement is a change, and the kinds of change are the three named above [Reference to 1067b1–7—Gy. T.], and of these those in the way of generation and destruction are not movements, and these are the changes from a thing to its contradictory, it follows that only the change from positive into positive is movement.[34]

---

[32] Cf.: Aristotle, *Metaphysica*. 1067b1–7.
[33] Aristotle, 1928. 15a14–24.
[34] Aristotle, *Metaphysica*. 1068a1–5,

Thus, in contradiction to what is written in the *Organon*, every motion is change but not every change is motion.

Inevitably Aristotle deserves credit for having raised this issue even if he could not solve it. Those coming after him failed to make progress in this field for a long time, or even to raise the question again. It was only when Descartes introduced variables that a major change occurred.

Another discovery of great importance as regards our subject was prepared by Descartes, too. He had proved that the quantity of motion is constant in the universe, and consequently, motion can be neither created nor destroyed. Later Lomonosov, relying on the strength of these findings, formulated the universal law of the conservation of motion.

Before proceeding to the discovery to which the above considerations also contributed additional material, let me briefly illustrate the way Kant tried to make use of these ideas in his philosophy.

> Coming to be and ceasing to be are not alterations of that which comes to be or ceases to be. Alteration is a way of existing which follows upon another way of existing of the same object. All that alters *persists*, and only its *state changes*. Since this change thus concerns only the determinations, which can cease to be or begin to be, we can say, using what may seem a somewhat paradoxical expression, that only the permanent (substance) is altered, and that the transitory suffers no alteration but only a *change*, inasmuch as certain determinations cease to be and others begin to be.[35]

This argument is an attempt to grasp the dialectic of change and permanence. It would be easy to point out its defects, e.g., the indefensibility of the proposition that the ephemeral does not change. Instead, I would like to call attention to the fact that these arguments were formulated at a time when metaphysical thought still flourished.

The major step which was to lead to the solution of this problem in due course was taken in the middle of the 19th century when the law of conservation of energy had been completed by, and combined with, that of the transformation of energy.

I shall try to describe the relation of change (becoming something else, transformation) and constancy (universality, permanence) in general. The

---

[35] Kant, *Kritik*. p. 281; Kemp Smith pp. 216–217.

views of those who recognized but one of the two to be valid have proved
to be one-sided. These two states exclude, but at the same time also
precondition, one another. The determining factor within their unity is
change. Change is an absolute state while permanence is a relative one.
The relation of change and motion can be summed up like this:

> Motion in the most general sense, conceived as the mode of existence, the
> inherent attribute, of matter, comprehends all changes and processes
> occurring in the universe, from mere change of place right up to thinking.[36]

The relation of rest and permanence is very easy to define formally:
every case of rest is permanence but not vice versa. The dialectical
approach, however, makes this relation appear more complicated and
subtle.

> Relative standstill is manifest not only in the fact that bodies stay in the
> same position in relation to one another, and neither in the fact only that a
> certain material body does not imply some concrete form of motion. One of
> the most important form of relative standstill is the constancy of processes,
> the conservation of motion typical of the bodies concerned, and the relative
> constancy of forms of motion occurring under the given circumstances.
> Specific living organisms, for example, exist only because always the same
> type of metabolism takes place in them, and they always have the same type
> of interrelation with their external environment. The relative immutability
> of life-functions, and the constancy of their types, refer to relative standstill
> and equilibrium. Nevertheless, what lies hidden behind this standstill and
> equilibrium is an endless motion, change, and self-regeneration of the
> organism.[37]

IV

The issue of development was also raised by the ancient Greeks, though in
a more primitive form than, e.g., motion. Anaximander looked for an
explanation for the development of living beings. He thought animals
evolved out of the moist under the influence of sunbeams. Originally

---

[36] Engels, *Dialektik der Natur* p. 61; Clemens Dutt tr., p. 92.
[37] *A marxista filozófia alapjai.* (Budapest 1961) pp. 148-149.

animals used to live in water. Some of them came out of it to live on the
land and adapted themselves to the new circumstances both in the
construction of their body, and in their way of life. Man, who originated
from fish, was one of them.—According to Heraclitus fire is the source of
all development.

Aristotle shared the view that there *is* development.[38] As regards
substance, i.e. the final instance, he, nevertheless, denied development.

> Obviously then the form also, or whatever we ought to call the shape
> present in the sensible thing, is not produced, nor is there any production of
> it, nor is the essence produced; for this is that which is made to be in
> something else either by art or by nature or by some faculty.[39]

For the two thousand years since Aristotle's time no remarkable
progress was made. In modern times Vico who called attention to the
development of society is among the first to be mentioned. In his opinion
every country passes through three stages, namely: a divine, a heroic, and
a human period. The human period being over, society returns to its initial
state, and the triple process repeats itself. So there is development only
within a definite cycle beyond which repetition prevails.

It was a generally accepted view in the first half of the 18th century that
nature is unchanged, and that everything is like it used to be at the very
beginning.

> The first breach in this petrified outlook on nature was made not by a
> natural scientist but by a philosopher. In 1755 appeared Kant's *Allgemeine
> Naturgeschichte und Theorie des Himmels*. The question of the first impulse
> was done away with; the earth and the whole solar system appeared as
> something that had *come into being* in the course of time. If the great
> majority of the natural scientists had had a little less of the repugnance to
> thinking that Newton expressed in the warning: Physics, beware of
> Metaphysics!, they would have been compelled from this single brilliant
> discovery of Kant's to draw conclusions that would have spared them
> endless deviations and immeasurable amounts of time and labour wasted in
> false directions. For Kant's discovery contained the point of departure for
> all further progress. If the earth was something that had come into being,

---

[38] Aristotle, *Metaphysica*. 994a41–43.
[39] *Op. cit.*, 1033b5–8.

then its present geological, geographical and climatic state, and its plants and animals likewise, must be something that had come into being; it must have had a history not only of co-existence in space but also of succession in time.[40]

The state of biology was similar to that of cosmogony. Just like the solar system, the species of animals and plants were seen as something unchanged from the very beginning. Linneas's artificial system contributed considerably to the firm establishment of this view.

The ideas of those propagating development could make a breach but very slowly. First among them, Buffon should be mentioned. He pointed out the transformation of living beings, in opposition to the defenders of the preformation theory. The idea of the variability of species was further developed by Lamarck who had demonstrated the effect of environment. Darwin, in turn, proved by evidence that the biological species are the products of an inevitable and normal process of development.

The achievements of special sciences served as a basis for philosophical generalization.

... we may safely say that experience is the real author of *growth* and *advance* in philosophy. For, firstly, the empirical sciences do not stop short at the mere observation of the individual features of a phenomenon. By the aid of thought, they are able to meet philosophy with materials prepared for it, in the shape of general uniformities, i.e. laws, and classifications of the phenomena. When this is done, the *particular* facts which they contain are ready to be received into philosophy. This, secondly, implies a certain compulsion on thought itself to proceed to these concrete specific truths. The reception into philosophy of these scientific materials, now that thought has removed their immediacy and made them cease to be mere data, forms at the same time a *development* of thought out of itself. Philosophy, then, owes its development to the empirical sciences. In return it gives their contents what is so vital to them, the *freedom* of thought,— gives them, in short, an *a priori* character. These contents are now *warranted necessary,* and no longer depend on the evidence of facts merely, that they were so found and so experienced. The fact as experienced thus becomes an illustration and a copy of the original and completely self-supporting activity of thought.[41]

[40] Engels, *Dialektik der Natur.* p. 14; Clemens Dutt tr., p. 37.
[41] Hegel, *Encyclopädie* pp. 45–46. Wallace tr., pp. 21—22.

The term 'development' is one of the central categories of Hegel's philosophy. He was the first to formulate such basic principles of development as the conversion of quantity into quality, contradiction as the source of development, and the negation of negation. He was not content, however, with the declaration of these basic principles, and assessed the history of society also as a regular process of development. He regarded logic as a science investigating the content and forms of thought in development. I think the following errors of the Hegelian approach should be mentioned here: (a) he interpreted the history of the world as the self-development of the spirit rather than that of material reality; (b) his system is opposed to the development of cognition; (c) he conceived nature as a process of (cyclic) repetition.

As contrasted with the Hegelian approach, Marxism asserts the following: (a) it is the development of the material world which determines the development of our notions and ideas; (b) the development of cognition is an endless process; (c) the development of nature must be accepted as proved by the evidence of the special sciences.

As far as the way of development is concerned, two conflicting views developed in the 19th century. This controversy was especially obvious in geology. Cuvier believed the changes on the surface of the earth to be the results of catastrophes that occur periodically.

> Cuvier's theory of the revolutions of the earth was revolutionary in phrase and reactionary in substance. In place of a *single* divine creation, he put a whole series of repeated acts of creation, making the miracle an essential natural agent.[42]

Lyell denied these revolutions in nature and maintained that there are only slow, gradual changes.

> The two basic (or two possible? or two historically observable?) conceptions of development (evolution) are: development as decrease and increase, as repetition, *and* development as a unity of opposites (the division of a unity into mutually exclusive opposites and their reciprocal relation)...

---

[42] Engels, *Dialektik der Natur.* p. 15; Clemens Dutt tr., p. 39.

The first conception is lifeless, pale and dry. The second is living. The second *alone* furnishes the key to the 'self-movement' of everything existing; it alone furnishes the key to the 'leaps', to the 'break in continuity', to the 'transformation into the opposite', to the destruction of the old and the emergence of the new.[43]

The views concerning the relation of development (Figure 41, circle 4) and repetition (2 *o* 3) can be classified, first of all, as follows:

(1)   neither of them exists, for motion is but appearance (e.g., the Eleatics);

(2)   there is repetition only, but no development (e.g., Saint-Simon called 'circular philosophers' those who maintained that the human mind comes full circle again and again for ever and always returns to the same point it started from);

(3)   everything is in permanent development (e.g., certain evolutionists);

(4)   there are both development and repetition (e.g., Vico).

Let us have a closer look at the fourth version. Formally, development and repetition exclude each other (as seen in Figure 41). The formal approach is justified within its own boundaries. The ringing of a bell, for example, must be taken as repetition only and not as development.

In a broader context, however, these boundaries turn out to be far too narrow. For in reality (1) development and repetition not only exclude, but are also imbued with, one another, and (2) development has a more decisive role than repetition, the latter being dependent on the former.

V

Naive materialists treated motion as one of the properties of matter. But what did they mean by matter and motion respectively? The former term denoted water, air, fire, etc., while the latter stood for such simple motions as, e.g., condensation and rarefaction. Thus, in the same way as only some concrete form of matter was accepted as substance, motion, too, was conceived only in its concrete forms.

---

[43] Lenin, *Philosophical Notebooks.* p. 360.

The following reasoning is very typical of Aristotle's approach:

> The question might be raised, why some things are produced spontaneously as well as by art, e.g. health, while others are not, e.g. a house. The reason is that in some cases the matter which governs the production in the making and producing of any work of art, and in which a part of the product is present,—some matter is such as to be set in motion by itself and some is not of this nature, and of the former kind some can move itself in the particular way required, while other matter is incapable of this; for many things can be set in motion by themselves but not in some particular way, e.g. that of dancing. The things, then, whose matter is of this sort, e.g. stones, cannot be moved in the particular way required, except by something else, but in another way they can move themselves—and so it is with fire. Therefore some things will not exist apart from some one who has the art of making them, while others will; for motion will be started by these things which have not the art but can themselves be moved by other things which have not the art or with a motion starting from a part of the product.[44]

Here Aristotle is concerned mainly with the existence of two kinds of matter: one of them is capable of self-movement while the other is not. Consequently not all matter has the quality of motion but only some. What does it mean from the point of view of motion? The above quotation contains only implications in this respect. Aristotle's argument reveal, however, that motion, too, is of two kinds. One of them is implied in matter and the other is immaterial. He calls the latter 'entelechia' i.e., pure activity.[45] According to this, not every, but just some, notion is material. In the language of logicians, the Aristotelean terms 'matter' and 'motion' are in a crossing relation.

Metaphysical thought was not content with the partial independence of these two categories any more but strove for their total separation. According to Newton, matter is an inert mass, and motion is just a force affecting it from outside and not a property inseparable from it. The separation of motion from matter inevitably led to the postulation of a prime mover.

---

[44] Aristotle, *Metaphysica*. 1034a8–22.
[45] *Op. cit.*, 1071b38.

Some, on the contrary, proceeded along the lines of naive materialism.

> *According to certain philosophers, the body in itself is without action or force*; this is a terrible falsehood, contrary to all sound physics and chemistry; for whether one considers it as molecules or as a mass, the body by itself, by the nature of its essential qualities, is full of action and force.[46]

Here matter was attached to such particular motions as, e.g., expansion and contraction, to a much smaller extent than previously, but the level of purely mechanical motion had not yet been surpassed.

In opposition to the mechanical approach, a theory of dynamism developed late in the 18th century. It proposed that bodies are composed of forces rather than atoms, and nature is the resultant of conflicting forces such as, e.g., attraction and repulsion. These forces were considered to be pure motions, independent of matter. Dynamism which had, among others, such adherents as Kant, Schlegel, and Hegel, was one of the forms of dialectical idealism.

Classical German idealism, having discovered the great variety of forms of motion and supposing their transition into one another, opened up new perspectives. It was, first of all, Hegel who postulated the evolutionary motion of matter. Mechanical materialism could not combine matter with the tendency of evolutional motion, however closely it connected motion with matter. This goal was first achieved in the Hegelian philosophy, but in an idealistic way. In Hegel's view matter develops merely as the alienated form of the idea, or as the manifestation of the *Weltgeist*.

As regards the relation concerned, Marxism proceeds from the supposition that matter is inevitably in motion:

> *Motion is the mode of existence of matter*, hence more than a mere property of it. There is no matter without motion, nor could there ever have been. Motion in cosmic space, mechanical motion of smaller masses on a single celestial body, the vibration of molecules as heat, electric tension, magnetic polarization, chemical decomposition and combination, organic life up to its highest product, thought—at each given moment each individual atom of matter is in one or other of these forms of motion. All equilibrium is either only relative rest or even motion in equilibrium, like

---

[46] Diderot, *Œuvres Philosophiques*. (Paris 1961). p. 394.

that of the planets. Absolute rest is only conceivable in the absence of matter. Neither motion as such nor any of its forms, such as mechanical force, can therefore be separated from matter nor opposed to it as something apart or alien, without leading to an absurdity.[47]

In contradiction to old views that motion is a property of matter in an abstract way, dialectical materialism argues that motion results from the internal contradiction of matter. The individual forms of motion can be detected in matters of certain kind. Motion is the manifestation of the internal contradictions of matter. Consequently, Marxism rejects both the idea of matter at rest and Aristotle's view that matter at least originally is supplied with motion from outside.

Dialectical materialism, having set the term 'motion' free from the idealistic shell of Hegelian philosophy, defined matter not merely as being in motion, but as being in motion in the course of its development.

> ...an extremely important feature of Marxist philosophy is that it recognizes not only the materiality of the universe but also the fact that matter and nature are in a state of eternal, permanent, and regular change and development. The materiality, change, and development of the universe are inseparable from one another... However, to accept the idea of development by itself is not sufficient. It is also important to form a true notion of the character and source of development, and universal laws of all kinds of motion and development. It is possible to accept motion and still to remain a metaphysician.[48]

Just as matter cannot exist without motion, motion cannot exist without matter.

> The concept matter expresses nothing more than the objective reality which is given us in sensation. Therefore, to divorce motion from matter is equivalent to divorcing thought from objective reality, or to divorcing my sensations from the external world—in a word, it is to go over to idealism. The trick which is usually performed in denying matter, and in assuming motion without matter, consists in ignoring the relation of matter to thought.[49]

---

[47] Engels, 'Vorarbeiten zum *Anti-Dühring*'.—*Anti-Dühring*. p. 421; Eng. tr. pp. 470–471.
[48] *A marxista filozófia alapjai.* (Budapest 1959). p. 259.
[49] Lenin, *Materialism and Empirio-Criticism.* (Moscow 1952). p. 276.

The relations of the categories of ontology and state (e.g.; what is matter is in motion, or what develops exists) are illustrated by the following figure. The designations are to be understood as follows: $m =$

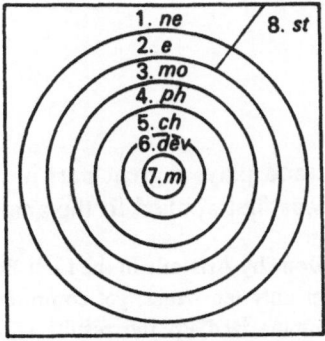

Fig. 42

matter, $dev =$ development, $ch =$ change, $ph =$ phenomenon, $mo =$ motion, $e =$ existent, $ne =$ non-existent, $st =$ standstill.

CONCLUSION

The combinatorial method plays a great part in the systematization of categories. In logic it was first applied to the analysis of syllogisms.

This research was done by Aristotle in the First Book of his *Prior Analytic* (Chapters 4–6). On only ten pages, yet comprehensively, and without a single error, he summarized up the results of a combinatorial logical research which had never occurred to anyone before. A similar combinatorial method had never been attempted previously in any science, not even in mathematics.

It might be worth noting that combinatorial analysis, now an important branch of mathematics, did not exist in the time of ancient Greece when numerous branches of mathematics had achieved perfection. The *use* of combinatorial operations, if at all present, was confined to a small number of combinations only and failed to develop into a *method*. It never happened previous to Aristotle's time—and not for a long time after it either—that someone tried to solve a scientific problem with the method of *combinatorial exploration*, i.e., examining a definite number of combinations composed of a given number of elements.[1]

Aristotle also made an attempt to apply the combinatorial method to categories. As is well known, he took ten categories and examined their interrelations. He did so, e.g., with the category 'to have something'. He concluded the analysis as follows:

Other senses of the word [i.e., to have something—Gy. T.] might perhaps be found, but the most ordinary ones have all been enumerated.[2]

[1] Aristotelés, *Organon.* (Budapest 1961). pp. LXVIII–LXIX.
[2] Aristotle. 1928. 15b31–33.

There was good reason for him to express himself so cautiously. For he failed to achieve in the field of categories what he had accomplished in syllogistic.

Theophrastus extended the use of the combinatorial method to conditional and disjunctive syllogism. The result, however, fell far behind what Aristotle had accomplished in categorical syllogism. Boethius, in turn, made a considerable progress in this field in his work *De syllogismo hypothetico*.[3]

The systematization of predicables was attempted by Porphyry. Let us take the following argumentation as a starting point:

> Now every proposition and every problem indicates either a genus or a peculiarity or an accident—for the differentia too, applying as it does to a class (or genus), should be ranked together with the genus. Since, however, of what is peculiar to anything part signifies its essence, while part does not, let us divide the 'peculiar' into both the aforesaid parts, and call that part which indicates the essence a 'definition', while of the remainder let us adopt the terminology which is generally current about these things, and speak of it as a 'property'. What we have said, then, makes it clear that according to our present division, the elements turn out to be four, all told, namely either property or definition or genus or accident.[4]

Theophrastus modified the above classification as follows. In his opinion difference should be regarded as an independent class, it therefore being an error to classify it as genus. Identity, however, was included in genus in his system. Accordingly, he distinguishes five versions: definition, genus, difference, property, and accident.

Porphyry identified definition with property, but considered species as an independent class of the same rank as the other four. Despite the fact that he had made minor modifications in the original classification he was credited for a long time with the statement that the predicate may contain five attributes *(quinque voces)*.

Although these attributes were to be interpreted in various ways later, the Porphyrian approach survived practically unchanged.

---

[3] See a sample of the versions discovered by Boethius in: Prantl, pp. 705–719.
[4] Aristotle. 1928. 101b17–25.

It is desirable that the reader, before proceeding further, should acquire an exact comprehension of the meaning of certain logical terms which are known as the Predicables, meaning the kinds of terms or attributes which can always be predicated of any subject. These terms are five in number; genus, species, difference, property, and accident; and when properly employed are of exceeding use and importance in logical science.[5]

In the Middle Ages, great efforts were made to discover new combinations. These endeavours aimed, first of all, at the extension of the variations of categorical syllogism, but also at the solution of other problems (e.g., Occam's modalities).

Lull found it needless to come to know the phenomena and connections of reality empirically for if one started with categories all problems could be solved with the aid of combinations. There were six groups in Lull's system, with nine categories in each. The first group contained what he called absolute properties: goodness, greatness, glory. Relations were contained in the second: difference, congruity, contradiction, fundamental principle, middle, end, majority, equality, minority. In addition, there still were nine questions, nine subjects, nine virtues, and nine sins.

Then he designed three circular sheets and wrote on them the above-mentioned 54 categories. He put these circular sheets with different diameters on one another, and by turning the sheets, he produced the combinations of the given concepts. Combinations with one component represented terms, those with two components stood for propositions, and those with three for syllogisms. Questions could be produced by the combination of interrogative words and categories.

Opinions were divided as to the evaluation of the system briefly outlined above. I quote here but one of them which I myself agree with.

> At first sight Lull's *Ars magna* as a whole seems to be a totally shallow and naive playing around with words collected at random. It would then as such not be worth serious attention. If there were, however, no sound ideas hidden behind the pompously presented empty words and senseless theses it would be impossible to understand why his thoughts have been thoroughly studied even by genuinely outstanding philosophers (e.g., Giordano Bruno, Leibniz) long after his death. Two things deserve notice, the importance of

[5] Jevons, 1872. p. 98.

which can be recognized even in this deformed *Ars magna*. One of them is the endeavour to construct all human knowledge out of some fundamental notions in a purely logical way, and the other is the idea of total formalization of logical operations which, of course, appears here in a totally distorted form. These two goals motivated also Leibniz's activity in logic. Lull's theory, however, lacks all logical foundations and therefore his method can by no means be said to be the first step towards some formalized system of reasoning or logical calculus. Thus, Lull deserves credit merely for being the first to develop the program of *ars inveniendi*—if that is to be regarded as a merit at all.[6]

Leibniz's approach essentially coincides with that of Lull. Both of them are aware of two interrelated tasks: to select adequate categories and to create special terms by their combinations. They differ significantly, however, in the way of realization of these tasks. Leibniz tried to select the fundamental terms on a scientific basis and to replace mystic relations by systematic analyses.

It was primarily the development of calculi which mathematical logic has taken over from Leibniz' ideas.

> A system of signs of a definite kind is called calculus if we define in which cases the assertions composed of the signs can be included into certain specific classes of assertion, and moreover, in what ways one or more assertions can be converted into other assertions. In Carnap's opinion every controlled language constructed according to formal rules can be regarded as such a calculus as far as only the formal aspect of the language is concerned, without the concrete meanings of the particular signs and expressions. Logical syntax deals with this calculus-like aspect of the language.[7]

Calculi have contributed a lot to the improvement of the technique of logic. Previously, the question of what are all the possible variations holding true was raised only in relation to categorical syllogism but now, in mathematical logic, it has been put into a much broader context, with numerous solutions suggested. The clumsy rules of traditional logic have been replaced by efficient decision-making operations, etc. The use of calculi, however, involves the risk of autotelism.

[6] Pozsonyi, p. 27.
[7] *Op. cit.*, p. 121.

> The modern mathematical logician has a strong support in the calculus; but the same calculus allows him all too often to absolve himself of mental labour—which perhaps might be just what was necessary.[8]

The development of calculus implied the use of signs. Letters as signs have been used by thinkers as early as Aristotle — as variables replaceable by concrete terms and propositions. It was the scholastic logic which developed the system of symbols of traditional logic ever since in use. The rise of mathematical logic, which uses symbols to denote not only variables but also operations, relations, etc., has brought about a decisive change in this field. In this way it has become possible to precisely and quickly solve such logical problems which previously could be solved only with hard work, or not at all.

How far is it justified to replace categories by symbols? Hegel wrote the following on this issue:

> There is as little to be said against the expression *power* when it is used only as a *symbol*, as there is against the use of numbers or any other kind of symbols for Notions — but also there is just as much to be said against them as against all symbolism whatever in which pure determinations of the Notion or of philosophy are supposed to be represented....... If numbers, powers, the mathematical infinite, and such-like are to be used not as symbols but as forms for philosophical determinations and hence themselves as philosophical forms, then it would be necessary first of all to demonstrate their philosophical meaning, i.e. the specific nature of their Notion. If this is done, then they themselves are superfluous designations; the determinateness of the Notion specifies its own self and its specification alone is the correct and fitting designation. The use of those forms is, therefore, nothing more than a convenient means of evading the task of grasping the determinations of the Notion, of specifying and of justifying them.[9]

In order to judge this opinion, the following should be taken into consideration. Leibniz regarded symbols as abstractions based on the objective. Boole, on the other hand, introduces and uses symbols for the

---

[8] Bochenski, p. 21.
[9] Hegel, *Wissenschaft* p. 404; A.v. Miller tr., p. 325.

elaboration of calculi before raising the question of their meaning. This constructive approach has come to be predominant in mathematical logic.

The following argument speaks in favour of abstraction as the principal approach.

> Mathematical logic is formalized much more perfectly and extensively than the traditional one and, as Russell and Whitehead have pointed out, it is in need of symbolic language so much because it works with concepts so abstract no adequate word of common usage can describe. It is not to say, however, that there is a *total* formalization in mathematical logic, or that the considerations of content are *everywhere* excluded from logic — it is just *localized* and *defined to a greater extent* than traditional formal logic. *Total* formalization is *inconceivable* even in mathematical logic (and in mathematics) for formalization, as said before, is a kind of abstraction, and abstraction aims at the cognition of reality (at the cost of disregarding *certain* connections of reality). Consequently, it would lose its meaning if it sacrificed all content relations with reality.[10]

The adherents of the constructive approach also disagree among themselves. Only the extremists are totally indifferent to the problems of content, or intension. The moderates find it necessary to study the meaning of symbols and the applicability of the calculus. As long as the introduction of symbols is seen as a working hypothesis rather than a triumphant breaking away from reality nothing can be said against the constructive method.

> The error in principle committed by Raymond Lull, Leibniz, and all of symbolic logic consists in the assumption that concepts instead of words can also be expressed by signs. For this position there are in fact historical examples: the Egyptian hieroglyphics and Chinese writing. The error is grounded in the assumption shared by all such approaches that the replacement of words by signs would make it possible *by purely mental operations, circumventing the reflection of reality, to achieve new discoveries, new knowledge.*[11]

Having considered what was said above, I have done the following. I have regarded categories as terms reflecting reality. In a critical analysis of

[10] Szalai, p. 386.
[11] Fogarasi, p. 91.

the various schools of thought developed historically, I have dwelt upon categories of the same type (e.g. modalities) and studied their interrelations. I have tried to make this investigation comprehensive by using the combinatorial method. In the course of analysis it was necessary to modify and supplement previous theories to a certain extent. Then, I have proceeded to the study of propositions which contain the categories concerned. First, I have assessed them by a selection of adequate criteria. Then I have used the tables of their values to check whether or not the conclusions hold true. In order to facilitate these investigations I have replaced categories by symbols.

In the present work I could analyse only a few categories, and even those not completely. As an excuse let me quote Fermat's words:

> ... and nevertheless, we do not regret that we have written this premature and immature work. Indeed, it is even of some use for science if we do not conceal from the next generation either the not wholly developed products of the mind or the way initially crude and immature thoughts get firmly established and multiply as a result of some new scientific discovery.[12]

Obviously there are two directions to follow. On the one hand, the categories concerned require a more thorough analysis. The study of the temporal, for example, should also be extended to propositions concerning the past, present, and future. On the other hand, the scope of research has to be widened by adding, e.g., the categories of deontic logic.

In the present work I have stayed practically within the framework of formal logic. I find it necessary to proceed towards the solution of such problems as the transitions and deducibility of categories, or of categories as the centres of historical development, etc.

> For rational thought the logical is purely subjective. As such it is *a priori* opposed to the historical. It might hold good, but it might as well be false, however consistent be its construction. For it is but an incidental and external possibility which might be more probable than others -- and that is all one can say about it. So, again, the historical alone seems to represent reality. Consequently, it is the historical alone that actually happens, demonstrating thereby *post festa* which of the incidental and external

---

[12] Quoted in Ribnikov, p. 138.

patterns that are equally possible on the ground of formal logic has indeed expressed truth and proved to be an inner necessity, allegedly ruling out anything else.—In fact, we should speak of the *organic* unity, i.e. dialectical relation, of the two even here. That is, it is the inner necessity manifest in the historical that is reflected in our mind as dialectical logic, and *vice versa*: the dialectical logical is but the historical summed up in its own inner necessity. Thus, the task of real, i.e. dialectical, logic is not simply to describe the historical as historical—for that would be shallow abstract speculation only. Real, or dialectical, logic has to define and reveal that inner law which is the reflection of inner historical necessity in our mind and which, in principle, allows us to decide in advance what kind of 'goals' a given historical process is striving after by its own innate nature.—The logical cannot be truly conceived without the historical; but as soon as it has been conceived, the historical has to be subordinated to the logical, and not the other way round. It is always to be performed dialectically, of course, that is, with a reflection up on itself.[13]

In principle, nothing prevents us from having a comprehensive knowledge of categories. We would be agnostic if we gave up this ambition. At the same time, however, we must realize that this goal will never be reached. One of the principal errors of Lull's *Ars magna* and Leibniz's *characteristica universalis* was their illusion to actually achieve such a totality. An absolute system of categories is just as impossible as an absolute system of axioms. Nevertheless, it would be a mistake to abandon the study of categories for this reason. On the contrary, the development of more and more comprehensive systems has to be striven for.

[13] Erdei, pp. 187–188.

INDEX OF SYMBOLS

a	= *all*	*Ic*	= it is contingent that some *S* is *P*
A	= all *S* is *P*		
Ac	= it is contingent that all *S* is *P*	*Inec*	= it is necessary that some *S* is *P*
Anec	= it is necessary that all *S* is *P*	*Ipo*	= it is possible that some *S* is *P*
Apo	= it is possible that all *S* is *P*	*ia*	= interaction
		id	= identity
al	= always	*ind*	= independence
bi	= biconditional	*ip*	= impossible
c	= contingent	*m*	= matter
ca	= causality	*M*	= middle term
ch	= change	*mo*	= motion
cj	= conjunction	*n*	= nothing
co	= conditional	*ne*	= nonexistent
d	= distributed	*nec*	= necessary
de	= dependence	*nev*	= never
dev	= development	*o*	= or
di	= disjunction	*O*	= some *S* is not *P*
dj	= disjunct	*Oc*	= it is contingent that some *S* is not *P*
e	= existent		
E	= no *S* is *P*	*Onec*	= it is necessary that some *S* is not *P*
Ec	= it is contingent that no *S* is *P*	*Opo*	= it is possible that some *S* is not *P*
Enec	= it is necessary that no *S* is *P*		
Epo	= it is possible that no *S* is *P*	*P*	= any term (traditionally: predicate)
f	= false	*p*	= any proposition
ft	= fact	*pc*	= partial coincidence
i	= inclusion	*ph*	= phenomenon
I	= some *S* is *P*	*po*	= possible
		q	= any proposition

R	= relation	*x*	= individual variable
S	= any term (traditionally: subject)	¬	= negation
s	= some	—	= negation
so	= sometimes	·	= conjunction
st	= standstill	∧	= conjunction
t	= true	∨	= disjunction
t/f	= it can be true, it can be false	→	= implication
th	= this	←	= converted implication
tt	= this time	↔	= equivalence
u	= universe	=	= identity
ud	= undistributed	∀	= universal quantifier
un	= union	∃	= existential quantifier

BIBLIOGRAPHY

A filozófia története (History of Philosophy) (Budapest, 1958), Vol. I.

A marxista filozófia alapjai (Fundamentals of Marxist Philosophy) (Budapest, 1959).

Alexander, 'Introduction' to *Aristotelis Analyticorum Priorum Librum* I *Commentarium* (Berolini, 1883).

Aquinas, as quoted in Bochenski, *Formale Logik*.

Aristotle, *Metaphysica* (Oxford, 1954).

Aristotle, *Metafizika* (Budapest, 1957).

Aristotle, *On the Soul* (Cambridge and London, 1957).

Aristotle, *Prior and Posterior Analytics*, ed. W. D. Ross (Oxford, 1949).

Aristotle, *Organon*, Vol. I (Budapest, 1961).

Aristotle, *Works* (Oxford, 1928), Vol. I.

Aristotle, *Topica cum Libro de Sophisticis Elenchis* (Leipzig, 1923).

Arnauld and Nicole, *La logique ou l'art de penser (Port Royal Logique)* (Paris, 1965); English translation: *The Art of Thinking*, tr. James Dickoff and Patricia James (Indianapolis, 1964).

Bacon, F., *A Selection of his Works* (New York, 1965).

Bacon, F., *The Works of Francis Bacon* (Stuttgart–Bad Canstatt, 1963) including the *Novum Organum*, etc. English translation by F. H. Anderson (Indianapolis, 1960).

Bakradze, K. S. *Logika* (Tbilisi, 1951).

Becker, A., *Die aristotelische Logik der Möglichkeitsschlüsse* (Berlin, 1933).

Berkeley, G., *The Works of George Berkeley*, ed. A. A. Luce and T. E. Jessop (London and New York, 1949), Vol. II. including 'Three Dialogues between Hylas and Philonous'.

Bochenski, J. M., *Formale Logik* (Freiburg and Munich, 1956); English translation: *A History of Formal Logic*, tr. Ivo Thomas (Notre Dame, Indiana, 1961).

Brentano, F., *Psychologie vom empirischen Standpunkt*, Vol. I (1911)

Bunge, M., *Causality* (Cambridge, Mass., 1959).

Buridan, *Tractatus consequentiarum magistri* as quoted in Bochenski, *Formale Logik*.

Burleigh, *De puritate artis logicae* as quoted in Bochenski, *Formale Logik*.

Church, A., 'Ontological Commitments', *J. Phil.* 55(1958): 1013.

Cornforth, M., *Science Versus Idealism* (London, 1946).

Couturat, L., *L'algèbre de la logique* (Paris, 1905).

Descartes, R., *Œuvres choisies*, Vol. I (Paris, n. d.).

Descartes, R., *Œuvres*, Vol. II (Paris, 1940), including 'Règles pour la direction de l'esprit'.

Descartes, R., *Válogatott filozófiai művek* (Selected Philosophical Works) (Budapest, 1961).

Descartes' Philosophical Writings, selected and translated by Norman Kemp Smith (London, 1952), including 'Rules for the Guidance of our Mental Powers' and 'Discourse on Method'.

Descartes, R., *Discourse on Method*, tr. F. E. Sutcliffe (Penguin ed., Middlesex and Baltimore, 1961).

Diderot, D., *Œuvres philosophiques* (Paris, 1961).

Drobisch, M., *Neue Darstellung der Logik (nach ihren einfachsten Verhältnissen mit Rücksicht auf Mathematik und Naturwissenschaften)*, 4th ed. (Leipzig, 1875).

Eisler, R., *Wörterbuch der philosophischen Begriffe (historisch-quellenmässig bearb.)* 3rd ed. in 3 volumes (Berlin, 1910).

Engels, F., *Anti-Dühring* (Berlin, 1948); English translation (Moscow, 1954).

Engels, F., *Dialektik der Natur* (Berlin, 1952); English translation: *Dialectics of Nature*, tr. Clements Dutt (Moscow, 1954).

Erdei, L., *Az ítélet dialektikus logikai elmélete* (The dialectical logical theory of the proposition) (Budapest, 1971).

Erdmann, B., *Logik*, Vol. I (Halle, 1892) (all published).

Fogarasi, B., *Logik* (Berlin, 1956) (Translated from the Hungarian by S. Szemere).

Frege, G., *Grundgesetze der Arithmetik*, Vols. I and II (Jena, 1893–1903); English translation: *The Basic Laws of Arithmetic*, tr. Montgomery Furth (Berkeley and Los Angeles, 1964).

Frege, G., 'Über Sinn und Bedeutung', *Z. f. Philosophie u. philosophische Kritik* 100(1892): 25–50; English translation in *Translations from the Philosophical Writings of Gottlob Frege*, ed. Peter Geach and Max Black (Oxford, 1952).

Galilei, G., quoted in Bunge, *Causality*.

Gergonne, J. D., 'Essai de dialectique rationnelle', *Annales de mathématique* 1816/17.

Halasy and Sólyom, *Út a modern algebrához* (A Way to Modern Algebra) (Budapest, 1972).

Händel, A. and Kneist, K., *Kurzer Abriss der Logik* (Berlin, 1960).

Hass, *Vadászok a tenger mélyén* (Hunters Deep in the Sea) (Budapest, 1965).

Hegel, G. W. F., *Encyclopädie der philosophischen Wissenschaften im Grundrisse* (Leipzig, 1905).

Hegel, G. W. F., *System der Philosophie*, ed. H. Glockner, Vol. 8/I: *Die Logik* (Stuttgart, 1929). (Note: In this so-called *Jubiläumausgabe*, Volume 8 of the *Sämtliche Werke* is the *System* which is the title given here to the *Encyclopädie*.) English translation: *The Logic of Hegel*, tr. William Wallace (London, 1931), which is a reprint of the Revised Second Edition of 1892.

Hegel, G. W. F., *Wissenschaft der Logik*, Vols. I and II (Stuttgart, 1958 and 1949); English translation: *Hegel's Science of Logic*, tr. A. V. Miller from Lasson's edition of 1923 (London and New York, 1969).

Herbart, J. F., *Lehrbuch zur Einleitung in die Philosophie* (Leipzig, 1812 and Königsberg, 1813).

Holbach, P. H. Th., *Système de la nature*, new ed. with notes and corrections by Diderot (Paris, 1821).

Hume, D., *Enquiries Concerning the Human Understanding*, ed. L. A. Selby-Bigge, 2nd ed. (Oxford, 1936).

Husserl, E., *Logische Untersuchungen*, Vol. I (Halle, 1928).

Huszár, *Logika* (Logic) (Budapest, 1937).

Jevons, W. S., *Elementary Lessons in Logic: Deductive and Inductive* (London, 1872).

Jevons, W. S., *The Principles of Science* (London, 1874).

Jourdain, P. E. B., *The Development of Mathematical Logic and the Principles of Mathematics* (London, 1912).

Kant, I., *Kritik der reinen Vernunft* (Leipzig, 1966); English translation: *Immanuel Kant's Critique of Pure Reason*, tr. Norman Kemp Smith (London, 1929 and New York, 1965).

Kant, I., *Schriften zur Metaphysik und Logik*, Vols. I and II (Frankfurt, 1968).

Kant, I., 'Die falsche Spitzfindigkeit der vier syllogistischen Figuren', in: *Frühschriften*, Vol. II (Berlin, 1961).

Klaus, G., *Einführung in die formale Logik* (Berlin, 1958). (Also published later as *Moderne Logik: Abriss der formalen Logik*.)

Kondakov, N. I., *Vvedenie v logiku* (Introduction to logic) (Moscow, 1967).

Krug, W. T., *System der theoretischen Philosophie* (Königsberg, 1825).

Lambert, J. H., *Neues Organon* (Leipzig, 1764).

Laplace, P., *Essai philosophique sur les probabilités* (Paris, various dates).

Leben und Meinungen berühmter Philosophen (Berlin, 1955).

Lenin, V. I., *Materialism and Empirio-criticism* (Moscow, 1952).

Lenin, V. I., *Philosophical Notebooks*, tr. Clements Dutt (Moscow, 1961); this is Vol. 38 of the *Collected Works*.

Lewis, C. I., *A Survey of Symbolic Logic* (Berkeley, 1918).

Logika (Logic) (Budapest, 1956).

Lotze, H., *System der Philosophie*, Vol. I (Leipzig, 1874).

Łukasiewicz, J., *Aristotle's Syllogistic from the Standpoint of Modern Formal Logic*, 2nd ed. (Oxford, 1957).

Maier, H., *Die Syllogistik des Aristoteles*, Vols. I–III (Tübingen, 1896–1900).

Markovic, *Logika* (Logic) (Novi Sad, 1967).

Marx, K., *Einleitung zur Kritik der Politischen Ökonomie*, in Volume 13 of *Marx–Engels Werke* (Berlin, 1961); English translation taken from the *Grundrisse*, tr. Martin Nicolaus (Penguin ed., London, 1973).

Marx, K. and Engels, F., *Die heilige Familie* (Berlin, 1953); English translation taken from *Marx–Engels Collected Works*, Vol. 4 (New York and London, 1978).

Menne, A., *Logik und Existenz: eine logistische Analyse der kategorischen Syllogismus-funktoren und das Problem der Nullklasse* (Meisenheim-Glan, 1954).

Menne, A., 'Zur Syllogistik strikt particulärer Urteile', in: *Contributions to Logic and Methodology in honor of J. M. Bochenski* (Amsterdam, 1965).

Mill, J. S., *A System of Logic, Ratiocinative and Inductive*, ed. J. M. Robson (Toronto, 1978), which is Volumes 7–8 of the *Collected Works*.

Occam, *Summa Logicae* (Louvain and Paderborn, 1954).

Patzig, G., *Die aristotelische Syllogistik* (Göttingen, 1959); English translation: *Aristotle's Theory of the Syllogism*, tr. Jonathan Barnes (Dordrecht and Boston, 1969).

Petrus Hispanus (Peter of Spain, and Pope John XXI), *Summulae Logicales*, ed. J. M. Bochenski (Turin, 1947); quoted from Bochenski, *Formale Logik*.

Pólya, G., *A gondolkodás iskolája* (School of Thinking) (Budapest, 1957); English original edition: *How to Solve It* (Princeton, 1945).

Porphyrius, *Isogoge et in Aristotelis Categories Commentarium* (Berolini, 1887).

Pozsonyi, *A logika tárgyköre és feladata* (The Subject Matter and Task of Logic) (Budapest. 1942).

Prantl, C. von, *Geschichte der Logik im Abendlande*, Vol. I–IV (Graz and Berlin, 1955).

Quine, W. V. O., *Methods of Logic*, rev. ed. (London and New York, 1959).

Ribnikov, *A matematika története* (The History of Mathematics) (Budapest, 1968).

Russell, B., *My Philosophical Development* (London, 1959).

Russell, B., 'Existence and Being', *Mind* 1901.

Russell, B., 'On Denoting' *Mind* n.s. 14 (1905).

Russell, B. and Whitehead, A. N., *Principia Mathematica* (Cambridge, 1910–1913).

Russell, B., *Our Knowledge of the External World* (London, 1914).

Schaff, A. *A marxista–leninista igazságelmélet néhány problémája* (Some Problems of the Marxist–Leninist Theory of Truth) (Budapest, 1955); see also the German ed.: *Zu einigen Fragen der Marxistischen Theorie der Wahrheit* (Berlin, 1954), tr. from the Polish original.

Scholz, H. and Hasenjaeger, G., *Grundzüge der mathematischen Logik* (Berlin, 1961).

Schröder, E., *Vorlesungen über die Algebra der Logik (Exakte Logik)* (Leipzig, 1890–1905), in 3 volumes.

Septulin, A. P. *A marxismus–leninizmus filozófiája* (The Philosophy of Marxism–Leninism) (Budapest, 1971).

Sextus Empiricus, *Adversos mathematicos*, in Vol. 3 of the *Opera*, ed. H. Mutschmann (Leipzig, 1914), as quoted in Prantl.

Sigwart, Ch., *Logik*, 4th ed., Vols. I and II (Tübingen, 1911).

Spinoza, B., *Spinoza's Ethics and 'De intellectus emendatione'* (Everyman's Library, London and New York, 1934), tr. A. Boyle.

Stegmüller, W., 'Das Universalien problem einst und jetzt', *Archiv f. Philosophie* 1956, No. 6 and 1957, No. 7.

Szalai, 'Az Organon keletkezésének és az arisztotelészi szillogisztika szerkezeti felépítésének főbb kérdései' (The major problems of the development of the *Organon* and the structure of the Aristotelian syllogistic) in: *Organon* (Hungarian ed., Budapest, Vol. I: 1961).

Szalai, 'Formális logika és matematikai logika' (Formal logic and mathematical logic), *Magyar Tudomány* 1957, Nos 9–10

Tavanets, *Suzhdenie i ego vidi* (The proposition and its kinds) (Moscow, 1953).

Tavanets, (On the truth value of terms), *Voprosy filozofii* 1959, No. 12.

Trendelenburg, F. A., *Logische Untersuchungen*, 3d ed. (Leipzig, 1870. Hildesheim, 1964).

Überweg, F. and Heinze, *Grundriss der Geschicnte der Philosophie*, Vol. I (1905).

Überweg, F., *System der Logik und Geschichte der logischen Lehre* (Bonn, 1865).

Uemov, A. I., *Dolgok, tulajdonságok, viszonyok* (Things, Properties, Relations) (Budapest, 1966), tr. from the Russian.

Uemov, A. I., (Empty classes and the Aristotelian logic), in: *Logicheskoi issledovanie* (Moscow, 1959).

Valla, *Dialecticae disputationes* (1543).

Voishvillo, E. K., *Poniatie* (The Notion) (Moscow, 1967).

Varga, T., *Matematikai logika,* (Mathematical logic) Vol. II. (Budapest, 1966).

Wittgenstein, L., *Tractatus logico-philosophicus* being Vol. I of *Schriften* (Frankfurt am Main, 1963).

INDEX OF NAMES

517

LIST OF FIGURES